Metallic Nanostructures

Yujie Xiong • Xianmao Lu
Editors

Metallic Nanostructures

From Controlled Synthesis to Applications

 Springer

Editors
Yujie Xiong
Chemistry
University of Science and Technology
of China
Anhui
China

Xianmao Lu
Chemical Engineering
National University of Singapore
Singapore
Singapore

ISBN 978-3-319-35577-1   ISBN 978-3-319-11304-3 (eBook)
DOI 10.1007/978-3-319-11304-3
Springer Cham Heidelberg New York Dordrecht London

Preface

Nanoscience is an intersection of chemistry, physics, materials science, biology, electrical engineering, and many other disciplines to study matters at nanometer scale. More specifically, this multidisciplinary field is focused on matters with at least one dimension sized from 1 to 100 nm. In parallel to quantum dots that were invented earlier, metallic nanostructures have attracted tremendous attention from the research community, and their research has become a fast-growing subfield of the nanoscience. Metals possess a range of wonderful properties, leading to extensive use in industrial applications including catalysis, electronics, photography, and magnetic information storage. As the dimensions of metallic materials shrink down to the nanoscale, they give birth to numerous new applications such as photonics, imaging, and medicine. Undoubtedly, a majority of these promising applications require the use of metals in a finely divided state, which not only brings about grand challenges to materials synthesis, but creates opportunities for chemists to collaborate with the scientists of various academic backgrounds.

The history of solution-phase synthesis—the most widely used approach to metallic nanoparticles can be traced back to the 1850s when Michael Faraday invented his now famous gold colloids. However, not until the recent decade had the controllability over size, shape, structure, and composition of metallic nanostructures reached the level usable for realistic applications. Now, the solution-phase methods can offer the quality, quantity, and reproducibility suitable for shape-property relationship investigations, which significantly blossoms the research and implementations of metallic nanostructures. Nevertheless, the people having insufficient chemistry training still suffer from the difficulty of transferring literature knowledge to bench work in wet labs. "Enabling anyone to make desired metal nanomaterials in a wet lab" is thus one of the major motivations of writing this book. Needless to say, understanding the designing rules for nanomaterials is also an obstacle to the chemists working on materials synthesis. Leveraging example applications, this book illustrates how to design a perfect metallic nanostructure for the application which the readers are specifically interested in. We do hope, by bridging controlled synthesis and applications, it will make it more clear how fascinating and versatile metal nanomaterials are.

This is the first book to address the fundamentals in the controlled synthesis of metallic nanostructures toward fulfilling specific functions, and is intended for the scientists and engineers making efforts in the fields related to nanotechnology. Here we just name a few examples: materials science, chemistry, physics, electronics, magnetic information storage, catalysis, biotechnology, optics, and photonics. Although it is not specially designed for a university textbook, it does not bother this book to become a reference for the college and graduate students studying the courses related to nanotechnology and materials science. Specifically, Chap. 1 first gives an overview on the fundamentals in the field of metallic nanostructures. Chapters 2 and 3 provide basic knowledge relative to metallic nanomaterial synthesis and allow the readers to have a clear picture for making desired metallic nanoproducts via direct synthesis or seeding growth. Chapters 4–9 illustrate the structure-dependent properties of metallic nanostructures for interfacial molecular interactions, biomedicine and sensing, electrocatalysis and catalysis, photonics and electronics, and magnetic applications case by case. They fully demonstrate what applications metallic nanomaterials can be implemented in, leading to the bright future of this research. We do hope that this book coherently integrates synthesis fundamentals with material functions and can unravel the puzzle that people frequently encounter when designing metal nanostructures for various applications.

Overall, the objective of this book is to strengthen research in metallic nanostructures by creating a coherent framework of fundamental understanding and case study for various disciplinary nanoscience communities. It represents a further step in the research of metal nanomaterials to achieve rational design materials. At the same time, it constitutes an updated overview of the state-of-the-art and a roadmap, reflecting frontier research trends in each subfield of metal nanomaterials.

University of Science and Technology of China   Yujie Xiong
National University of Singapore   Xianmao Lu

Contents

Contributors

Pedro H. C. Camargo Departamento de Química Fundamental, Instituto de Química, Universidade de São Paulo, São Paulo, Brazil

Jingyi Chen Department of Chemistry and Biochemistry, University of Arkansas, Fayetteville, AR, USA

Joseph S. DuChene Department of Chemistry and Center for Nanostructured Electronic Materials, University of Florida, Gainesville, FL, USA

Yanglong Hou College of Engineering, Peking University, Beijing, China

Samir V. Jenkins Department of Chemistry and Biochemistry, University of Arkansas, Fayetteville, AR, USA

Mingshang Jin Center for Materials Chemistry, Frontier Institute of Science and Technology, Xi'an Jiaotong University, Shaanxi, China

Ran Long School of Chemistry and Materials Science, University of Science and Technology of China, Hefei, Anhui, P. R. China

Xianmao Lu Department of Chemical and Biomolecular Engineering, National University of Singapore, Singapore, Singapore

Timothy J. Muldoon Department of Biomedical Engineering, University of Arkansas, Fayetteville, AR, USA

Wenxin Niu Department of Chemical and Biomolecular Engineering, National University of Singapore, Singapore, Singapore

Zhenmeng Peng Department of Chemical and Biomolecular Engineering, University of Akron, Akron, OH, USA

Jingjing Qiu Department of Chemistry and Center for Nanostructured Electronic Materials, University of Florida, Gainesville, FL, USA

Thenner S. Rodrigues Departamento de Química Fundamental, Instituto de Química, Universidade de São Paulo, São Paulo, Brazil

Anderson G. M. da Silva Departamento de Química Fundamental, Instituto de Química, Universidade de São Paulo, São Paulo, Brazil

Brendan C. Sweeny Department of Chemistry and Center for Nanostructured Electronic Materials, University of Florida, Gainesville, FL, USA

Jiale Wang Departamento de Química Fundamental, Instituto de Química, Universidade de São Paulo, São Paulo, Brazil

Zhenni Wang Center for Materials Chemistry, Frontier Institute of Science and Technology, Xi'an Jiaotong University, Shaanxi, China

W. David Wei Department of Chemistry and Center for Nanostructured Electronic Materials, University of Florida, Gainesville, FL, USA

Yujie Xiong School of Chemistry and Materials Science, University of Science and Technology of China, Hefei, Anhui, P. R. China

Wenlong Yang College of Engineering, Peking University, Beijing, China

Jing Yu College of Engineering, Peking University, Beijing, China

Shan Zhou School of Chemistry and Materials Science, University of Science and Technology of China, Hefei, Anhui, P. R. China

Editors Biography

Yujie Xiong   received his B.S. in chemical physics in 2000 and Ph.D. in inorganic chemistry in 2004 (with Professor Yi Xie), both from the University of Science and Technology of China (USTC). After four-year training with Professors Younan Xia and John A. Rogers, he joined the National Nanotechnology Infrastructure Network (NSF-NNIN) at Washington University in St. Louis as the Principal Scientist and Lab Manager. Starting from 2011, he is a Professor of Chemistry at the USTC. He has published 88 papers with over 8,000 citation (H-index 46). His research interests include synthesis, fabrication and assembly of inorganic materials for energy and environmental applications.

Xianmao Lu   is an assistant professor in the Department of Chemical & Biomolecular Engineering at National University of Singapore (NUS). Before he joined NUS, he was a postdoctoral research fellow at University of Washington and Washington University. He received his PhD in Chemical Engineering from the University of Texas at Austin, where he started his research in nanomaterials. His current research interest is mainly on shape-selective growth of noble metal nanocrystals and understanding of their fundamental properties.

Chapter 1
Metallic Nanostructures: Fundamentals

Wenxin Niu and Xianmao Lu

Abstract This chapter focuses on the fundamental aspects of metallic nanostructures. We firstly introduce the definition, classification, and historical background of metallic nanostructures. A summary of their novel optical, catalytic, and magnetic properties is provided in the next section. General methods for the synthesis of metallic nanostructures are then outlined, followed by the discussion of their morphological, structural, and compositional characterization methods. Finally, representative examples for controlled syntheses of metallic nanostructures are highlighted.

1.1 Metallic Nanostructures: General Introduction and Historical Background

1.1.1 General Introduction

Nanoscience has emerged as one of the most exciting areas of modern science and technology. The development of nanoscience has revolutionized many applications ranging from catalysis to electronics, photonics, information storage, biological imaging and sensing, as well as energy conversion and storage [1–5]. Among these advances, metallic nanostructures have played a major role in many directions and have become one of the most studied subjects in nanoscience [6]. Metallic nanostructures are defined as metallic objects with at least one dimension in the range of one to a few hundred of nanometers. They exhibit many remarkable chemical and physical properties that are different from both individual metal atoms and bulk metals. The emergence of these novel properties is the driving force for the rapid development of research in metallic nanostructures, which has been on the forefront of scientific disciplines including chemistry, physics, materials science, medicine, and biology.

X. Lu (✉) · W. Niu
Department of Chemical and Biomolecular Engineering, National University of Singapore,
4 Engineering Drive 4, Singapore 117585, Singapore
e-mail: chelxm@nus.edu.sg

© Springer International Publishing Switzerland 2015
Y. Xiong, X. Lu (eds.), *Metallic Nanostructures,* DOI 10.1007/978-3-319-11304-3_1

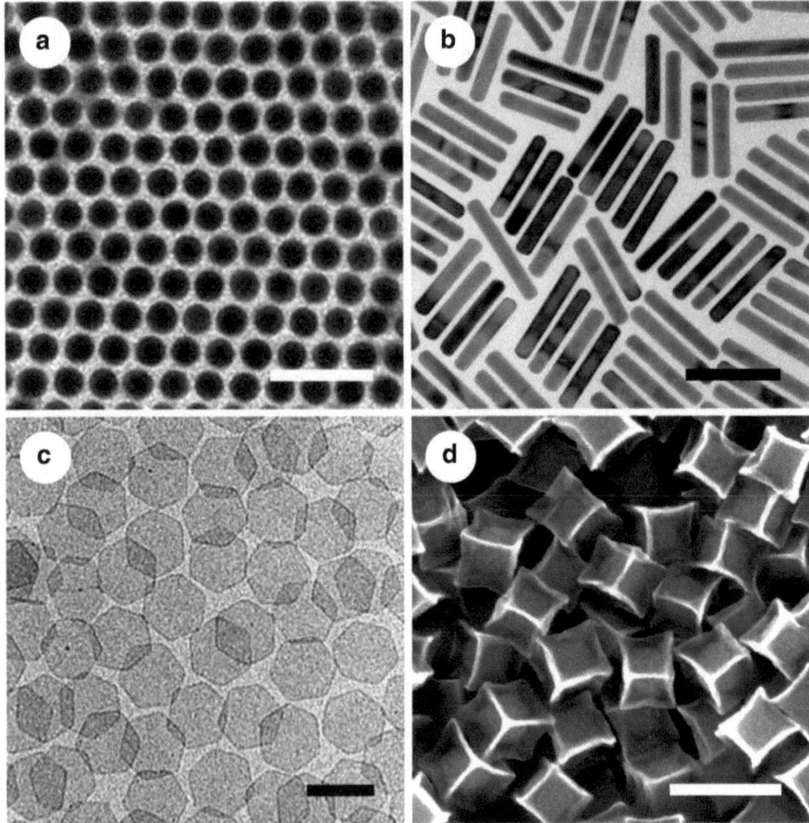

Fig. 1.1 Metallic nanostructures of different dimensions: **a** 0D Pd spherical nanoparticles (scale bar: 20 nm, reproduced with permission from reference [7], Copyright 2010 Wiley-VCH), **b** 1D Au nanorods (scale bar: 100 nm, reproduced with permission from reference [8], Copyright 2012 American Chemical Society), **c** 2D Pd ultrathin nanosheets (scale bar: 100 nm, reproduced with permission from reference [9], Copyright 2011 Nature Publishing Group), and **d** 3D Pd concave nanocubes (scale bar: 200 nm, reproduced with permission from reference [10], Copyright 2014 American Chemical Society)

1.1.2 Classification of Metallic Nanostructures

A typical way of classifying metallic nanostructures of different morphologies is to identify them according to their dimensions. Based on the number of dimensions, metallic nanostructures can be roughly classified as zero-dimensional (0D), one-dimensional (1D), two-dimensional (2D), and three-dimensional (3D). 0D metallic nanomaterials include small clusters composed of a few to roughly a hundred metal atoms, and common spherical metal nanoparticles. Figure 1.1a shows an example of 0D Pd spherical nanoparticles [7]. 1D metallic nanomaterials are nanostructures with large aspect ratios such as nanorods, nanowires, nanobelts (nanoribbons), and

Fig. 1.2 The dichroism of the Lycurgus cup: **a** viewed in transmitted light, **b** viewed in reflected light (from reference [20], Copyright Trustees of the British Museum), **c** medieval stained glass from Chartres Cathedral (from reference [21], Copyright Dr Stuart Whatling)

nanotubes. Figure 1.1b shows an example of 1D Au nanorods [8]. 2D nanomaterials are thin films with nanometer thickness such as nanosheets, nanoplates, and nanoprisms. Figure 1.1c shows an example of Pd ultrathin nanosheets with a thickness of 1.8 nm [9]. 3D nanomaterials include more complicated structures such as various polyhedra, as well as assemblies of 0D, 1D, and 2D nanostructures. Figure 1.1d shows an example of 3D Pd concave nanocubes with sub-10 nm sharp edges [10].

1.1.3 Historical Background of Metallic Nanostructures

Metallic nanostructures have been unknowingly used for a few purposes dating back for millennia. Gold and silver nanoparticles have been incorporated into glasses as colorant for over 2000 years [3]. One of the most well-known examples is the Lycurgus cup from the Roman times [11]. The Lycurgus cup has a remarkable characteristic of dichroism: the glass appears a color of green jade under normal external lighting conditions, while it gives a deep ruby red color when it is lighted from within (Fig. 1.2a, b). Modern transmission electron microscopy (TEM) studies proved that the fascinating colors originate from alloy nanoparticles of Ag and Au with sizes of 50–100 nm embedded in the glass [12]. When viewed in reflected light, the green color of the cup is due to the scattering contribution of the alloy nanoparticles [13]. When it is illuminated from inside, the color is resulted from the absorption contribution: the green wavelength is absorbed and the light that goes through the glass appears with the complementary red color.

In the Middle Ages, Au, Ag, and Cu nanoparticles have been frequently employed as colorants, particularly for church windows and pottery [14, 15]. In the seventeenth century, Andreas Cassius et al. described a procedure that can produce Au nanoparticles with an intense purple color [16]. This so-called "Purple of Cassius" was used to color glasses and produce the so-called "gold ruby glass" [17]. The preparation of "Purple of Cassius" involved the dissolution of gold metal in aqua regia followed by the reduction of Au(III) by a mixture of stannous chloride, although the true chemical reaction mechanisms were not understood at that time [18]. Figure 1.2c shows a medieval-stained glass from Chartres Cathedral, in which the red color originates from Au nanoparticles. Ag and Cu nanoparticles were also dispersed in the glassy matrix of the ceramic glaze for luster decoration of medieval and Renaissance pottery [19].

The first documented scientific study of metal nanoparticles was performed by Michael Faraday [22, 23]. In 1857, Faraday reported that colloidal Au nanoparticles can be prepared by the reduction of gold salts with reagents including organic compounds or phosphorus [24]. Faraday's Au nanoparticles, with size between 3 and 30 nm, are still stable now [25]. Faraday correlated the red color of Au colloid with the small size of the Au particles in metallic form. He concluded that it is gold present in solution in a "finely divided metallic state" smaller than the wavelength of visible light that shows colors different from the original color [24]. Faraday also examined the synthesis of other metal nanoparticles such as Pt, Pd, Rh, and Ag [24].

Zsigmondy and Svedberg also made significant contributions to the development of metallic nanostructures [26]. Zsigmondy employed an ultramicroscope to study the optical properties of Au nanoparticles [27, 28]. He developed a method for preparing colloidal Au nanoparticles by boiling gold chloride with formaldehyde [27]. He even used small Au particles as nuclei for the growth of large Au nanoparticles, which is possibly one of the first seed-mediated growth methods for metallic nanostructures [27]. Svedberg also studied the synthesis of metal nanoparticles extensively [29]. He explored many reducing agents to produce Au nanoparticles from chloroauric acid, ranging from gas phase reductant hydrogen, hydrogen sulphide, and carbon monoxide (CO), to various organic and inorganic reducing agents [29]. In addition, he invented the ultracentrifuge, which allowed size separation of nanoparticles [30].

The revolutionary development of TEM enabled researchers to directly explore the morphology of metallic nanostructures [31]. In 1937, Beischer et al. first studied the shapes of Au nanoparticles using a TEM [32]. Since then, researchers have been able to correlate the sizes and shapes of metallic nanostructures with their synthetic conditions. In 1951, Turkevich et al. developed a method for producing spherical Au nanoparticles via citrate reduction of tetrachloroauric acid [33]. The citrate reduction method is very simple and can give Au nanoparticles with narrow size distribution. In 1973, Frens refined Turkevich's work and discovered that the size of Au nanoparticles can be simply controlled by tuning the ratio between sodium citrate and gold salt [34]. The citrate reduction method is still one of the most simple and reproducible approaches in synthesizing Au nanoparticles with controlled size. Besides TEM, the invention of numerous characterization and analytical technologies has greatly simulated the developments of modern research in metallic nanostructures [35]. Based on these high-resolution chemical and physical analyses, elucidation of structure–function relationships in metallic nanostructures has become possible. Morphology-controlled synthesis of metallic nanostructures has achieved unprecedented progress, even at atomic precision [36]. Many rational synthetic methods for metallic nanostructures have been developed and "materials by design" has become a goal of research in this area.

Table 1.1 Calculated sizes of cuboctahedral Pt nanoparticles and their corresponding surface atom percentages (reproduced with permission from reference [37], Copyright 2012 Wiley-VCH)

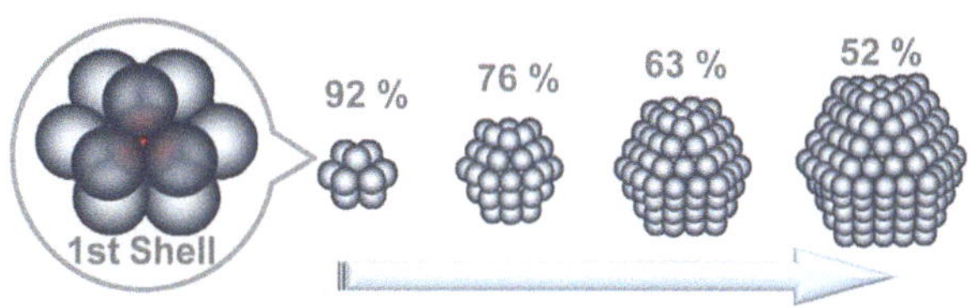

Number of shells	Number of surface atoms	Number of total atoms	Percentage of surface atoms (%)	Size of Pt nanoparticles (nm)
1	12	13	92	0.8
2	42	55	76	1.4
3	92	147	63	1.9
4	162	309	52	2.4
5	252	561	45	3.0
6	362	923	39	3.5
7	492	1415	35	4.1
8	642	2057	31	4.6
9	812	2869	28	5.1

1.2   Fundamental Properties of Metallic Nanostructures

The fast development in the synthesis of metallic nanostructures with controllable morphologies has enabled the discovery of their new properties [38]. Now it is well-known that size and shape of metallic nanostructures can profoundly alter their properties. One of the most well-known size-dependent properties of metallic nanostructures is their high surface area to volume ratio. Table 1.1 shows the relationships between particle size versus the percentage of surface atoms for Pt nanoparticles with cuboctahedral shapes [37]. It can be clearly seen that as the size of Pt nanoparticles decreases, a dramatic increase in the surface atom percentages can be obtained. The large surface area to volume ratio of metallic nanostructures is the origin of a number of unique applications, especially in catalysis [39]. Besides their sizes, the shape of metallic nanostructures also strongly affects their optical, catalytic, and magnetic properties. In this section, the correlation between these properties and the morphology of metallic nanostructures is briefly introduced.

1.2.1   Optical Properties

The brilliant colors of Au and Ag nanostructures have fascinated people for thousands of years. These phenomena arise from localized surface plasmon resonance (LSPR). LSPR is the resonant collective oscillation of free electrons of metallic

Fig. 1.3 Metallic nanostructures with tunable localized surface plasmon resonance (LSPR): **a** Au nanorods with different aspect ratios, **b** Au nanoshells with different shell thickness, and **c** Au/Ag nanocages with different composition and shell thickness (reproduced with permission from reference [40], Copyright 2012 The Royal Society of Chemistry)

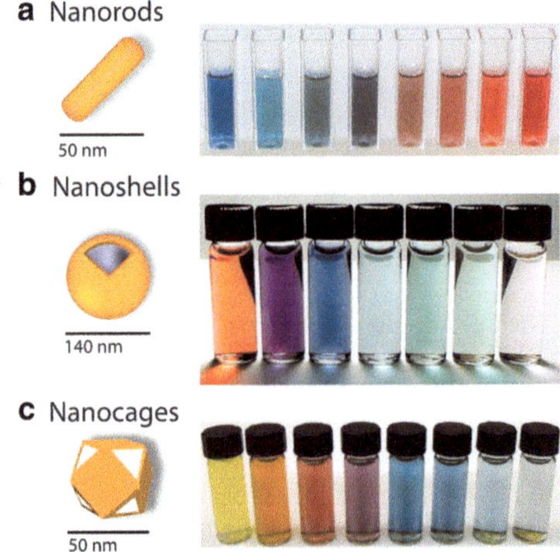

nanostructures under light irradiation [41–43]. LSPR can result in strong light absorption, scattering, and enhanced local electromagnetic field. Based on these characteristics, plasmonic nanostructures have found broad application in biological sensing and imaging, photothermal therapy, and solar energy harvesting [43–46]. Metallic nanostructures possessing LSPR in visible wavelength range are usually metals with a negative real and small positive imaginary dielectric constant such as gold, silver, and copper [41]. One remarkable feature of LSPR is that its frequency and intensity are strongly dependent on the size, shape, and composition of the plasmonic metallic nanostructures. This provides an important method to tailor the LSPR of metallic nanostructures by synthetically tuning their structural parameters. For example, Fig. 1.3 shows three types of metallic nanostructures with tunable LSPR properties: nanorods, nanoshells, and nanocages [40, 47, 48]. Their optical properties can be controlled by tuning the aspect ratio, shell thickness, and composition. Another feature of LSPR is that it is very sensitive to the refractive index of the surrounding medium [41]. Pronounced red shifts in the LSPR spectral position can be observed when the refractive index of the surrounding medium increases. Based on this property, LSPR has been used for ultra-sensitive sensing [49].

1.2.2 Catalytic Properties

The rapid development of metallic nanostructures has impacted the field of heterogeneous catalysis considerably [51–54]. Metallic nanostructures with well-defined sizes and shapes can be a new family of model systems for establishing structure–function relationships in heterogeneous catalysis [55]. In conventional catalysis, the detailed crystal structure and crystallographic facets of metallic nanocatalysts are usually not well resolved. However, these structural parameters can have a dramatic effect on their catalytic activity. A typical example is the discovery of Au catalysts in low-temperature CO oxidation [56, 57]. Although bulk Au is very unreactive for

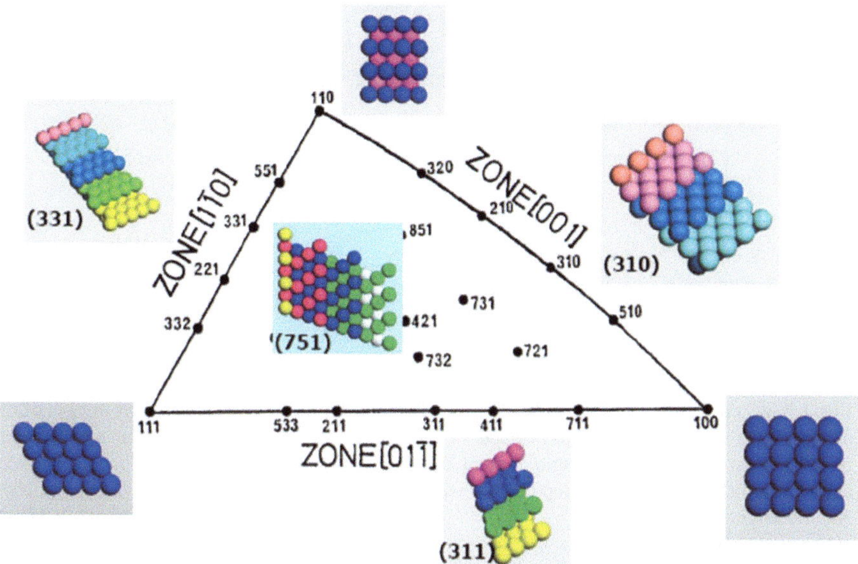

Fig. 1.4 Unit stereographic triangle of face-centered cubic (fcc) metal single-crystal and models of surface atomic arrangement (reproduced with permission from reference [50], Copyright 2008 American Chemical Society)

low-temperature CO oxidation, Au nanoparticles in sizes of 3–4 nm are catalytically active for this reaction. The advance in metallic nanostructures now allows one to control their size, shape, and composition precisely, offering new opportunities to discover cost-effective and active metallic catalysts [58]. For instance, by controlling the surface crystal facets of Pt nanocrystals, their catalytic activity can be significantly increased [59]. Figure 1.4 shows the surface atomic arrangements of various crystal facets of face-centered cubic (fcc) metals [50]. Pt nanostructures with high-index facets can exhibit up to 400 % increase in catalytic activity for the electrooxidation of small organic fuels compared with common Pt catalysts [60]. The activity is attributed to the open structure with a high density of unsaturated atoms on the high-index facets. The selectivity of catalytic reaction can also be control by the shape of metallic nanostructures [37]. Somorjai et al. showed that during the benzene hydrogenation reactions, cyclohexane and cyclohexene products were formed on cuboctahedral Pt nanocrystals with both {111} and {100} facets, whereas, only cyclohexane was produced on Pt nanocubes enclosed by {100} facets [61].

1.2.3 Magnetic Properties

The size and shape of ferromagnetic metal nanostructures have a significant impact on their magnetic properties [62]. Superparamagnetism is one of the most interesting phenomena arising from the shrinking size of metallic nanostructures. In metallic nanostructures of a critical size, the thermal energy is sufficient to invert

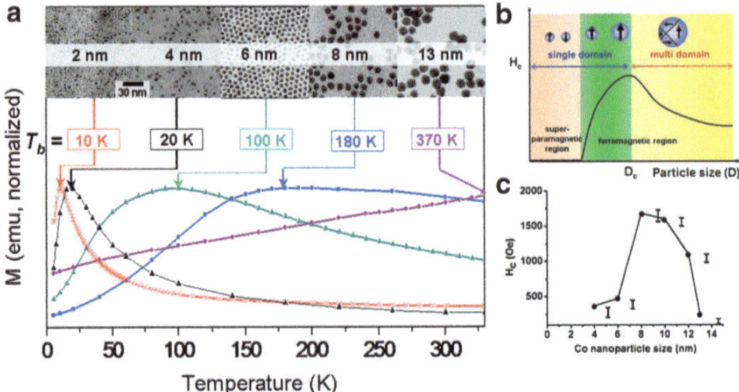

Fig. 1.5 **a** Zero-field cooling curves and transmission electron microscopy (TEM) images of Co nanoparticles with sizes of 2, 4, 6, 8, and 13 nm. **b** Size-dependent magnetic domain structures from superparamagnetism to single domain and multidomain ferromagnetism. **c** Size-dependent coercivity of Co nanoparticles (reproduced with permission from reference [62], Copyright 2008 American Chemical Society)

the magnetic spin direction and the magnetic fluctuation results in a net magnetization of zero. This phenomenon is called superparamagnetism [63]. Apparently, the transition from ferromagnetism to superparamagnetism is temperature dependent and this transition temperature is referred as the blocking temperature [64]. The blocking temperature of metallic nanostructures is strongly dependent on their sizes. In the case of cobalt nanoparticles, ferromagnetic to superparamagnetic transitions of nanoparticles with sizes of 2, 4, 6, 8, and 13 nm occur at 10, 20, 100, 180, and 370 K, respectively (Fig. 1.5a) [65]. The magnetic coercivity of cobalt nanoparticles is also dependent on their sizes. For Co nanoparticles with single magnetic domain structures, the magnetic coercivity increases with the size of the nanoparticles. For Co nanoparticles with multiple magnetic domains, the magnetic coercivity decreases as the size of the Co nanoparticles increases (Fig. 1.5b and c) [66]. The magnetic coercivity of metallic nanostructure is also dependent on their shape and composition [67–69]. For example, Co nanowires with high aspect ratios have a much higher coercivity than those with small aspect ratios [67].

1.3 General Methods for the Synthesis of Metallic Nanostructures

The quest for methods of producing metallic nanostructures with controllable sizes and morphologies has been always a challenging subject. The interest in rationally synthesizing metallic nanostructures with controllable sizes and morphologies is not based solely on their esthetic appeal. The sizes and morphologies of metallic nanostructures will play a pivotal role in determining their properties [42, 52, 70–72]. Therefore, the ability to generate metallic nanomaterials with well-controlled

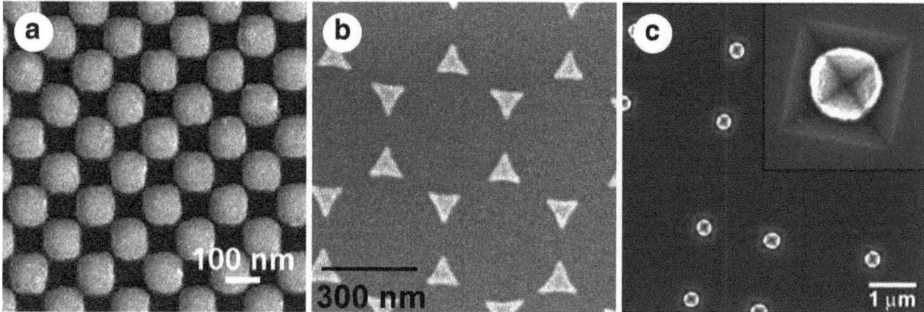

Fig. 1.6 Metallic nanostructures fabricated by combining physical vapor deposition and nanofabrication methods: **a** A checkerboard cluster of Au nanoblocks separated by nanogaps through electron beam lithography (reproduced with permission from reference [79], Copyright 2011 American Chemical Society), **b** An array of triangular copper nanostructures by nanosphere lithography (reproduced with permission from reference [88], Copyright 2007 American Chemical Society), **c** Pyramidal Au nanostructures supported on Si pedestals fabricated by templating Si wafer with etched holes (reproduced with permission from reference [89], Copyright 2005 American Chemical Society)

sizes, shapes, and compositions is central to unraveling the chemical and physical properties of metal nanostructures. Since the time of Faraday, numerous approaches have been developed to control the morphology and composition of metallic nanostructures. Besides simple chemical reduction or thermal decomposition routes, various techniques such as thermal evaporation, laser ablation, and electrochemical, photochemical, sonochemical processes have been developed. These methods could be realized either through chemical or physical processes in gas, liquid, or solid phases.

1.3.1 Gas- and Solid-Phase Methods

Many gas-phase synthesis approaches are based on physical vapor deposition (PVD), in which metallic nanostructures are generated through homogeneous or heterogeneous nucleation and growth through a vapor of metal atoms. A few techniques can be used to evaporate metals, such as arc discharge [73], laser ablation [74], ion sputtering [75], electron beam [76], or simply thermal evaporation [77]. However, most of these methods can only produce metallic nanomaterials with broad size distribution and usually have limited control on their morphology. Combined PVD with nanofabrication methods such as electron beam lithography can generate 2D metallic nanostructures with arbitrary sizes and shapes, although the throughput may be low (Fig. 1.6a) [78, 79]. Besides lithographic methods, templates can also be used to produce nanostructures with anisotropic shapes [80, 81]. For example, Van dyne et al. has developed a technique called nanosphere lithography to fabricate arrays of metallic nanostructures (Fig. 1.6b) [82]. This method uses close-packed sphere arrays of monodispersed polystyrene or silica nanospheres as templates for the deposition of truncated triangular prisms of different metals. By templating against a single-crystal Si wafer with an array of etched holes, Odom et al. were able to generate free-standing pyramidal metallic nanostructures with diameters down to 80 nm and curvatures of tip radii as small as 2 nm (Fig. 1.6c) [81].

An alternative gas phase synthesis approach is chemical vapor deposition (CVD). In a typical CVD process, a substrate is exposed to one or more volatile precursors of metallic nanostructures, then the precursors may react or decompose on the substrate surface to produce corresponding metal nanostructures [83]. A popular variation of CVD is atomic layer deposition (ALD) [84]. ALD has emerged as an important technique for depositing metallic thin films and supported metallic nanoparticles with controllable size, composition, and structures [85]. During ALD processes, metallic nanostructures are grown via sequential, self-limiting chemical reactions between precursor vapors and the substrate surface [86]. There are only a finite number of surface sites on the substrate surface, therefore, only a finite number of surface species can be deposited [87]. Precise control over metal nanoparticle size, composition, and structure can be achieved by manipulating the combination of ALD sequence, surface treatment, and deposition temperature [84].

Most PVD and CVD processes start at the molecular level to build up metallic nanostructures. Conversely, metallic nanostructures can be produced from bulk metals through top-down approaches. For example, solid-phase mechanical processes such as grinding and milling have been used to reduce the size of metal materials into nanoscale [90, 91]. Nanoscale metallic powders are produced by high-energy ball milling of bulk materials. Colloidal stabilizers or supports are commonly added during the grinding and milling processes to avoid the aggregation of metal nanoparticles during the size reduction processes.

1.3.2 Wet Chemical (Liquid Phase) Methods

Gas- and solid-phase methods have advantages in obtaining nanomaterials with high purity and some of the processes are readily applicable to large-scale production. However, size and morphology control of metallic nanostructures are difficult to achieve with these methods if they are not combined with nanofabrication techniques. In addition, some of the techniques require high vacuum and expensive instruments. In contrast, synthetic methods based on wet chemistry have received intense attention and exhibit a few advantages over gas- and solid-phase methods [72, 92–94]. Wet chemical methods have greater flexibility and versatility in synthesizing metal nanostructures with controlled size, shape, and composition. For wet chemical methods, different reaction parameters such as precursor, surfactant, shape-directing agent, and reaction temperature can be independently manipulated. Moreover, wet chemical methods are easier to scale up and there is no need for expensive equipment. Over the past 30 years, many wet chemical methods have been developed for metallic nanostructures [72]. Most of these methods inevitably involve the precipitation of metals into solid phase in solution through chemical reduction or decomposition. In the following sections, a few wet chemical methods are introduced. These methods vary from common chemical reduction and thermal decomposition, to hydrothermal and solvothermal reactions at high pressure, synthetic processes involving microwave heating, radiation, electrochemistry, sonochemistry, syntheses at interfaces and in reversed micelles, and syntheses directed by templates.

1.3.2.1 Chemical Reduction and Thermal Decomposition

Chemical reduction is the most traditional method for metallic nanostructures [95, 96]. It has been successfully applied in synthesizing noble metal nanostructures such as Au, Ag, Pd, Rh, and Pt [72]. In a typical chemical reduction process, several regents are essential in addition to metal salts. The first one is the reducing agent, which donates electrons and reduces the metal salts to metal atoms. Important reducing agents include sodium borohydride, hydrazine, hydroxylamine, hydrogen, CO, and organic compounds such as sodium citrate, various sugars, alcohols, aldehydes, amides, and amines [26, 96]. To avoid the aggregation of metal nanostructures, a stabilizing agent is generally needed [97]. Stabilizing agents are usually surfactants or polymers that can be absorbed to the surface of metal nanostructures. To achieve morphology control, some shape-directing agents might be added. These shape-directing agents may have specific interactions with certain facets of the metal nanostructures or can change the reduction kinetics and induce the anisotropic growth of the nanostructures [92]. These agents include atomic/ionic species such halides [98, 99] and metal ions [100], surfactants [97], polymers [101], or biomolecules such as peptides [102–106].

Salts of reactive metals such as Fe, Co, and Ni have relatively low reduction potentials and cannot be reduced with common reducing agents. In these cases, thermal decomposition of their organometallic compounds becomes an attractive route for the synthesis of these metallic nanostructures [107, 108]. A typical thermal decomposition process usually involves the hot injection of an organometallic precursor into a heated solvent in the presence of suitable stabilizing agent [109]. Various metal nanoparticles including Fe, Ni, Co, as well as their alloys have been prepared through thermal decomposition [109–111]. It should be noted that CO generated from the decomposition of metal carbonyl precursors also has a strong effect on the shape evolution of metal nanostructures [112].

The synthesis of metallic nanostructures with narrow size distribution and uniform morphology has long been a great challenge [72, 92, 93, 113, 114]. To achieve this goal, the nucleation step must be separated from the growth step during the synthesis of metallic nanostructures, thus simultaneous secondary nucleation and growth can be avoided. Therefore, it is necessary to induce a single nucleation event in the nucleation step. This could be realized through a hot-injection method, in which a quick injection of metal precursor into hot reducing solvent could yield a fast nucleation process [115, 116]. An alternative and possibly better approach is seed-mediated growth [117, 118]. Seed-mediated growth is a typical heterogeneous nucleation process, which involves the synthesis of metal nanoseeds and the growth of the seeds in another growth solution. The seeds can catalyze the reduction of metal salts by reducing agents, therefore, they can continually grow into large nano-structures and the secondary nucleation can be efficiently discouraged. Therefore, a better control over size and shape of metallic nanostructures can be attained. More-over, seed-mediated method is promising in identifying shape-directing agents and providing mechanistic insight into the growth mechanisms of metal nanostructure due to its high controllability [119, 120].

1.3.2.2 Hydrothermal and Solvothermal Method

For common chemical reduction or thermal decomposition methods in solution phase, the reaction temperatures generally cannot exceed the boiling points of the solvents. Under hydrothermal and solvothermal conditions, the reactions are proceeded in a sealed container and water or solvents can be heated to tempera-tures above their boiling points [127–129]. In the meantime, heating can increase the autogenous pressure in the reaction container, causing significantly increased reactivity of the reducing agents and precursors [130, 131]. There have been many successful examples using hydrothermal and solvothermal processes to synthesize metallic nanostructures. For example, different Ag nanocrystals including nano-cubes, right bipyramids, nanorods, nanospheres, and nanowires can be produced through cetyltrimethylammonium bromide (CTAB)-modified silver mirror reac-tions at hydrothermal conditions [132, 133]. Pd nanowires have been synthesized in high yields through an iodide-assisted hydrothermal process (Fig. 1.7a) [121]. Solvothermal methods have been especially successful to synthesize noble met-als such as Pt and Pd (Fig. 1.7b). A few Pd, Pt, and their alloy nanostructures with high surface energy have been synthesized [134–139]. Under hydrothermal and solvothermal conditions, small adsorbates with low boiling point can be effectively introduced to control the growth of metal nanostructures [140].

1.3.2.3 Microwave Method

Microwave can be used as an energy source to heat reaction solutions and induce the reduction of metal salts or the decomposition of metal complexes into metallic nanostructures [141, 142]. Under microwave irradiation, polar molecules such as H_2O and polyols tend to orientate with the external electric field from microwave. When dipolar molecules try to re-orientate with alternating high-frequency elec-tric fields, molecular friction between the polar molecules will generate heat [143]. Compared with normal heating methods such as heating using oil bath and heating mantle, microwave methods provide rapid, uniform, and efficient heating of reagents and solvents [144, 145]. The rapid and uniform heating can promote the reduction of metal precursors and the nucleation of metal clusters, leading to the formation of nanostructures with uniform sizes. Besides the common heating effect, hot spots may be created on the solid–liquid surfaces when metal nanostructures are heated by microwave [141]. Upon microwave heating, the adsorption of surfactant with a

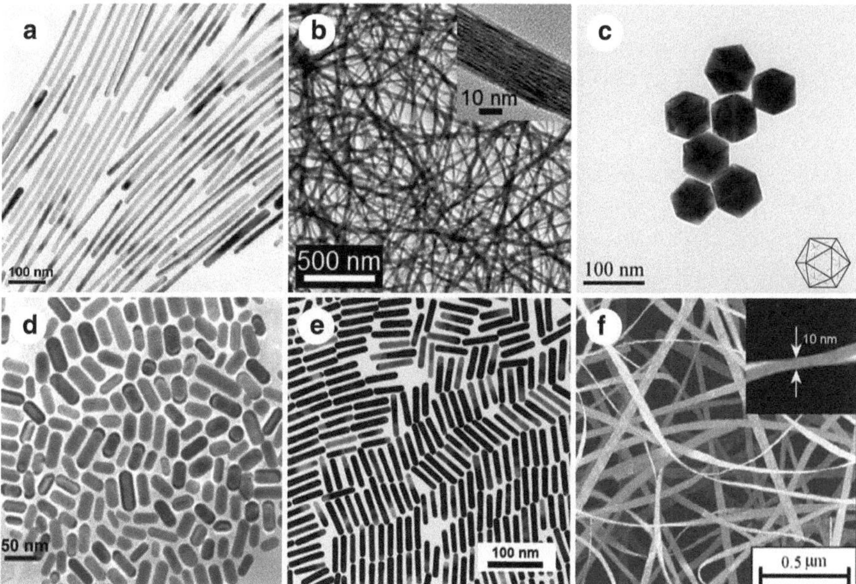

Fig. 1.7 Metallic nanostructures synthesized with different wet chemical methods: **a** Pd nanowires via hydrothermal method (reproduced with permission from reference [121], Copyright 2009 American Chemical Society), **b** Pt nanowire bundles via solvothermal method (reproduced with permission from reference [122], Copyright 2013 American Chemical Society), **c** Pd icosahedral nanocrystals via microwave method (reproduced with permission from reference [123], Copyright 2009 International Union of Pure and Applied Chemistry), **d** Au nanorods via photochemical method (reproduced with permission from reference [124], Copyright 2002 American Chemical Society), **e** Au nanorods via electrochemical method (reproduced with permission from reference [125], Copyright 1998 American Chemical Society), **f** Au nanobelts via sonochemical method (reproduced with permission from reference [126], Copyright 2006 Wiley-VCH)

large dielectric loss constant on metal nanostructures can create hot surfaces on solid metals, and this could be a specific shape-control strategy not attainable by other methods. For example, Liu and Huang et al. have applied microwave irradiation to synthesize uniform Pd icosahedral and cubic nanocrystals (Fig. 1.7c) [123, 146, 147].

1.3.2.4   Radiolytic and Photochemical Method

Radiolytic reduction can be employed as a powerful tool to produce metallic nanostructures [148]. Normal ionization radiations such as electron beam, X-ray, gamma-ray can cause the formation of hydrated electrons, OH^*, and H^* radicals during radiolysis of aqueous solutions [149, 150]. These species can cause the reduction of metal salts. Ultra-violet (UV) light, on the other hand, can also cause the formation of radicals through photolysis reactions. For example, Au nanorods were obtained after a solution containing acetone irradiated with a 254-nm UV light. Radicals are generated via the excitation of acetone by UV light and cause the reduction of gold salts (Fig. 1.7d) [124]. The aspect ratio of the final rods can be simply tuned by the amount of silver ions added to the system [151].

Visible light can also be involved in the synthesis of metallic nanostructures [152]. This process is mainly based on the plasmonic properties of noble metal nanostructures. Upon the excitation the plasmonic modes of Au or Ag nanostructures, visible light can induce the shape transformation of metal nanocrystals. Mirkin, Xia et al. discovered a photo-induced method for converting small Ag nanoparticles into nanoprisms, nanoplates, and nanobelts [153–156]. For Ag nanoprisms, edge lengths in the 30–120 nm range can be controlled by using dual-beam illumination of the Ag nanoparticles [155]. The process is believed to be driven by surface plasmon excitations. The plasmon excitations may cause the photo-oxidation of citrate by hot "holes" from plasmon dephasing on the surface of Ag nanostructures, oxidative etching of Ag in the presence of oxygen, and selective reduction of aqueous Ag ions to form larger nanoprisms [157].

1.3.2.5 Electrochemical Method

Electrochemical process was employed as a synthesis method of metallic nanostructures almost 100 years ago by Svedberg [29]. Reetz et al. further developed this method in recent years to synthesize metal nanoparticles [158, 159]. Their process generally is based on a two-electrode setup. The sacrificial anode consists of a bulk metal is transformed into metal nanoparticles upon electrochemical treatment. In 1997, the electrochemical method was successfully used to synthesize single-crystalline Au nanorods with tunable aspect ratios (Fig. 1.7e) [125, 160]. Recently, electrochemical methods have also been applied to synthesize metal nanostructures with defined crystal facets [50]. Sun et al. pioneered the synthesis of platinum, palladium, rhodium, and iron nanocrystals with high-index facets and unconventional shapes using electrochemical method [50]. Repetitive adsorption/desorption of oxygen generated by square-wave potential are believed to play a key role in the formation of high-index facets. Oxygen atoms preferentially adsorb at high-index facets because high-index facets contain many step atoms with low coordination numbers. Therefore, high-index facets are preserved during the electrochemical treatment [59].

1.3.2.6 Sonochemical Method

Ultrasound has also been used as a unique tool for the synthesis of metallic nanostructures in recent years [161, 162]. A few metallic nanostructures, such as Au clusters, Ag nanoprisms, Au nanorods, dodecahedra, octahedral, and belt-like nanostructures (Fig. 1.7f) have been synthesized by using sonochemical methods [126, 163–168]. The effect of ultrasonic radiation on chemical reactions arises from a phenomenon called acoustic cavitation. Acoustic cavitation involves the formation, growth, and implosive collapse of bubbles in a liquid [169]. The transient temperature and pressure in and around the collapsing bubbles can reach as high as

5000 K and 1800 atm. Moreover, the heating and cooling rates are extremely high, which is in the order of 10^{10} K/s [170]. The extreme temperature and pressure and high energy produced during acoustic cavitation can cause the reduction or decomposition of metal precursors and lead to the formation of metallic nanostructures in a room-temperature liquid phase, eliminating the requirement of high temperatures, high pressures, or long reaction times [171]. Ultrasonic irradiation treatment of volatile precursor and non-volatile precursors follows different mechanisms. For volatile precursors such as organometallic compounds, free metal atoms will be generated by bond dissociation due to the high temperatures created during bubble collapse [170]. In the case of non-volatile precursors, ultrasonic irradiation of water can generate highly reactive H^* and OH^* radicals, which are strong reducing agents for the chemical reduction of metal salts. Moreover, these reactive radicals can further react with organic species in the solution and generate secondary radicals (R^*) [172]. The secondary radicals can dramatically promote the reduction rate of metal salts. Metallic nanoparticles synthesized with volatile organometallic compounds are generally amorphous, while metallic nanostructures produced from non-volatile precursors are usually well crystallized [161].

1.3.2.7 Reversed Micelle Method

Reverse micelles are defined as globular aggregates formed through the self-assembly of surfactants in apolar solvents [173]. A reverse micelle has a polar core and an apolar shell, which are formed by the assembly of the hydrophilic heads and the hydrophobic chains of the surfactants, respectively [174]. Water confined in the polar cores of reverse micelles are separated as many water-in-oil droplets, which act as effective nanoreactors for the growth of metallic nanostructure with tailored size, shape, and composition [175]. By modulating water to surfactant ratio, the droplet size of reverse micelles can be readily tuned in the nanometer range. An important feature of reverse micelles is that the water-in-oil droplets are not kinetically stable and a dynamic exchange process is occurring between different colliding droplets [176]. Based on this characteristic, the synthesis of metallic nanostructure in reverse micelles can be achieved by mixing two reverse micelle solutions containing hydrophilic metal salts and reducing agents, respectively [177]. After mixing, the droplets collide and exchange the metal salts and reducing agents, leading to the nucleation and growth metallic nanostructures in the water-in-oil droplets. The synthesis of metallic nanostructure in reverse micelle systems has proved to be effective for monodisperse metal nanoparticles [178, 179].

1.3.2.8 Multiphase Process

Interfaces play an important role in the nucleation and growth of metallic nanostructures [180]. By carefully designing the chemical reactions taking place at the interfaces of different phases, better control over the size and shape of metallic nanostructures can be achieved. A well-known example is the broadly used Brust–Schiffrin method [181]. The Brust–Schiffrin method is especially useful for the synthesis of ultra-small gold clusters and nanoparticles with small diameters [182]. Because of the utilization of thiols as stabilizing agents, the as-formed Au nanoparticles can be selectively functionalized and have a much better stability [183]. The nanoparticles can be repeatedly isolated and redispersed in common organic solvents without aggregation. The typical Brust–Schiffrin procedure involves three major steps: (1) phase transfer of gold species from an aqueous solution to the organic phase assisted by a long-chain quaternary ammonium surfactant; (2) reduction of Au(III) to Au(I) by thiols; and (3) reduction of Au(I) to Au(0) by a strong reducing agent such as sodium borohydride [184].

Li et al. developed a liquid–solid–solution phase transfer and separation strategy to synthesize monodispersed metal nanoparticles [185]. When aqueous solutions of noble metal salts, sodium linoleate, linoleic acid, and ethanol were mixed, three phases would form: an upper liquid phase of ethanol and linoleic acid, a middle solid phase of sodium linoleate, and a bottom solution phase of noble metal salts in water and ethanol mixed solvent. During the reaction, metal ions will enter the solid phase and form metal linoleate through ion exchange. Then ethanol could reduce the noble metal ions at the liquid–solid or solution–solid interfaces. The formed metal nanoparticles can be stabilized by in situ generated linoleic acid. Finally, the hydrophobic metal nanoparticles could be collected at the bottom of the reaction container. This method can be generalized to synthesize various noble metal nanoparticles, including Ag, Ru, Rh, Ir, Au, Pd, and Pt.

1.3.2.9 Template-Based Syntheses

Many of the previously introduced methods can only produce 0D or polyhedral metallic nanostructures with isotropic shapes. Forming anisotropic metallic nanostructures through symmetry breaking is especially challenging for metallic nanostructures because most metals possess highly symmetric crystal lattices [186]. Therefore, there are only limited strategies to control the morphology of anisotropic metallic nanostructures, such as judiciously control over the reaction conditions, selection of shape-direction agents, manipulation of growth kinetics, or formation of specifically twinned structures [187–191]. Synthesis using a pre-existing nanostructured template is an alternative and effective strategy that can achieve high degree of morphology control, especially for highly anisotropic metallic nanostructures [192, 193]. Many types of nanomaterials can be considered as templates to direct the growth of metallic nanostructures. Most of them can be classified into three major categories: soft templates, hard templates, and sacrificial templates.

Soft matters such as polymers and biological molecular assemblies can be used as templates for the growth of metallic nanostructures. For example, Crooks et al. developed a method using dendrimers as templates for the synthesis of metal nanoparticles [200]. After sequestering metal ions within dendrimers, corresponding metal nanoparticles can be synthesized through chemical reduction.

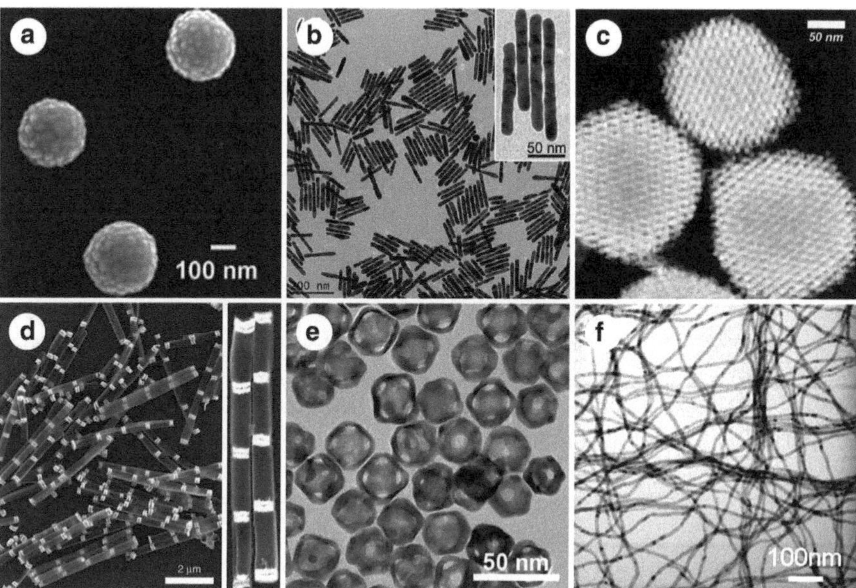

Fig. 1.8 Metallic nanostructures synthesized with template-based methods: **a** Au nanoshells with silica spheres as templates (reproduced with permission from reference [194], Copyright 2005 American Chemical Society), **b** Au nanorods with silica nanotubes as templates (reproduced with permission from reference [195], Copyright 2011 American Chemical Society), **c** mesoporous Pt nanostructures with 3D mesoporous silica as templates (reproduced with permission from reference [196], Copyright 2011 American Chemical Society), **d** metal nanostructures with controllable gaps via on-wire lithography (reproduced with permission from reference [197], Copyright 2009 Nature Publishing Group), **e** Au/Ag nanocages with Ag nanocubes as sacrificial templates (reproduced with permission from reference [198], Copyright 2013 The Royal Society of Chemistry), **f** Pt nanowires with tellurium nanowires as sacrificial templates (reproduced with permission from reference [199], Copyright 2009 Wiley-VCH)

The uniform structure of dendrimer templates allows for the generation of metal nanoparticles with uniform sizes. The dendrimer-encapsulated metal nanoparticles have a substantial fraction of unpassivated surface and may have a higher catalytic activity than common metal nanoparticles. By using lamellar bilayer membranes as 2D templates, ultrathin single-crystalline Au nanosheets were obtained in confined water layers [201, 202].

Soft templates usually have a limited effect on the shapes of the metallic nanostructures. In contrast, hard templates with specific structures have a profound effect of the as-obtained metallic nanostructures. By using silica spheres as templates, Halas et al. were able to fabricate Au nanoshells with tunable optical properties [194, 203]. To coat the silica spheres with Au nanoshells, small Au nanoparticles were first attached to the silica cores by using linker molecules. Then more gold was deposited onto the seeds until the seed nanoparticles coalesced into a complete shell (Fig. 1.8a) [204]. By varying the diameters of the silica sphere templates and the thickness of the nanoshell, the optical resonance of Au nanoshells can be systematically tuned over a broad region ranging from visible to the mid-infrared [205]. By using a similar seeding technique, Yin et al. synthesized various noble metal nanorods in silica nanotube templates [195]. Silica nanotubes can confine the growth of metal nanostructures and form a metal nanorod@silica core-shell structure. Silica can be easily etched to release dispersed metal nanorods (Fig. 1.8b). 3D mesoporous silica has also been used as templates for the growth of mesoporous metallic nanostructures (Fig. 1.8c) [196, 206, 207].

Channels in porous membranes were also employed as 1D hard templates for the synthesis of metallic nanostructures [208]. By using chemical or electrochemical methods, metal can be deposited into channels of the porous templates, after removal of the templates, nanotubes, or nanowires of a wide range of materials can be obtained [209]. Two types of porous membranes are commercially available: ion-track-etched membranes and anodic aluminum oxide (AAO) templates [210, 211]. Based on AAO templates, Mirkin et al. further developed a technique called on-wire lithography for synthesizing metallic nanostructures with controllable gaps (Fig. 1.8d) [197, 212]. They first used electrochemical deposition to prepare segmented nanowires composed of a noble metal and a sacrificial metal. After the removal of the template and deposition of a backing material on the segmented nanowires, the sacrificial metal layers can be selectively etched, leaving noble metal nanostructures with controllable gaps. The gap size is controlled by the thickness of the sacrificial metal layer [213].

Besides serving solely as templates, sacrificial templates can be also involved in chemical transformations during the synthesis of metallic nanostructures. Templates can function as reducing agents and lead to the deposition of metal on their surfaces. For example, Xia et al. employed galvanic replacement reactions to synthesize various hollow metal nanostructures (Fig. 1.8e) [214]. Ag nanostructures can be used as sacrificial templates for generating hollow nanostructures of Au, Pt, Pd, and Ag alloys [198]. Many pre-synthesized nanostructures can act as reducing agents for metal ions. For example, selenium and tellurium nanowires have been used as both reducing agent and sacrificial templates for noble metal and their alloy nanowires/nanotube (Fig. 1.8f) [199, 215]. Hollow metal nanocages have been synthesized with Cu_2O nanocrystals as templates, in which Cu_2O was oxidized into Cu(II) species while metal salts were reduced to metallic cages [216, 217].

1.4 Characterizations of Metallic Nanostructures

1.4.1 Techniques for Morphological Analysis

1.4.1.1 Transmission Electron Microscope

TEM is one of the most useful and straightforward microscopic methods to characterize the morphologies of metallic nanostructures [222]. It can be directly utilized to visualize the size, shape, structure, and dispersity of metallic nanomaterials.

TEM operates on similar working principles as the light microscope, except that TEM uses an electron beam instead of light to probe the samples. Electron beams can exhibit wavelengths that are 10,000 times smaller than those of visible light, which makes it possible to get a much better resolution using TEM [223]. In TEM, the high energy electron beam is collimated and focused by electrostatic and electromagnetic lenses, and transmitted through the sample [224]. The transmitted electron beam is magnified and focused by an objective lens and appears on the imaging screen. A more detailed schematic comparison between the structures of TEM and the light microscope can be found in the book by Bozzola and Russell [225]. High-resolution transmission electron microscope (HRTEM) has allowed the imaging of samples with resolutions below 0.5 Å (with magnifications more than 50 million times) [226]. Therefore, HRTEM can determine the position of atoms and defects of the samples, and the locations of atoms and grain boundaries can be rigorously interrogated and examined [227]. Figure 1.9a–d shows HRTEM of metal

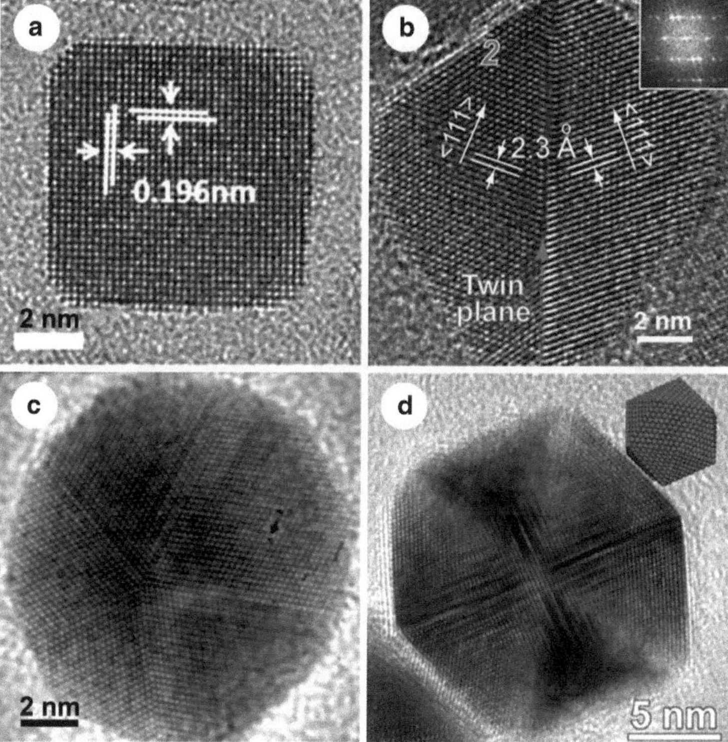

Fig. 1.9 High-resolution transmission electron microscope (HRTEM) images of metal nanoparticles with different crystal structures: **a** single-crystalline Pt nanocubes (reproduced with permission from reference [218], Copyright 2012 American Chemical Society), **b** single-twinned Pd bipyramids (reproduced with permission from reference [219], Copyright 2013 American Chemical Society), **c** penta-twinned Au nanoparticles (reproduced with permission from reference [220], Copyright 2006 The Royal Society of Chemistry), **d** icosahedral multiply twinned Pt nanoparticles (reproduced with permission from reference [221], Copyright 2013 American Chemical Society)

nanostructures with four typical crystal structures: single-crystalline Pt nanocubes [218], single-twinned Pd bipyramids [219], penta-twinned Au nanoparticles [220], and icosahedral multiply-twinned Pt nanoparticles [221], respectively.

1.4.1.2   Scanning Electron Microscope (SEM)

SEM is one of the most versatile instruments available for the examination and analysis the 3D morphology of metallic nanostructures [228]. It produces images of a sample by scanning it with a focused beam of electrons. The electrons interact with the atoms in the sample and produce various signals that contain information about the sample's surface morphology and composition [229]. To study the topographic feature of a sample, the secondary electron operation mode of SEM is commonly used, which collects the inelastically scattered electrons. When a sample is scanned with an electron beam, the secondary electrons can only escape from a thin layer near surface of the sample, therefore, the surface topography information of the sample can be provided [230]. Although the resolution of the SEM is lower than that of the TEM, it has very large depth of field and can image a bulk sample of several centimeters in size with 3D representation. By using electron beams generated with a field emission gun, it is possible to achieve up to 900 K magnification for SEM imaging [231]. Figure 1.10 shows several examples using SEM to study the assembly behaviors of Ag and Au nanostructures [8, 232, 233].

1.4.1.3   Atomic Force Microscope (AFM)

AFM is a type of scanning probe microscopy techniques for examining the surface of materials [234]. AFM can provide a direct 3D visualization of materials. Qualitative and quantitative information on the size, morphology, and surface roughness can be obtained accordingly. AFM provides the information of the topography of the sample by measuring the atomic force between the atoms at the surface of the sample and the AFM tip [235]. During the operation of AFM, a tip at the end of a cantilever is scanned across the sample surface, and the forces between the tip and the surface cause the cantilever to deflect, providing data to form images for the surface of the sample [236]. AFM can provide measurements of ultrathin nanomaterials with a vertical resolution of less than 0.1 nm with XY resolution of approximately 1 nm [237]. Figure 1.11b shows an AFM image of ultrathin Rh nanosheets [137]. Based on AFM measurement, the thickness of the Rh nanosheets is only 0.4 nm. Compared with SEM, AFM can be operated in ambient air and liquid environment. A major drawback of AFM is the limited area that it can scan: typically, AMF can only scan samples on the order of micrometer [238].

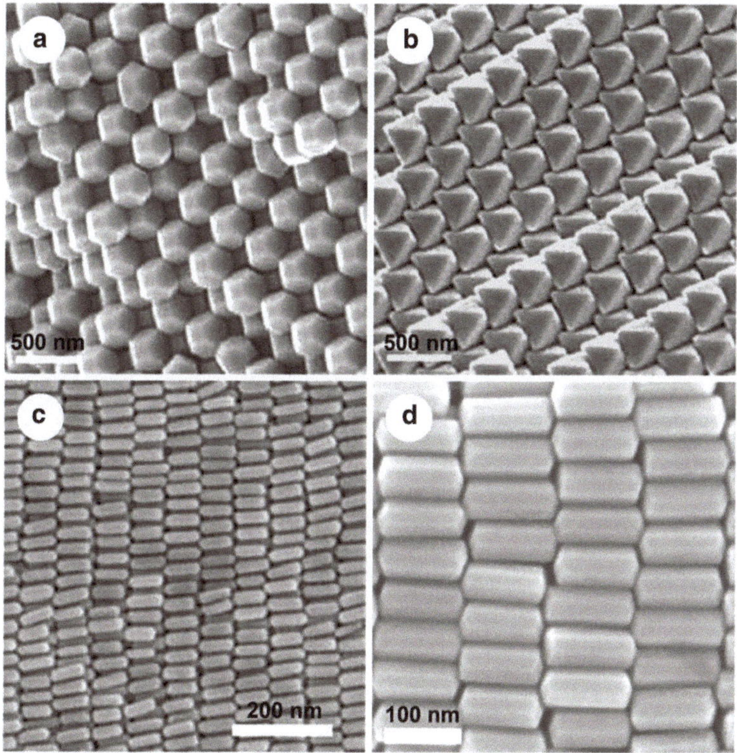

Fig. 1.10 Scanning electron microscopy (SEM) images of assemblies of metal nanostructures: **a** and **b** Ag truncated octahedral nanocrystals and Ag octahedral nanocrystals, respectively (reproduced with permission from reference [232], Copyright 2012 Nature Publishing Group); **c** and **d** Au nanorods with different aspect ratios (reproduced with permission from reference [8] and [233], Copyright 2012 and 2013 American Chemical Society)

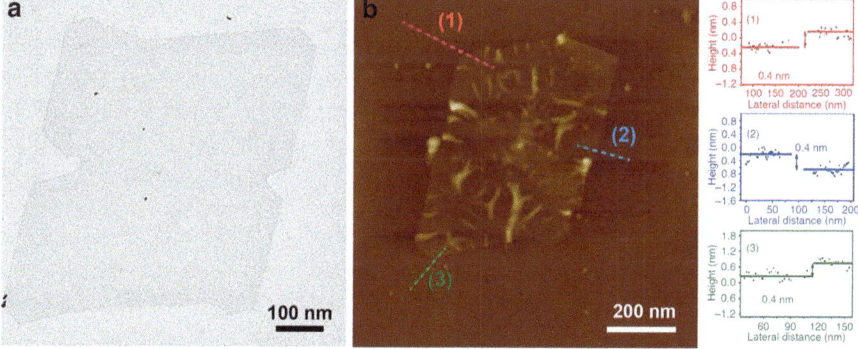

Fig. 1.11 Transmission electron microscopy (TEM) image, atomic force microscopy (AFM) image, and the corresponding height profiles of ultrathin Rh nanosheets (reproduced with permission from reference [137], Copyright 2014 Nature Publishing Group)

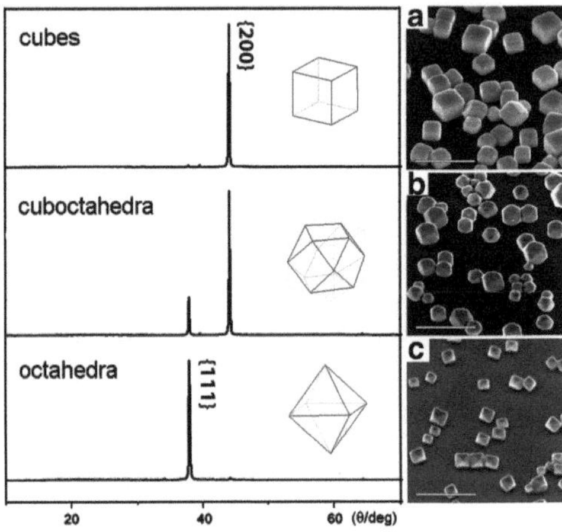

1.4.2 Techniques for Crystal Structural Analysis

1.4.2.1 X-Ray Diffraction (XRD)

XRD is a non-destructive technique to characterize the crystallographic structure, grain size, and preferred orientation in solid samples [239]. Powder diffraction is commonly used to identify unknown crystalline substances. It may also be used to characterize heterogeneous solid mixtures to determine the relative abundance of crystalline materials [238]. During XRD testing, a monochromatic X-ray directed onto a sample and the interaction between X-rays and different crystal planes of the sample will lead to the diffraction of X-rays [240]. By recording the scattered intensity of the X-ray beam as a function of incident and scattered angle, a spectrum will be obtained. Two types of fingerprint information of a particular crystalline material are provided in the spectrum: the peak positions (corresponding to lattice spacing according to Bragg's law) and the relative intensity of the peaks [231]. XRD is also useful in size measurement of structural nanocrystalline materials using Scherrer equation. The Scherrer equation is a formula that relates the size of sub-micrometer crystallites to the broadening of a peak in a diffraction pattern [241]. It can determine the size of crystalline nanoparticles in powder samples. XRD is capable of investigating the structural information of metal nanocrystals. For example, Fig. 1.12 shows the XRD patterns and corresponding SEM images of Au nanocrystals with different shapes [242]. When the shapes of Au nanocrystals gradually evolved from cubes with {100} facets to octahedra with {111} facets, their XRD patterns also show the decrease of {200} peaks and the increase of {111} peaks.

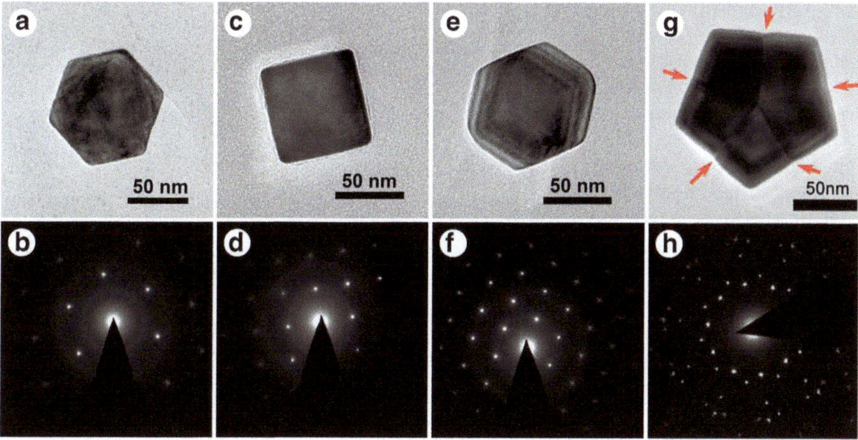

Fig. 1.13 Transmission electron microscopy images and corresponding selected area electron diffraction patterns of different metallic nanostructures: **a, b** an octahedral Pd nanocrystal recorded along the [$\bar{1}$11] zone axis; **c, d** a cubic Pd nanocrystal recorded along the [001] zone axis; **e, f** a rhombic dodecahedral Pd nanocrystal recorded along the [011] zone axis (reproduced with permission from reference [247], Copyright 2010 American Chemical Society); and **g, h** a penta-twinned Au nanocrystal recorded along the [011] zone axis (reproduced with permission from reference [248], Copyright 2013 The Royal Society of Chemistry)

1.4.2.2 Selected Area Electron Diffraction (SAED)

SAED is a crystallographic experimental technique that usually coupled with TEM microscopes for microanalysis [243]. Unlike XRD, which can only provide crystallographic structure of bulk samples, SAED can provide crystallographic information on the single nanostructure level. SAED analysis can reveal whether a material is polycrystalline, single-crystalline, or amorphous. A crystalline material will produce a diffraction pattern composed of several spots. The center spot is associated with the non-diffracted transmitted electron beam, whereas all the other spots are diffracted spots [244]. These diffraction spots represent the beams diffracted from different sets of planes. The distance between the diffraction spots and center spot is $1/d_{hkl}$, where d_{hkl} is the interplanar spacing between (hkl) planes [245]. If the sample is polycrystalline, in which individual crystal domains are oriented in different directions, the diffraction pattern is the overlap of diffraction spots from each individual crystal domain. Small crystal size and large numbers of crystals will lead to the formation of continuous diffraction rings [246]. Figure 1.13a–f show the TEM images and corresponding SAED patterns of single-crystalline octahedral, cubic, and rhombic dodecahedral Pd nanocrystals, respectively [247]. Their characteristic diffraction patterns are indicative of their crystal facets of {111}, {100}, and {110}, respectively. In Fig. 1.13g and h, a SAED pattern of a penta-twinned Au nanocrystal along the [110] direction is shown [248]. The penta-twinned Au nanocrystal is formed by five single crystals with {111} twin planes as indicated in Fig. 1.13g. The diffraction pattern can be interpreted by superimposing the [110] SAED pattern of each crystal rotated by 72° with respect to each other [220].

Fig. 1.14 X-ray photoelectron spectroscopy (XPS) analysis of Ag adsorption on Au nanocrystals with different crystal facets: **a** models of Ag atoms adsorbed on different Au facets (*Blue spheres* represent surface atoms), **b** Ag/Au ratio for each facets obtained from XPS data (*black*), and theoretical values for monolayer coverage of Ag based on models in (**a**). (reproduced with permission from reference [253], Copyright 2011 American Chemical Society)

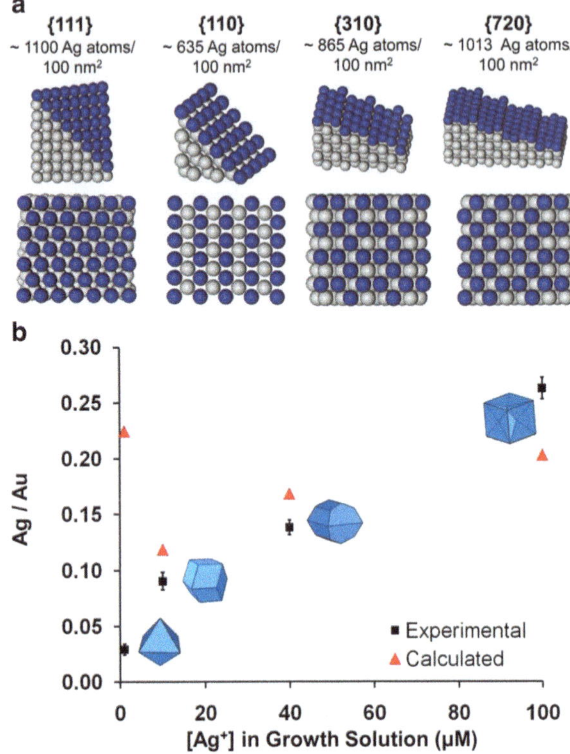

1.4.3 Techniques for Composition Analysis

1.4.3.1 X-ray Photoelectron Spectroscopy (XPS)

XPS is a spectroscopic technique to analyze the materials within several nanometers from the surface [249]. XPS is an important analytical technique that can probe the chemical state and chemical composition of a sample. XPS can detect elements with atomic number larger than 3, with parts per million detection limit [226]. During XPS measurement, the system is irradiated with a monochromatic X-ray beam to promote the emission of photoelectrons of the surface atoms [2]. The emitted photoelectrons are then collected and analyzed using an electron spectrometer. XPS spectra are recorded by plotting the number of photoelectrons escaped from the surface of the sample versus the binding energy. The binding energy of the emitted photoelectrons can be calculated from the energy of the incident beam and the measured kinetic energy of the electrons [250]. The chemical shifts in the binding energies of the photoelectrons are indicative of the chemical states of the detected elements [251]. The short mean free path of the emitted electrons in solids makes XPS a surface-sensitive method to analyze the surface composition of metallic nanostructures [252]. For example, Mirkin et al. have used XPS to investigate the surface coverage of Ag atoms on different Au crystal facets and illustrate the function of Ag underpotential deposition in the synthesis of Au nanocrystals (Fig. 1.14) [253].

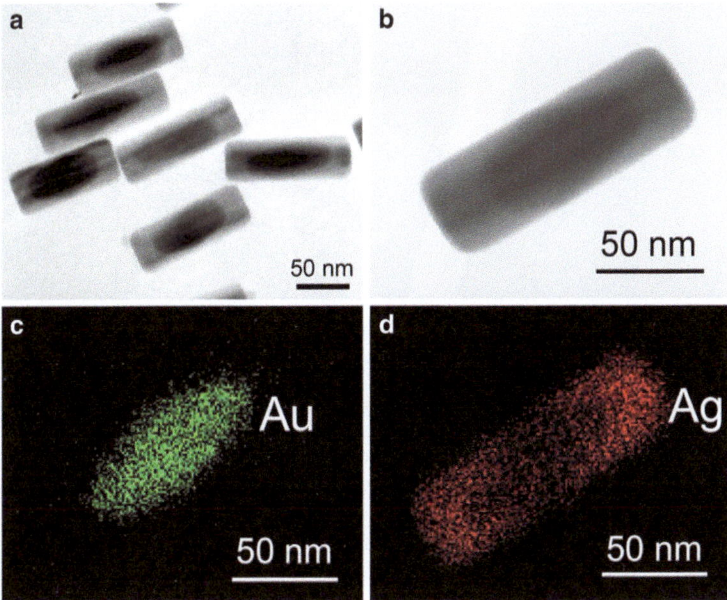

Fig. 1.15 **a**, **b** Transmission electron microscopy images of Au@Ag core-shell nanorods and **c**, **d** Energy-dispersive X-ray spectroscopy elemental mapping of Ag and Au for the Au@Ag nanorod shown in (**b**) (reproduced with permission from reference [257], Copyright 2013 Wiley-VCH)

1.4.3.2   Energy-Dispersive X-Ray Spectroscopy

Energy-dispersive X-ray spectroscopy (EDX, EDS, or EDAX) is an analytical technique used for the elemental analysis of a sample [254]. It is based on the analysis of characteristic X-ray lines from the sample when exposed with a beam of charged particles such as electron beams. Electron beams can remove an inner shell electron from the sample and create a vacancy. Then an electron of higher states fill the resulted vacancy and release the energy difference between these two states of electrons in the form of X-rays [255]. The number and energy of the X-rays emitted from a specimen are measured by an energy dispersive spectrometer. An EDX spectrum is a plot of intensity versus X-ray energy. The energy of emitted X-rays is distinctive characteristic of each element, therefore, the EDX spectra can be used to identify the composition and measure the quantity of elements in the sample [256]. EDX systems are most commonly found as a feature of SEM or TEM. Therefore, the composition analysis can be coupled with the morphology of the nanomaterials through different scanning modes such as spot, linear, or mapping scan. For example, EDX mapping has been used to study the structure of Au@Ag core-shell nanorods (Fig. 1.15) [257].

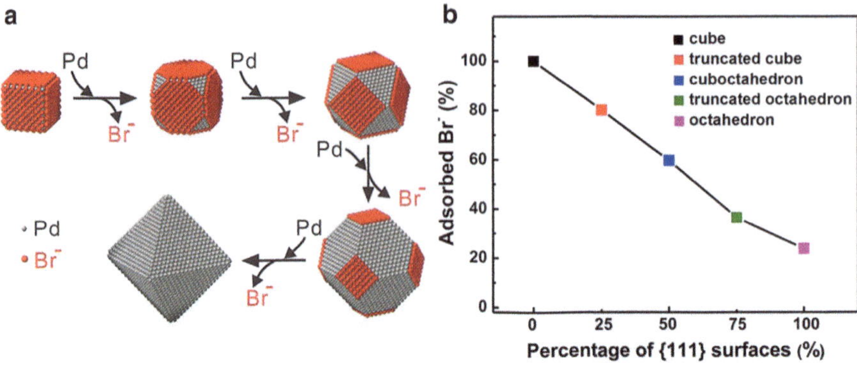

Fig. 1.16 **a** Schematic illustration for the processes of bromide ion desorption during the Pd overgrowth on Pd {100} surfaces, **b** inductively coupled plasma mass spectrometry analysis of the residual bromide ions at different growth stages, indicating that bromide ions desorbed from the Pd {100} surfaces during the Pd overgrowth (reproduced with permission from reference [264], Copyright 2013 American Chemical Society)

1.4.3.3 Inductively Coupled Plasma Atomic Emission Spectroscopy/Mass Spectrometry

Inductively coupled plasma atomic emission spectroscopy (ICP-AES, also called inductively coupled plasma optical emission spectrometry—ICP-OES) [258] and inductively coupled plasma mass spectrometry (ICP-MS) [259] are two of the most popular methods for quantitative elemental analysis of trace metals. Both techniques are based on inductively coupled plasma (ICP) techniques. ICP is a high temperature excitation source that can efficiently excite and ionize atoms [260]. Liquid samples are dried, vaporized, atomized, ionized, and finally transformed into a fine liquid aerosol after it travels through the plasma torch [261]. The excited atoms and ions that represent the elemental composition of the sample are ready to be quantified by atoms emission spectroscopy (AES) or mass spectrometry (MS). In the case of AES, the excited atoms and ions for a particular element emit electromagnetic radiation at characteristic wavelengths. The intensity of the radiation measured at specific wavelengths is proportional to the concentration of the corresponding element in the solution. Most elements can be quantitatively measured using ICP-AES with sensitivity down to parts per billion [262]. In ICP-MS, the sample is ionized with ICP and then the formed ions are separated and quantified by a mass spectrometer. The intensity of a specific peak in the mass spectrum is proportional to the amount of the element in the original sample. ICP-MS is capable of detecting metals and several non-metals at concentrations as low as part per trillion [263]. The ICP methods offer multi-element analytical capability within a broad dynamic range and good precision [261]. Xia et al. studied the bromide ion-assisted growth of Pd nanocrystals with ICP-MS. During the growth of Pd cubic seeds into cuboctahedra and then octahedra, the bromide ions are gradually released from the surface of Pd nanocrystals, as illustrated by the ICP-MS results in Fig. 1.16 [264]. This quantitative information can be used to estimate the amount of bromide ions needed to generate Pd nanocrystals with {100} and {111} facets of desired ratios.

1.5 Representative Examples for Shape-Controlled Synthesis of Metallic Nanostructures

1.5.1 Seed-Mediated Growth Methods

Seed-mediated growth methods have been broadly used to control the sizes and morphologies of metallic nanostructures [117, 265]. In seed-mediated growth methods, the nucleation and growth stages of metallic nanostructures are well separated, therefore, a better control over the size distribution and morphology evolution of metallic nanostructures can be readily achieved [118–120]. The seeding strategy has been used to synthesize Au nanoparticles since the time of Zsigmondy [28]. In the 1990s, Brown and Natan developed a method to synthesize monodisperse Au nanoparticles based on Au nanoparticle-catalyzed reduction of Au salts by hydroxylamine [266–268]. They have described the observation of some rod-like Au nanostructures in their reports. However, it is until the year of 2001 when seed-mediated growth methods were rediscovered as an efficient strategy for controlling the morphologies of metallic nanostructures [269, 270]. The Murphy group first reported the synthesis of Ag and Au nanorods using seed-mediated growth methods with CTAB as a surfactant and ascorbic acid as a reducing agent. In the case of Au nanorods, they developed a three-step protocol that yielded long Au nanorods with penta-twinned structures (Fig. 1.17a) [270–272]. Although a high yield of Au nanorods could be achieved after shape separation, the actual purity of nanorods in the original products was only 5–10 % [273]. In 2002, the El-Sayed group developed an alternative seed-mediated approach to synthesize Au nanorods. By using single-crystalline Au nanoparticles as seeds and silver nitrate as a shape-directing agent, single-crystalline Au nanorods with high yields approaching 100 % were synthesized [274]. Moreover, by changing the amount of silver nitrate, the aspect ratio of Au nanorods could be tuned and this leads to tunable plasmonic properties. Subsequent studies have made a few variations of the initial reaction conditions of Au nanorods to achieve more control and better monodispersity (Fig. 1.17b) [275–279]. For example, the Wang group explored the effect of chain lengths of alkyl group of surfactants and was able to synthesize Au nanorods with very high aspect ratios [280, 281]. The Murray group has explored aromatic additives [8], binary surfactant mixture composed of CTAB and sodium oleate [282], and bromide-free surfactant composed of alkyltrimethylammonium chloride and sodium oleate for the synthesis of Au nanorods [233], respectively. These studies have achieved large-scale synthesis of Au nanorods with ultrahigh monodispersity and purity.

In 2004, another report from the Murphy group described the application of seed-mediated growth methods in synthesizing multiple shapes of Au nanostructures, such as rectangle-, hexagon-, cube-, triangle-, and star-like morphologies [283]. Shape control by manipulation of the kinetic and thermodynamic parameters of Au

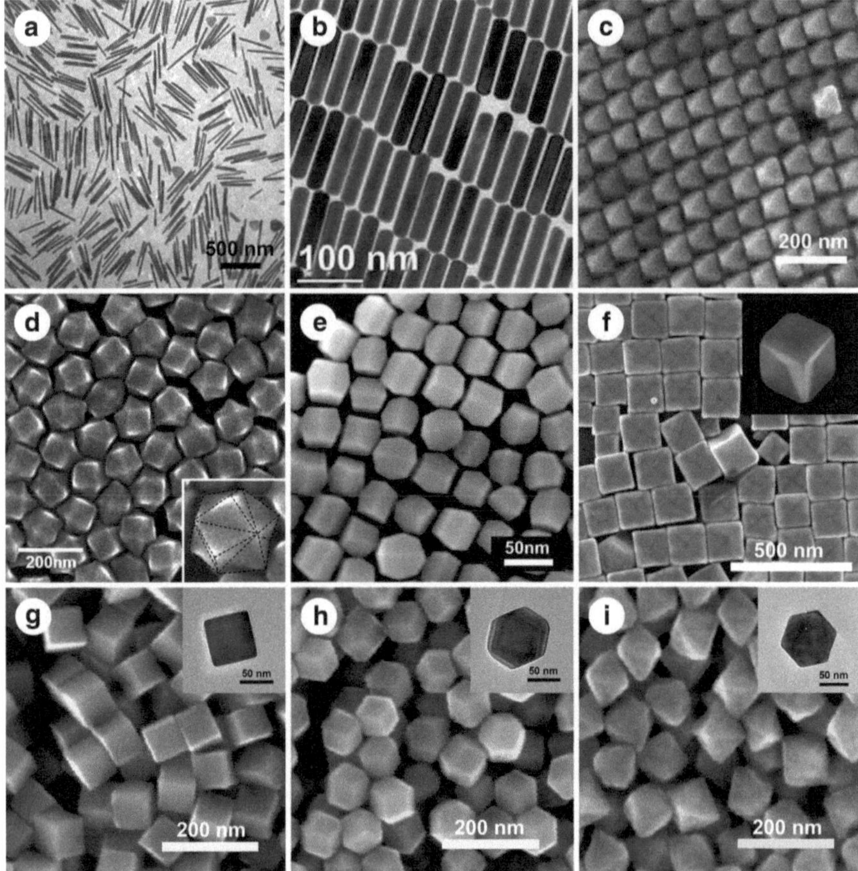

Fig. 1.17 Metallic nanostructures synthesized with seed-mediated growth methods: **a** penta-twinned Au nanorods via three-step growth (reproduced with permission from reference [272], Copyright 2004 American Chemical Society), **b** single-crystalline Au nanorods via Ag(I)-assisted growth (reproduced with permission from reference [282], Copyright 2013 American Chemical Society), **c** Au octahedra (reproduced with permission from reference [284], Copyright 2008 American Chemical Society), **d** Au trisoctahedral nanocrystals (reproduced with permission from reference [287], Copyright 2010 American Chemical Society), **e** Au truncated ditetragonal nano-prisms (reproduced with permission from reference [293], Copyright 2011 American Chemical Society), **f** concave Au nanocubes (reproduced with permission from reference [295], Copyright 2010 American Chemical Society), **g–i** Pd cubic, rhombic dodecahedral, and octahedra nano-crystals (reproduced with permission from reference [247], Copyright 2010 American Chemical Society)

nanostructures has received considerable attention since this study. The synthesis of a variety of Au nanostructures such as cubes, octahedra (Fig. 1.17c), rhombic do-decahedra, and trisoctahedra (Fig. 1.17d) has been rationalized by correlating their shapes and growth kinetics and thermodynamics [284–287]. In 2005, the function of silver nitrate in the synthesis of Au nanorods was revealed as an underpotential deposition (UPD) mechanism [279]. The UPD of foreign metal atoms on metal nanostructures can stabilize high-energy facets and has been an important strategy to synthesize metallic nanostructures with high-energy facets [253, 288, 289]. Elongated tetrahexahedra with {730} facets [290], tetrahexahedra with {520} facets [291], rhombic dodecahedra with {110} facets [292, 293], truncated ditetragonal prisms with {310} facets (Fig. 1.17e) [293], hexagonal bipyramids with {210} facets [294], and concave cubes with {720} facets (Fig. 1.17f) [295] have been synthesized based on the UPD mechanism. UPD was also employed to control the architectural diversity of heterogeneous metallic nanostructures [296].

Seed-mediated growth methods have been successfully applied in synthesizing Pd and various core-shell metallic nanostructures [10, 247, 257, 297–306]. For example, rhombic dodecahedral, cubic, and octahedral Pd nanocrystals, as well as their derivatives with varying degrees of edge- and corner-truncation were synthesized through manipulation of the concentration of KI and the reaction temperature (Fig. 1.17g–i) [247]. Polyhedral Au@Pd core-shell nanocrystals including concave trisoctahedra with {hhl} facets, concave hexoctahedra NCs with {hkl} facets, and tetrahexahedra NCs with {hk0} facets were synthesized under careful control of their growth kinetics [303]. Au@Ag core-shell nanocrystals with cubic, truncated cubic, cuboctahedral, truncated octahedral, and octahedral shapes were synthesized by regulating the growth rates of different crystal facets [300].

1.5.2 The Polyol Process

The polyol process is one of the most versatile methods to synthesize metal nanostructures with controllable morphologies [72, 307]. Nanostructures of many different metals including Ag, Au, Pt, Pd, and Rh have been synthesized with polyol method [72]. In polyol processes, various polyols such as ethylene glycol, 1,5-pentanediol, di(ethylene glycol), and triethylene glycol are used as both the solvent and the reducing agent for the reduction of metallic precursors [187, 308, 309]. Polyvinylpyrrolidone (PVP) is commonly used as a stabilizing agent and sometimes a shape-directing agent for the growth of metallic nanostructures. The rapid development of the polyol process in the synthesis of metallic nanostructures dates back to the seminal report of polyol synthesis of Ag nanocubes by the Xia group in 2002 [310]. In the report, monodispersed Ag nanocubes were synthesized by reducing silver nitrate with ethylene glycol in the presence of PVP (Fig. 1.18a). In their subsequent studies, rational synthesis of Ag nanostructures was demonstrated by a mechanism called oxidative etching [311]. During the growth of Ag nanostructures, oxygen and halides can selectively remove Ag seeds with certain crystal structures. In the case of Ag nanocubes, twinned seeds can be removed by oxygen and chloride due to their high activity [312, 313]. Thus, a high yield of Ag nanocubes can be obtained. If the reaction was performed without oxidative etching, penta-twinned Ag nanowires can be synthesized (Fig. 1.18b) [314]. By using bromide and oxygen as a weaker etchant, single-twinned Ag right bipyramids were synthesized (Fig. 1.18c)

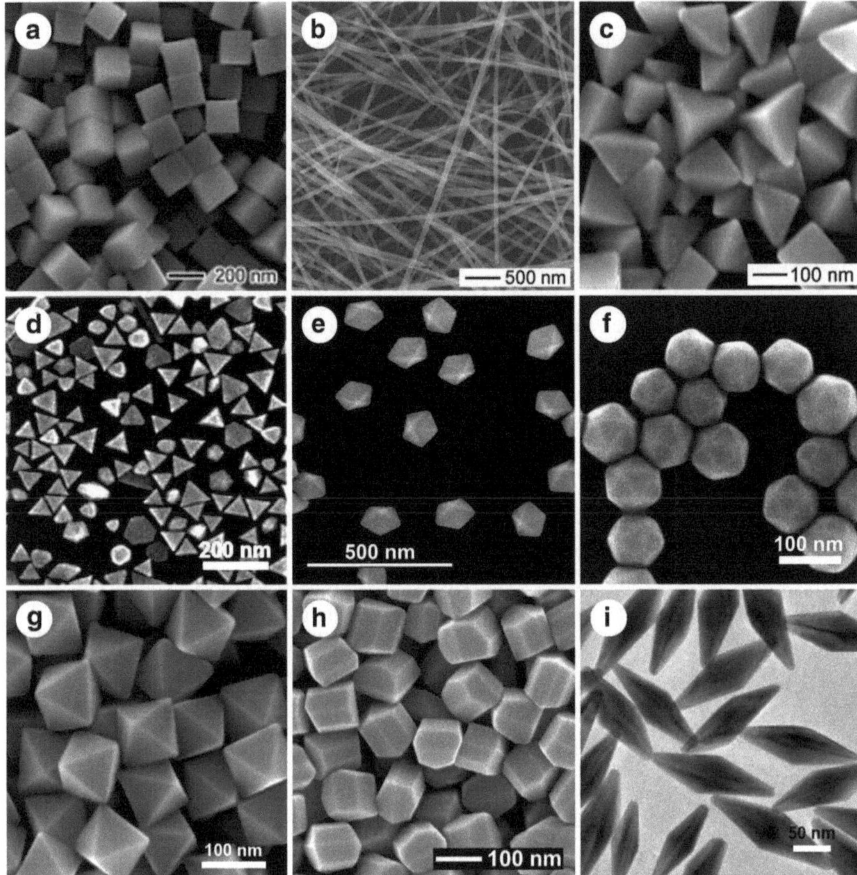

Fig. 1.18 Metallic nanostructures synthesized via polyol processes: **a–c** Ag nanocubes (reproduced with permission from reference [328], Copyright 2005 Wiley-VCH), nanowires (reproduced with permission from reference [312], Copyright 2004 American Chemical Society), and right bipyramids (reproduced with permission from reference [315], Copyright 2006 American Chemical Society), **d–f** Au plate-like, decahedral (reproduced with permission from reference [308], Copyright 2008 American Chemical Society), and icosahedral nanocrystals (reproduced with permission from reference [329], Copyright 2014 American Chemical Society), **g–h** Au octahedra (reproduced with permission from reference [323], Copyright 2008 American Chemical Society), truncated ditetragonal prisms (reproduced with permission from reference [324], Copyright 2011 American Chemical Society), and **i** multiple-twinned bipyramids (reproduced with permission from reference [325], Copyright 2013 The Royal Society of Chemistry)

[315]. A similar strategy is also applied in the synthesis of Pd nanostructure such as Pd nanocubes, nanorods, and nanobars [316–318]. By blocking the oxidative etching of Pd nanostructure with citric acid, multiple-twinned Pd icosahedra were obtained [319]. The Song group showed that by tuning the oxidative etching rate of Au seeds, Au nanocrystals with decahedral, icosahedral, and truncated tetrahedral shapes can be synthesized (Fig. 1.18d–f) [308].

The growth of metallic nanostructures through the polyol process is very sensitive to growth kinetics and inorganic species. For example, the Yang group have shown that the synthesis of Ag nanocrystals with shapes from cubes, to cuboctahedra and octahedra with the assistance of copper(II) chloride [309]. Different shapes can be obtained at different stages of the reaction. By introducing silver nitrate in the growth of Au nanocrystals, Au nanocrystals with continually tunable shapes including octahedra, truncated octahedra, cuboctahedra, cubes, and higher polygons were synthesized, which is attributed to the UPD of silver on the surface of Au nanostructures [242, 320, 321]. By introducing different amounts of silver nitrate, the shapes of platinum nanocrystals could also be tunable across octahedra, cuboctahedra, and cubes [322].

An alternative stabilizing agent for the polyol process is poly(diallyl dimethylammonium chloride) (PDDA). Li et al. developed a PDDA-mediated polyol route to synthesize Au octahedral nanocrystals with high yields and tunable sizes (Fig. 1.18g) [323]. PDDA was believed to selectively stabilize the {111} facets of the Au octahedra. By introducing silver and palladium ions into the growth of Au nanostructures, truncated tetrahexahedra enclosed by {310} and {111} facets, truncated ditetragonal prisms enclosed by {310} facets (Fig. 1.18h), and penta-twinned bipyramids could be obtained [324, 325]. Penta-twinned Au decahedra, Au@Ag core-shell nanorods, and penta-twinned bipyramids (Fig. 1.18i) were also synthesized through Ag(I)-assisted PDDA-mediated polyol processes [325–327].

1.5.3 N,N-Dimethylformamide-Mediated Syntheses

Similar to polyols, N,N-dimethylformamide (DMF) can also be used as a solvent and reducing agent for the synthesis of metallic nanostructures [335]. DMF was first applied as a reaction medium for the synthesis of metallic nanostructures by the Liz-Marzán group in 1999 [336]. In a later work, they showed that Ag nanoplates could be synthesized by using a high silver nitrate concentration and a low PVP concentration (Fig. 1.19a) [330]. Gao et al. synthesized Ag decahedra in DMF with PVP of high molecular weights as a stabilizing agent (Fig. 1.19b) [331]. By increasing the amount of PVP, tetrahedra Ag nanocrystals were obtained. The Xie group reported the synthesis of decahedral Au nanostructures in DMF [337]. By introducing sodium hydroxide, rhombic dodecahedral Au nanocrystals were obtained. Further increasing of the reaction temperature leads to the formation of Au octahedra. Moreover, hexagonal nanosheets were synthesized at a high concentration of Au salts. Subsequent studies have synthesized a few Au nanocrystal enclosed by {110} facets in DMF, including rhombic dodecahedra (Fig. 1.19c), squashed dodecahedra, truncated pentagonal bipyramids (Fig. 1.19d), and Platonic dodecahedra (Fig. 1.19e) [248, 329, 332, 338]. DMF or one of its oxidation products during the reactions is responsible to stabilize the high-energy {110} facets. Xia et al. showed that in an ethylene glycol and DMF mixture, Pt nanowire assemblies composed of ultrathin Pt nanowires were synthesized under solvothermal conditions [122].

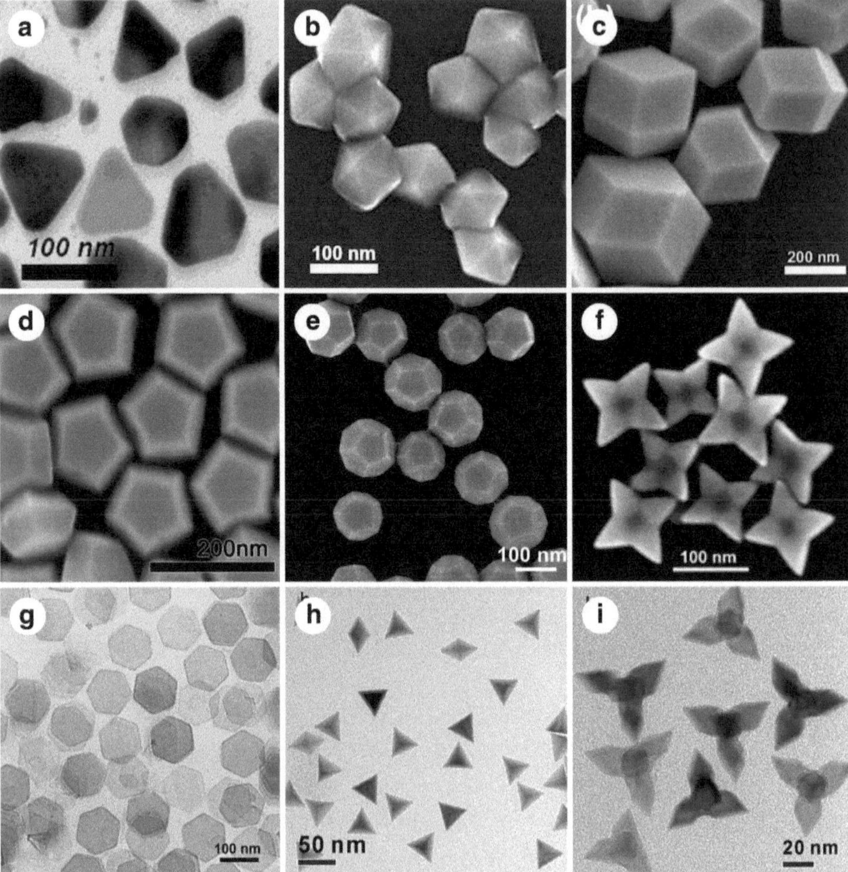

Fig. 1.19 Metallic nanostructures synthesized via N,N-dimethylformamide (DMF)-mediated syntheses: **a** Ag nanoplates (reproduced with permission from reference [330], Copyright 2002 American Chemical Society), **b** Ag decahedral nanocrystals (reproduced with permission from reference [331], Copyright 2006 Elsevier B.V.), **c–e** Au rhombic dodecahedra (reproduced with permission from reference [332], Copyright 2009 American Chemical Society), truncated pentagonal bipyramids (reproduced with permission from reference [248], Copyright 2013 The Royal Society of Chemistry), and Platonic dodecahedra (reproduced with permission from reference [329], Copyright 2014 American Chemical Society), **f** concave Pt nanocrystals (reproduced with permission from reference [333], Copyright 2011 American Chemical Society), **g–i** Pd nanosheets (reproduced with permission from reference [9], Copyright 2011 Nature Publishing Group), tetrahedra, and tetrapods (reproduced with permission from reference [334], Copyright 2012 American Chemical Society)

A few small molecular adsorbates have been introduced into DMF-mediated reduction to control the morphologies of metallic nanostructures [339]. The Zheng group reported the synthesis of Pt NCs with {411} high-index facets with the addition of short chain amines (Fig. 1.19f). Amines are believed to stabilize the open structures on the {411} surface [333]. By introducing CO, palladium nanosheets with less than 10-atomic-layer thickness were obtained, due to the growth confinement effect of CO (Fig. 1.19g) [9]. If hydrogen is introduced in addition to CO, single-crystalline Pd tetrapod and tetrahedral nanocrystals enclosed by (111) surfaces can be obtained (Fig. 1.19h–i) [334]. This is due to the fact that CO can stabilize the {111} facets of β-PdH$_x$ nanocrystals, which formed in the presence of H$_2$ but transformed into pure Pd nanocrystals in air. Cui et al. also reported that CO could be used to confine the growth of Pt–Ni alloy nanoparticles [340].

1.5.4 Oleylamine-Mediated Syntheses

Oleylamine (cis-1-amino-9-octadecene) is a long-chain primary alkylamine broadly used in the synthesis of metallic nanostructures [341]. A variety of metallic nanostructures including Au, Ag, Pt, Pd, and their alloy has been synthesized in the presence of oleylamine. It has several important functions in the synthesis of metallic nanostructure. Oleylamine can simultaneously function as a solvent for the reaction mixture and stabilizing agent for metallic nanostructures. In addition, oleylamine has a boiling point of 364 °C; therefore, it can be used to synthesize metallic nanostructures at high temperatures. Peng et al. used Fourier transform infrared spectroscopy to study the presence of oleylamine on Ag nanoparticles and observed a new peak of Ag–N bonds [342]. Oleylamine-stabilized metallic nanostructures can be easily dispersed in non-polar solvents assisted by the long hydrocarbon chains of the oleylamine molecules. The Sun group has done extensive research on the oleylamine-mediated syntheses of metal nanostructures. For example, monodispersed FePt nanoparticles were synthesized by simultaneous reduction of platinum acetylacetonate and decomposition of iron pentacarbonyl in the presence of oleic acid and oleylamine (Fig. 1.20a) [343]. The composition of the as-synthesized FePt nanoparticles can be controlled by changing the ratio of iron and platinum precursors. Based on oleylamine, they have developed the synthesis protocol for many metallic nanostructures, such Pt, Pd, AgPd, CoPt, NiPt, CuPt, ZnPt, FePtAu alloy nanoparticles, various core-shell structures such as Pd/Au and Pd/Au/FePt core/shell nanoparticles, and Pt-Au heterostructures [344–351].

A few strategies for the shape control of metallic nanostructures have been reported in the oleylamine-mediated syntheses. A typical example is the synthesis of ultrathin Au nanowires in the presence of oleylamine (Fig. 1.20b). Au nanowires can be synthesized in the presence of oleylamine under mild reducing conditions. 1D mesostructures formed between HAuCl$_4$ and oleylamine are believed to act as templates for the growth of Au nanowires [353, 356–360]. Mild reducing conditions can preserve the 1D mesostructure and convert Au salts into metallic nanowires. The Sun group discovered that oleylamine can induce the 1D growth of FePt and CoPt nanostructures (Fig. 1.20c) [354, 361]. Based on this process, 1D FePt, CoPt, FePtCu, FePtNi, and FePtPd nanostructures have been synthesized [354, 361–364]. Ag nanocubes were also synthesized by oleylamine-mediated processes based on oxidative etching process [365–367]. For example, dimethyldistearylammonium

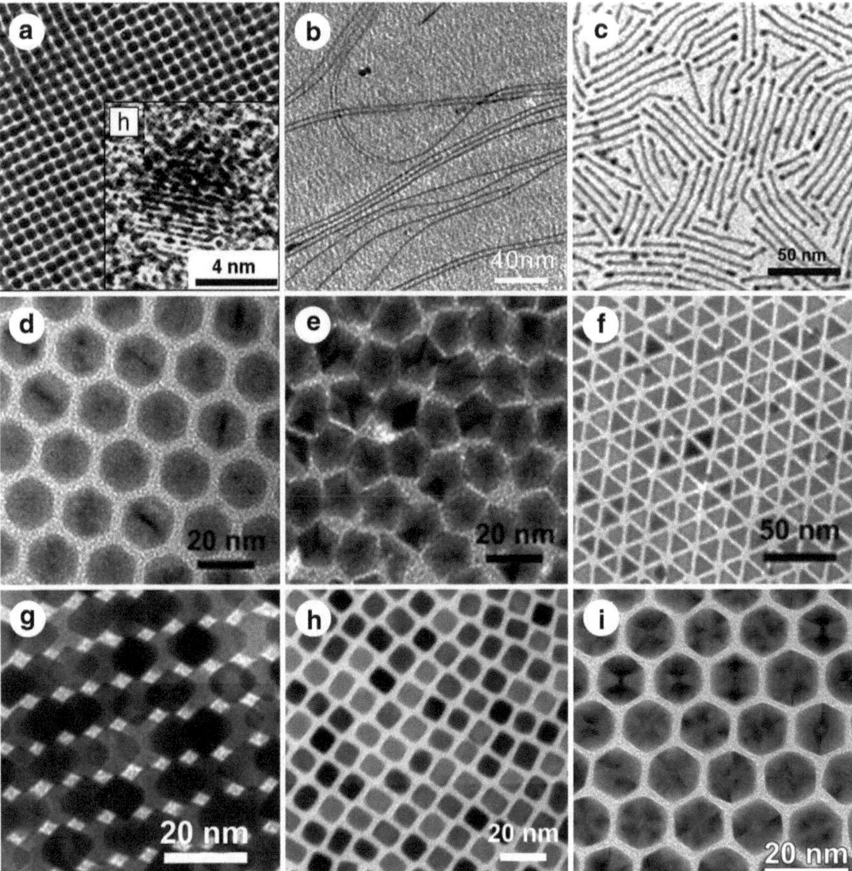

Fig. 1.20 Metallic nanostructures synthesized via oleylamine-mediated synthesis: **a** FePt nanoparticles (reproduced with permission from reference [352], Copyright 2001 Materials Research Society), **b** ultrathin Au nanowires (reproduced with permission from reference [353], Copyright 2008 American Chemical Society), **c** ultrathin FePt nanorods (reproduced with permission from reference [354], Copyright 2007 Wiley-VCH), **d–f** Pd icosahedra, decahedra, and triangular plates (reproduced with permission from reference [355], Copyright 2011 Wiley-VCH), **g–i** Pt octahedra, cubes (reproduced with permission from reference [218], Copyright 2012 American Chemical Society), and icosahedra (reproduced with permission from reference [221], Copyright 2013 American Chemical Society)

chloride and Fe(III) species were introduced to remove twinned seeds and facilitate the formation of exclusively single-crystalline Ag nanocubes [365, 367]. Pd icosahedral, decahedral, octahedral, tetrahedral, and triangular plate-like Pd NCs were selectively synthesized by introducing formaldehyde and varying the quantities of oleylamine (Fig. 1.20d–f) [355].

Metal carbonyls are an important class of shape-directing agents in oleylamine-mediated syntheses of metal nanostructures. Tungsten hexacarbonyl, iron pentacarbonyl, chromium hexacarbonyl, and dimanganese decacarbonyl have been explored to control the shapes of metallic nanostructures [218, 368–375]. The Fang group has developed a general strategy to synthesize Pt_3Fe, Pt_3Co, Pt_3Ni nanocubes, and Pt_3Ni nanoctahedra based on tungsten hexacarbonyl-assisted growth in the oleylamine-mediated syntheses [368–370, 374]. The Murray group reported the synthesis of multiple Pt nanostructures with the assistance of dimanganese decacarbonyl (Fig. 1.20g–h) [218]. These structures include truncated cubes, cuboctahedra, spheres, tetrapods, star-shaped octapods, multipods, and hyper-branched structure. Wang and Loukrakpam et al. show that Pt nanocubes could be synthesized in the presence of iron pentacarbonyl and chromium hexacarbonyl, respectively [372, 373]. In these studies, the zero-valence metals from the metal carbonyls were considered to play an essential role for the shape control of Pt nano-cubes. Wu et al. proposed that CO from the decomposition of metal carbonyls also plays a major role and promotes the formation of Pt nanocubes [112]. The Murray and Yang groups also directly introduced CO in the oleylamine-mediated syntheses of metallic nanocrystals. A variety of metallic nanostructures have been synthesized through this versatile method, including Pt nanocube, Au nanowires, Pd nanopar-ticles, icosahedral nanocrystals of Pt (Fig. 1.20i), PtAu, Pt_3Ni, Pt_3Pd, PtPd, nanocu-bes of Pt_3Ni, PtNi, and $PtNi_3$, octahedral nanocrystals of Pt_3Ni, PtNi, and $PtNi_3$ [7, 221, 376, 377]. This method is named by the Yang group as a gas-reducing agent in liquid solution method. The Tilley group used hydrogen instead of CO as a reducing agent in similar conditions and synthesized monodispersed Pt concave nanocubes and highly branched Pd nanostructures [378–380].

1.5.5 *Plasmon-Mediated Syntheses*

The interaction of light and plasmonic metallic nanostructures can be employed as a novel tool to synthesize metallic nanostructures with controllable morpholo-gies, especially in the case of Ag nanostructure [152]. In 2001, the Mirkin group first reported that plasmonic photo-excitation of Ag nanostructures could be used to synthesize monodispersed Ag nanoprisms [156]. In their synthesis, spherical Ag nanoparticles were first prepared by the reduction of silver nitrate by sodium boro-hydride in the presence of trisodium citrate and bis(p-sulfonatophenyl) phenylphos-phine dihydrate dipotassium salt (BSPP). Irradiating the Ag nanoparticle solution with visible light results the gradual conversion of spherical nanoparticles into tri-angular nanoprisms. A plasmon-induced electron transfer mechanism is believed to be responsible for the synthesis of triangular Ag nanoprisms [157, 389–391]. The plasmon excitation of spherical Ag nanoparticles will result in the generation of en-ergetic electron-hole pairs from plasmon decay, which could catalyze the reduction of Ag^+ by citrate, leading to the conversion of spherical nanoparticles into triangular nanoprisms. In the meantime, smaller Ag nanoparticles were oxidized into Ag^+ for the continuous growth of Ag nanoprisms. By using dual-beam illumination, they can improve the synthesis of monodisperse Ag nanoprisms with desired edge lengths in the range of 30–120 nm (Fig. 1.21a) [155]. They also found that edge lengths of the nanoprisms can also be controlled by the pH of the growth solutions [392].

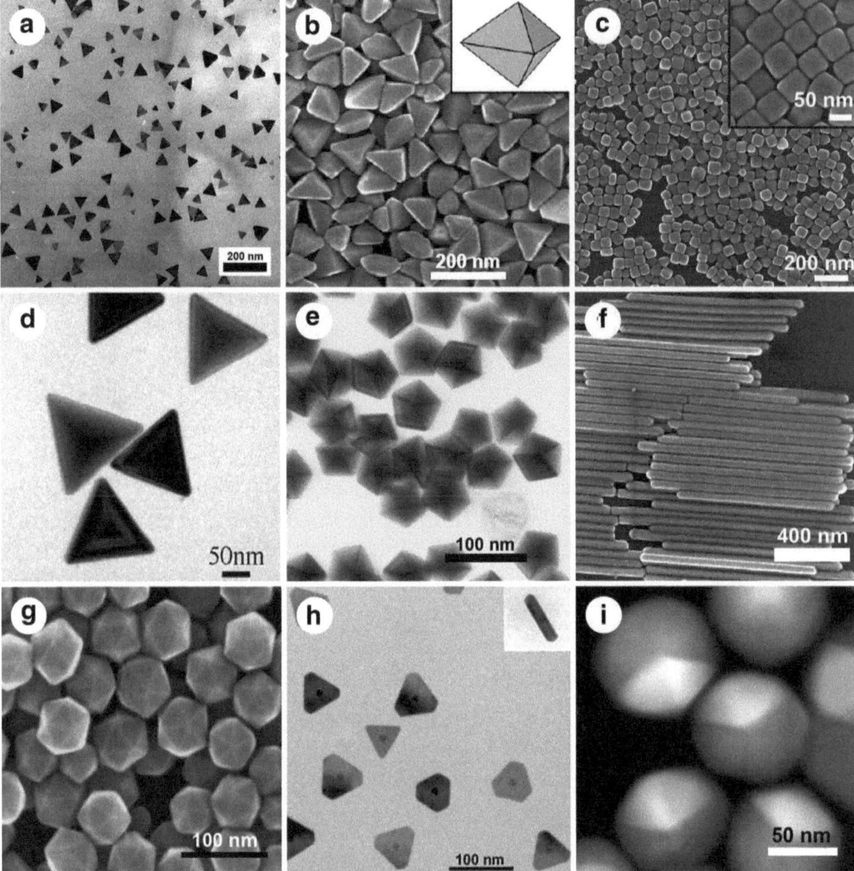

Fig. 1.21 Metallic nanostructures synthesized via plasmon-mediated synthesis: **a** Ag nanoprisms (reproduced with permission from reference [155], Copyright 2003 Nature Publishing Group), **b** Ag right bipyramids (reproduced with permission from reference [381], Copyright 2010 American Chemical Society), **c** twinned Ag nanocubes (reproduced with permission from reference [382], Copyright 2013 Wiley-VCH), **d** Ag tetrahedra (reproduced with permission from reference [383], Copyright 2008 American Chemical Society), **e** Ag decahedra (reproduced with permission from reference [384], Copyright 2008 American Chemical Society), **f** Ag nanorods (reproduced with permission from reference [385], Copyright 2011 American Chemical Society), **g** Ag icosahedra (reproduced with permission from reference [386], Copyright 2014 The Royal Society of Chemistry), **h** Au@Ag core-shell nanoprisms (reproduced with permission from reference [387], Copyright 2007 Wiley-VCH), **i** Au@Ag core-shell icosahedra (reproduced with permission from reference [388], Copyright 2011 Wiley-VCH)

The Ag nanoprisms synthesized through plasmon-mediated syntheses have a multiply planar-twinned structure and are primarily enclosed by {111} facets [152]. By using silver nitrate as a precursor and performing the light-radiation reaction at a higher pH, Ag right triangular bipyramids enclosed by {100} facets were obtained (Fig. 1.21b) [393]. These bipyramids have similar planar-twinned structures. The selective conversion of these two shapes was realized in a later study [381]. The {100}-faceted right triangular bipyramids are favored at a high pH (> 10) and a [BSPP]/[Ag$^+$] ratio close to 1. Triangular prisms with mostly {111} facets are favored at a lower pH or a higher [BSPP]/[Ag$^+$] ratio. Ag nanocubes with multiply planar-twinned structures were also synthesized by fine tuning the irradiation wavelength of light during plasmon-mediated syntheses (Fig. 1.21c) [382].

The plasmon-mediated syntheses are capable of producing metallic nanostructures with other crystal structures. For example, with disodium tartrate as a structural-directing agent, tetrahedral Ag nanocrystals enclosed by 4 {111} facets were synthesized with relatively high yields (Fig. 1.21d) [383]. By using arginine as a photochemical promoter, the Kitaev group reported that decahedral Ag nanocrystals with penta-twinned structures were synthesized with a high shape selectivity (Fig. 1.21e) [384]. The Mirkin group reported that at low-energy irradiation wavelengths, Ag penta-twinned nanorods were obtained (Fig. 1.21f) [385]. By using a combination of Cu^{2+} additive, H$_2$O$_2$ etchant, and violet light irradiation, the Kitaev group synthesized Ag icosahedral nanoparticles with multiply twinned structures (Fig. 1.21g) [386]. Besides Ag nanostructures, the plasmon-mediated syntheses were also applied in synthesizing Au@Ag core-shell nanostructures by exploring plasmonic Au seeds (Fig. 1.21h–i) [387, 388, 394].

1.5.6 Electrochemical Square-Wave-Potential Methods

Metallic nanostructures with high-energy facets have received considerable attention recently due to their importance in catalytic applications [400, 401]. High-energy facets are usually referred to the {110} and high-index {hkl} facets of metallic nanostructures, which possess a high density of atoms with low coordination numbers and a high density of atomic steps, ledges, and kinks [402]. However, in most common synthetic conditions, the high-energy facets will easily disappear due to their low stability. Electrochemical square-wave-potential (SWP) methods have emerged as a versatile method to synthesize metallic nanostructures with high-energy facets [50, 59]. The Sun pioneered the synthesis of metallic nanostructures with high-energy facets using the electrochemical SWP methods. In 2007, the Sun group reported the synthesis of single-crystalline tetrahexahedral Pt nanocrystals with {730} high-index facets by the electrochemical SWP method [60]. In their report, tetrahexahedral Pt nanocrystals were synthesized by treating electrode-posited ~750 nm Pt nanospheres with SWP of certain lower (E$_L$) and upper (E$_U$) potentials in a solution of H$_2$SO$_4$ and ascorbic acid. The formation of the high-index {730} facets was attributed to the repetitive adsorption/desorption of oxygen species (OH$_{ad}$ and O$_{ad}$) generated by SWP processes. At the E$_U$ potential, some of the oxygen species may invade into Pt and squeeze out some Pt atoms. At the E$_L$, all oxygen species will be desorbed and lead to the formation of Pt surface with open structures. Repeating the SWP process for thousands of times will lead to the formation of Pt nanocrystals with well-defined high-index facets.

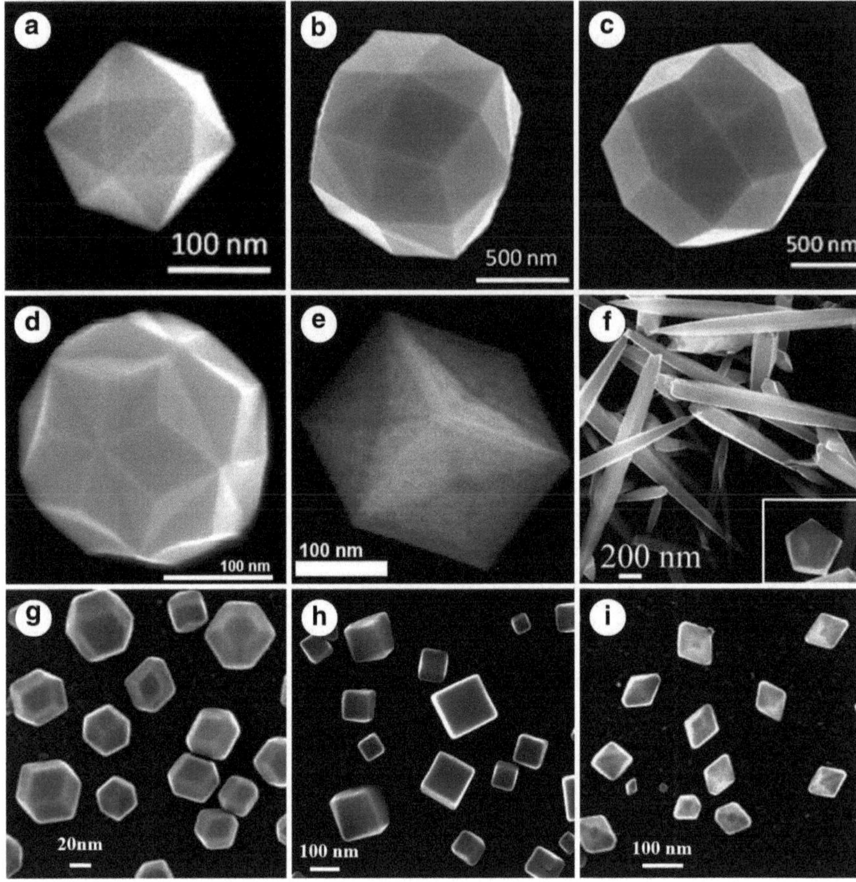

Fig. 1.22 Metallic nanostructures synthesized via electrochemical square-wave-potential (SWP) methods: **a** Pt tetrahexahedra, **b** Pt hexoctahedra, **c** trapezohedra (reproduced with permission from reference [395], Copyright 2013 American Chemical Society), **d** triambic icosahedral Pt nanocrystals (reproduced with permission from reference [396], Copyright 2013 American Chemical Society), **e** concave Pt tetrahexahedra (reproduced with permission from reference [397], Copyright 2011 American Chemical Society), **f** Pd nanorods (reproduced with permission from reference [398], Copyright 2009 The Royal Society of Chemistry), **g–i** Fe rhombic dodecahedra, cubes, and tetragonal bipyramids (reproduced with permission from reference [399], Copyright 2009 American Chemical Society)

The electrochemical SWP methods can be generalized in synthesis various Pt nanocrystals with high-index facets. In a recent report from the Sun group, they described a process that can selectively synthesize tetrahexahedral, hexoctahedral, and trapezohedral Pt nanocrystals, which are enclosed by {hk0}, {hkk}, and {hkl} high-index facets, respectively (Fig. 1.22a–c) [395]. The transition among these different shapes was realized by increasing either the upper or lower potential of the SWP process. Concave hexoctahedral, triambic icosahedral (Fig. 1.22d), concave tetrahexahedral Pt nanocrystals (Fig. 1.22e) were also synthesized through similar electrochemical SWP techniques [50, 396, 397, 403]. The electrochemical SWP method can also be extended to synthesize other metallic nanostructures, such as tetrahexahedral and trapezohedral Pd nanocrystals, five-fold twinned Pd nanorods (Fig. 1.22f), tetrahexahedral Rh nanocrystals, tetrahexahedral Pd–Pt alloyed nanocrystals, and Fe nanocrystals with different shapes ranging from rhombic dodecahedra to tetragonal bipyramids and cubes (Fig. 1.22g–i) [50, 398, 399, 404–406].

Acknowledgment We thank the Ministry of Education Singapore (Grant# R279-000-391-112) for the financial support of this work.

References

1. C. Burda, X. Chen, R. Narayanan et al., Chem. Rev. **105**, 1025 (2005)
2. R. Vajtai, Springer Handbook of Nanomaterials. (Springer, 2013)
3. K.J. Klabunde, R. Richards, Nanoscale materials in chemistry. (Wiley Online Library, 2001)
4. P. Yang, The chemistry of nanostructured materials. (World Scientific, 2011)
5. R. Nagarajan, T.A. Hatton, Nanoparticles: synthesis, stabilization, passivation, and functionalization. (Oxford University Press, USA, 2008)
6. T.K. Sau, A.L. Rogach, Complex-shaped metal nanoparticles: bottom-up syntheses and applications. (John Wiley & Sons, 2012)
7. Y. Kang, X. Ye, C.B. Murray, Angew. Chem. Int. Ed. **49**, 6156 (2010)
8. X. Ye, L. Jin, H. Caglayan et al., ACS nano **6**, 2804 (2012)
9. X. Huang, S. Tang, X. Mu et al., Nat. Nanotechnol. **6**, 28 (2011)
10. W. Niu, W. Zhang, S. Firdoz et al., Chem. Mater. **26**, 2180 (2014)
11. I. Freestone, N. Meeks, M. Sax et al., Gold Bull. **40**, 270 (2007)
12. D. Barber, I. Freestone, Archaeometry **32**, 33 (1990)
13. S.A. Love, B.J. Marquis, C.L. Haynes, Appl. Spectrosc. **62**, 346A (2008)
14. O. Schalm, V. Van Der Linden, P. Frederickx et al., Spectrochimica Acta B **64**, 812 (2009)
15. S. Padovani, D. Puzzovio, C. Sada et al., Appl. Phys. A **83**, 521 (2006)
16. L. Hunt, Gold Bull. **9**, 134 (1976)
17. B.T. Bell, Rev. Prog. Color. Relat. Top. **9**, 48 (1978)
18. J. Carbert, Gold Bull. **13**, 144 (1980)
19. S. Padovani, I. Borgia, B. Brunetti et al., Appl. Phys. A **79**, 229 (2004)
20. Lycurgus cup, British Museum. http://www.britishmuseum.org/explore/highlights/highlight_objects/pe_mla/t/the_lycurgus_cup.aspx
21. Chartres Cathedral - the Medieval Stained Glass, Whatling Stuart http://www.medievalart.org.uk/index.html
22. P.P. Edwards, J.M. Thomas, Angew. Chem. Int. Ed. **46**, 5480 (2007)
23. R.D. Tweney, Perspect. Sci. **14**, 97 (2006)
24. M. Faraday, Phil. Trans. R. Soc. Lond. 145 (1857)
25. J.M. Thomas, Pure Appl. Chem. **60**, 1517 (1988)
26. V. Sharma, K. Park, M. Srinivasarao, Mat. Sci. Eng. R. **65**, 1 (2009)
27. R. Zsigmondy, Colloids and the ultramicroscope: a manual of colloid chemistry and ultramicroscopy. (J. Wiley & Sons, 1909)
28. R. Zsigmondy, J.F. Norton, E.B. Spear, The chemistry of colloids. (John Wiley & sons, Incorporated, 1917)
29. T. Svedberg, The formation of colloids. (D. Van Nostrand, 1921)
30. T. Svedberg, K.O. Pedersen, The Ultracentrifuge. (1940)
31. J. Turkevich, J. Hillier, Anal. Chem. **21**, 475 (1949)
32. D. Beischer, And Krause, F. , Naturwissenschaften **25**, 825 (1937)
33. P. Cooperástevenson, Discuss. Faraday Soc. **11**, 55 (1951)

34. G. Frens, Nature **241**, 20 (1973)
35. Z.L. Wang, Characterization of nanophase materials. (Wiley Online Library, 2000)
36. Y. Zhu, R. Jin, Y. Sun, Catalysts **1**, 3 (2011)
37. K. An, G.A. Somorjai, ChemCatChem **4**, 1512 (2012)
38. P. Mulvaney, MRS Bull. **26**, 1009 (2001)
39. L.N. Lewis, Chem. Rev. **93**, 2693 (1993)
40. E.C. Dreaden, A.M. Alkilany, X. Huang et al., Chem. Soc. Rev. **41**, 2740 (2012)
41. K.A. Willets, R.P. Van Duyne, Annu. Rev. Phys. Chem. **58**, 267 (2007)
42. H. Chen, L. Shao, Q. Li et al., Chem. Soc. Rev. **42**, 2679 (2013)
43. J. Zhao, X.Y. Zhang, C.R. Yonzon et al., Nanomedicine **1**, 219 (2006)
44. H.A. Atwater, A. Polman, Nat. Mater. **9**, 205 (2010)
45. A.J. Haes, R.P. Van Duyne, Anal. Bioanal. Chem. **379**, 920 (2004)
46. K. Ueno, H. Misawa, J. Photochem. Photobiol., C **15**, 31 (2013)
47. J.L. West, N.J. Halas, Annu. Rev. Biomed. Eng. **5**, 285 (2003)
48. S.E. Skrabalak, L. Au, X. Li et al., Nat. Protoc. **2**, 2182 (2007)
49. B. Sepúlveda, P.C. Angelomé, L.M. Lechuga et al., Nano Today **4**, 244 (2009)
50. N. Tian, Z.-Y. Zhou, S.-G. Sun, J. Phys. Chem. C **112**, 19801 (2008)
51. F. Zaera, ChemSusChem **6**, 1797 (2013)
52. B. Roldan Cuenya, Acc. Chem. Res. **46**, 1682 (2012)
53. S. Cheong, J.D. Watt, R.D. Tilley, Nanoscale **2**, 2045 (2010)
54. J. Wu, P. Li, Y.-T.F. Pan et al., Chem. Soc. Rev. **41**, 8066 (2012)
55. Y. Li, G.A. Somorjai, Nano Lett. **10**, 2289 (2010)
56. M. Haruta, Nature **437**, 1098 (2005)
57. M. Haruta, N. Yamada, T. Kobayashi et al., J. Catal. **115**, 301 (1989)
58. Y. Xiong, B.J. Wiley, Y. Xia, Angew. Chem. Int. Ed. **46**, 7157 (2007)
59. Z.-Y. Zhou, N. Tian, Z.-Z. Huang et al., Faraday Discuss. **140**, 81 (2009)
60. N. Tian, Z.-Y. Zhou, S.-G. Sun et al., science **316**, 732 (2007)
61. K.M. Bratlie, H. Lee, K. Komvopoulos et al., Nano Lett. **7**, 3097 (2007)
62. Y.-W. Jun, J.-W. Seo, J. Cheon, Acc. Chem. Res. **41**, 179 (2008)
63. S.P. Gubin, Magnetic nanoparticles. (John Wiley & Sons, 2009)
64. A.P. Guimarães, Principles of nanomagnetism. (Springer, 2009)
65. Y.-W. Jun, J.-S. Choi, J. Cheon, Chem . Commun. 1203 (2007)
66. J.I. Park, N.J. Kang, Y.W. Jun et al., ChemPhysChem **3**, 543 (2002)
67. J.-R. Choi, S.J. Oh, H. Ju et al., Nano Lett. **5**, 2179 (2005)
68. J.-I. Park, M.G. Kim, Y.-W. Jun et al., J. Am. Chem. Soc. **126**, 9072 (2004)
69. J.-I. Park, J. Cheon, J. Am. Chem. Soc. **123**, 5743 (2001)
70. M.A. El-Sayed, Acc. Chem. Res. **34**, 257 (2001)
71. D. Wang, Y. Li, Adv. Mater. **23**, 1044 (2011)
72. Y. Xia, Y. Xiong, B. Lim et al., Angew. Chem. Int. Ed. **48**, 60 (2009)
73. A.A. Ashkarran, J. Cluster Sci. **22**, 233 (2011)
74. N.G. Semaltianos, Crit . Rev. Solid State Mater. Sci. **35**, 105 (2010)
75. F.B. De Mongeot, U. Valbusa, J. Phys.. Condens. Matter. **21**, (2009)
76. A. Botman, J.J.L. Mulders, C.W. Hagen, Nanotechnology **20**, (2009)
77. Z.R. Dai, Z.W. Pan, Z.L. Wang, Adv. Funct. Mater. **13**, 9 (2003)
78. M. Altissimo, Biomicrofluidics **4**, 026503 (2010)
79. S. Gao, K. Ueno, H. Misawa, Acc. Chem. Res. **44**, 251 (2011)
80. J. Henzie, J. Lee, M.H. Lee et al., Annu. Rev. Phys. Chem. **60**, 147 (2009)
81. X. Ye, L. Qi, Nano Today **6**, 608 (2011)
82. C.L. Haynes, R.P. Van Duyne, J. Phys. Chem. B **105**, 5599 (2001)
83. M. Seipenbusch, A. Binder, J. Phys. Chem. C **113**, 20606 (2009)
84. J. Lu, J.W. Elam, P.C. Stair, Acc. Chem. Res. **46**, 1806 (2013)
85. J. Lu, K.-B. Low, Y. Lei et al., Nat. Commun. **5**, 3264 (2014)
86. M. Knez, K. Nielsch, L. Niinistö, Adv. Mater. **19**, 3425 (2007)
87. S.M. George, Chem. Rev. **110**, 111 (2009)

88. G.H. Chan, J. Zhao, E.M. Hicks et al., Nano Lett. **7**, 1947 (2007)
89. J. Henzie, E.-S. Kwak, T.W. Odom, Nano Lett. **5**, 1199 (2005)
90. C.N.J. Wagner, J. Non-Cryst. Solids **150**, 1 (1992)
91. S. Orimo, H. Fujii, Intermetallics **6**, 185 (1998)
92. A.R. Tao, S. Habas, P. Yang, Small **4**, 310 (2008)
93. W.X. Niu, G.B. Xu, Nano Today **6**, 265 (2011)
94. D. Wang, T. Xie, Y. Li, Nano Research **2**, 30 (2009)
95. Q.A. Yuan, X. Wang, Nanoscale **2**, 2328 (2010)
96. M.C. Daniel, D. Astruc, Chem. Rev. **104**, 293 (2004)
97. J.Y. Xiao, L.M. Qi, Nanoscale **3**, 1383 (2011)
98. T.H. Ha, H.-J. Koo, B.H. Chung, J. Phys. Chem. C **111**, 1123 (2006)
99. S.E. Lohse, N.D. Burrows, L. Scarabelli et al., Chem. Mater. **26**, 34 (2013)
100. X. Lu, T.T. Tran, W. Zhang, Chem. Eng. Process. Tech. **1**, 1009 (2013)
101. J.U. Kim, S.H. Cha, K. Shin et al., Adv. Mater. **16**, 459 (2004)
102. C.-Y. Chiu, Y. Li, L. Ruan et al., Nature chemistry **3**, 393 (2011)
103. L. Ruan, C.-Y. Chiu, Y. Li et al., Nano Lett. **11**, 3040 (2011)
104. J. Xie, J.Y. Lee, D.I. Wang et al., ACS nano **1**, 429 (2007)
105. J. Xie, J.Y. Lee, D.I. Wang et al., Small **3**, 672 (2007)
106. C.-Y. Chiu, L. Ruan, Y. Huang, Chem. Soc. Rev. **42**, 2512 (2013)
107. J.R. Thomas, J. Appl. Phys. **37**, 2914 (1966)
108. D. Decaro, T. Ould-Ely, A. Mari et al., Chem. Mater. **8**, 1987 (1996)
109. V.F. Puntes, K.M. Krishnan, A.P. Alivisatos, Science **291**, 2115 (2001)
110. S.J. Park, S. Kim, S. Lee et al., J. Am. Chem. Soc. **122**, 8581 (2000)
111. T.O. Ely, C. Amiens, B. Chaudret et al., Chem. Mater. **11**, 526 (1999)
112. B. Wu, N. Zheng, G. Fu, Chem. Commun. **47**, 1039 (2011)
113. M. Grzelczak, J. Pérez-Juste, P. Mulvaney et al., Chem. Soc. Rev. **37**, 1783 (2008)
114. Y. Sun, Chem. Soc. Rev. **42**, 2497 (2013)
115. C.D. Donega, P. Liljeroth, D. Vanmaekelbergh, Small **1**, 1152 (2005)
116. S.G. Kwon, T. Hyeon, Small **7**, 2685 (2011)
117. C.J. Murphy, T.K. Sau, A.M. Gole et al., J. Phys. Chem. B **109**, 13857 (2005)
118. W. Niu, L. Zhang, G. Xu, Nanoscale **5**, 3172 (2013)
119. W. Niu, L. Zhang, G. Xu, Sci. China Chem. **55**, 2311 (2012)
120. C.J. Murphy, A.M. Gole, S.E. Hunyadi et al., Inorg. Chem. **45**, 7544 (2006)
121. X. Huang, N. Zheng, J. Am. Chem. Soc. **131**, 4602 (2009)
122. B.Y. Xia, H.B. Wu, Y. Yan et al., J. Am. Chem. Soc. **135**, 9480 (2013)
123. Y. Yu, Y. Zhao, T. Huang et al., Pure Appl. Chem. **81**, (2009)
124. F. Kim, J.H. Song, P. Yang, J. Am. Chem. Soc. **124**, 14316 (2002)
125. S.-S. Chang, C.-W. Shih, C.-D. Chen et al., Langmuir **15**, 701 (1998)
126. J. Zhang, J. Du, B. Han et al., Angew. Chem. **118**, 1134 (2006)
127. W.T. Yao, S.H. Yu, Int. J. Nanotechnol. **4**, 129 (2007)
128. Y.J. Li, H. Zhu, C. Hon et al., Prog. Chem. **25**, 276 (2013)
129. W.D. Shi, S.Y. Song, H.J. Zhang, Chem. Soc. Rev. **42**, 5714 (2013)
130. S. Feng, R. Xu, Acc. Chem. Res. **34**, 239 (2001)
131. B.A. Xu, X. Wang, Dalton Trans. **41**, 4719 (2012)
132. D. Yu, V.W.-W. Yam, J. Am. Chem. Soc. **126**, 13200 (2004)
133. D. Yu, V.W.-W. Yam, J. Phys. Chem. B **109**, 5497 (2005)
134. X. Sun, Y. Li, Adv. Mater. **17**, 2626 (2005)
135. Y. Jia, Y. Jiang, J. Zhang et al., J. Am. Chem. Soc. **136**, 3748 (2014)
136. F. Saleem, Z. Zhang, B. Xu et al., J. Am. Chem. Soc. **135**, 18304 (2013)
137. H. Duan, N. Yan, R. Yu et al., Nat. Commun. **5**, 3093 (2014)
138. X. Huang, S. Tang, H. Zhang et al., J. Am. Chem. Soc. **131**, 13916 (2009)
139. Z.-C. Zhang, J.-F. Hui, Z.-C. Liu et al., Langmuir **28**, 14845 (2012)
140. M. Chen, B.H. Wu, J. Yang et al., Adv. Mater. **24**, 862 (2012)
141. M. Tsuji, M. Hashimoto, Y. Nishizawa et al., Chem. Eur. J **11**, 440 (2005)

142. M. Baghbanzadeh, L. Carbone, P.D. Cozzoli et al., Angew. Chem. Int. Ed. **50**, 11312 (2011)
143. T. Yamamoto, Y. Wada, T. Sakata et al., Chem. Lett. **33**, 158 (2004)
144. N. Elander, J.R. Jones, S.-Y. Lu et al., Chem. Soc. Rev. **29**, 239 (2000)
145. W. Tu, H. Liu, J. Mater. Chem. **10**, 2207 (2000)
146. Y. Yu, Y. Zhao, T. Huang et al., Mater. Res. Bull. **45**, 159 (2010)
147. X. Tong, Y. Zhao, T. Huang et al., Appl. Surf. Sci. **255**, 9463 (2009)
148. A. Abedini, A.R. Daud, M.a.A. Hamid et al. , Nanoscale Res. Lett. **8**, 1 (2013)
149. J. Marignier, J. Belloni, M. Delcourt et al., (1985)
150. J. Belloni, Catal. Today **113**, 141 (2006)
151. Y. Niidome, K. Nishioka, H. Kawasaki et al., Chem. Commun. 2376 (2003)
152. M.R. Langille, M.L. Personick, C.A. Mirkin, Angew. Chem. Int. Ed. **52**, 13910 (2013)
153. Y. Sun, Y. Xia, Adv. Mater. **15**, 695 (2003)
154. Y. Sun, B. Mayers, Y. Xia, Nano Lett. **3**, 675 (2003)
155. R.C. Jin, Y.C. Cao, E.C. Hao et al., Nature **425**, 487 (2003)
156. R.C. Jin, Y.W. Cao, C.A. Mirkin et al., Science **294**, 1901 (2001)
157. X. Wu, P.L. Redmond, H. Liu et al., J. Am. Chem. Soc. **130**, 9500 (2008)
158. M.T. Reetz, W. Helbig, S.A. Quaiser et al., Science **267**, 367 (1995)
159. M.T. Reetz, W. Helbig, J. Am. Chem. Soc. **116**, 7401 (1994)
160. Y.-Y. Yu, S.-S. Chang, C.-L. Lee et al., J. Phys. Chem. B **101**, 6661 (1997)
161. H. Xu, B.W. Zeiger, K.S. Suslick, Chem. Soc. Rev. **42**, 2555 (2013)
162. Y.-J. Zhu, F. Chen, Chem. Rev. **114**, 6462 (2014)
163. E. Carbó-Argibay, B. Rodríguez-González, J. Pacifico et al., Angew. Chem. **119**, 9141 (2007)
164. A. Sánchez-Iglesias, I. Pastoriza-Santos, J. Pérez-Juste et al., Adv. Mater. **18**, 2529 (2006)
165. L.-P. Jiang, S. Xu, J.-M. Zhu et al., Inorg. Chem. **43**, 5877 (2004)
166. K. Okitsu, K. Sharyo, R. Nishimura, Langmuir **25**, 7786 (2009)
167. H. Xu, K.S. Suslick, ACS nano **4**, 3209 (2010)
168. H. Liu, X. Zhang, X. Wu et al., Chem. Commun. **47**, 4237 (2011)
169. K.S. Suslick, Y. Didenko, M.M. Fang et al., Philos. Trans. A Math. Phys. Eng. Sci. **357**, 335 (1999)
170. K.S. Suslick, S.-B. Choe, A.A. Cichowlas et al., Nature **353**, 414 (1991)
171. K.S. Suslick, G.J. Price, Annu. Rev. Mater. Sci. **29**, 295 (1999)
172. R.A. Caruso, M. Ashokkumar, F. Grieser, Langmuir **18**, 7831 (2002)
173. L. Qi, Encyclopedia of Surface and Colloid Science. Second Edition. New York: Taylor & Francis 6183 (2006)
174. V.T. Liveri, Reversed micelles as nanometer-size solvent media. (Marcel Dekker: New York, 2002)
175. I. Lisiecki, J. Phys. Chem. B **109**, 12231 (2005)
176. J. Klier, C.J. Tucker, T.H. Kalantar et al., Adv. Mater. **12**, 1751 (2000)
177. J. Eastoe, M.J. Hollamby, L. Hudson, Adv. Colloid Interface Sci. **128**, 5 (2006)
178. M. Pileni, J. Phys. Chem. **97**, 6961 (1993)
179. M.P. Pileni, Langmuir **13**, 3266 (1997)
180. X. Wang, Q. Peng, Y. Li, Acc. Chem. Res. **40**, 635 (2007)
181. L.M. Liz-Marzan, Chem. Commun. **49**, 16 (2013)
182. Q. Yao, Y. Yu, X. Yuan et al., Small **9**, 2696 (2013)
183. M. Brust, J. Fink, D. Bethell et al., J. Chem. Soc., Chem. Commun. 1655 (1995)
184. M. Brust, M. Walker, D. Bethell et al., J. Chem. Soc., Chem. Commun. 801 (1994)
185. X. Wang, J. Zhuang, Q. Peng et al., Nature **437**, 121 (2005)
186. N.R. Jana, Small **1**, 875 (2005)
187. Y.J. Xiong, H.G. Cai, B.J. Wiley et al., J. Am. Chem. Soc. **129**, 3665 (2007)
188. B.J. Wiley, Y. Chen, J.M. Mclellan et al., Nano Lett. **7**, 1032 (2007)
189. Y. Sun, Y. Xia, Adv. Mater. **14**, 833 (2002)
190. B.J. Wiley, Z. Wang, J. Wei et al., Nano Lett. **6**, 2273 (2006)
191. X. Liu, N. Wu, B.H. Wunsch et al., Small **2**, 1046 (2006)

192. M.R. Jones, K.D. Osberg, R.J. Macfarlane et al., Chem. Rev. **111**, 3736 (2011)
193. Y. Liu, J. Goebl, Y. Yin, Chem. Soc. Rev. **42**, 2610 (2013)
194. H. Wang, G.P. Goodrich, F. Tam et al., J. Phys. Chem. B **109**, 11083 (2005)
195. C. Gao, Q. Zhang, Z. Lu et al., J. Am. Chem. Soc. **133**, 19706 (2011)
196. H. Wang, H.Y. Jeong, M. Imura et al., J. Am. Chem. Soc. **133**, 14526 (2011)
197. M.J. Banholzer, L. Qin, J.E. Millstone et al., Nat. Protoc. **4**, 838 (2009)
198. L. Polavarapu, L.M. Liz-Marzan, Nanoscale **5**, 4355 (2013)
199. H.W. Liang, S. Liu, J.Y. Gong et al., Adv. Mater. **21**, 1850 (2009)
200. R.M. Crooks, M. Zhao, L. Sun et al., Acc. Chem. Res. **34**, 181 (2001)
201. H.L. Qin, D. Wang, Z.L. Huang et al., J. Am. Chem. Soc. **135**, 12544 (2013)
202. J. Niu, D. Wang, H. Qin et al., Nat. Commun. **5**, 3313 (2014)
203. H. Wang, D.W. Brandl, P. Nordlander et al., Acc. Chem. Res. **40**, 53 (2007)
204. C. Loo, A. Lin, L. Hirsch et al., Technol. Cancer. Res. Treat. **3**, 33 (2004)
205. S.J. Oldenburg, R.D. Averitt, S.L. Westcott et al., Chem. Phys. Lett. **288**, 243 (1998)
206. P. Karthika, H. Ataee-Esfahani, H.J. Wang et al., Chem. Asian J. **8**, 902 (2013)
207. H.J. Wang, M. Imura, Y. Nemoto et al., Chem. Asian J. **7**, 802 (2012)
208. C. Martin, Science **266**, 1961 (1994)
209. Y.Z. Piao, H. Kim, J. nanosci. nanotechnol. **9**, 2215 (2009)
210. M.E. Toimil-Molares, Beilstein J. Nanotechnol. **3**, 860 (2012)
211. T.R. Kline, M. Tian, J. Wang et al., Inorg. Chem. **45**, 7555 (2006)
212. L.D. Qin, S. Park, L. Huang et al., Science **309**, 113 (2005)
213. A.B. Braunschweig, A.L. Schmucker, W.D. Wei et al., Chem. Phys. Lett. **486**, 89 (2010)
214. S.E. Skrabalak, J. Chen, Y. Sun et al., Acc. Chem. Res. **41**, 1587 (2008)
215. B. Mayers, X. Jiang, D. Sunderland et al., J. Am. Chem. Soc. **125**, 13364 (2003)
216. Y. Qin, R. Che, C. Liang et al., J. Mater. Chem. **21**, 3960 (2011)
217. J. Fang, S. Lebedkin, S. Yang et al., Chem. Commun. **47**, 5157 (2011)
218. Y. Kang, J.B. Pyo, X. Ye et al., ACS nano **7**, 645 (2012)
219. X. Xia, S.-I. Choi, J.A. Herron et al., J. Am. Chem. Soc. **135**, 15706 (2013)
220. J.L. Elechiguerra, J. Reyes-Gasga, M.J. Yacaman, J. Mater. Chem. **16**, 3906 (2006)
221. W. Zhou, J. Wu, H. Yang, Nano Lett. **13**, 2870 (2013)
222. D.B. Williams, C.B. Carter, Micron **28**, 75 (1997)
223. P.W. Hawkes, Advances in imaging and electron physics. (Academic Press, 2003)
224. B. Schmidt, K. Wetzig, Ion beams in materials processing and analysis. (Springer, 2012)
225. J.J. Bozzola, L.D. Russell, Electron microscopy: principles and techniques for biologists. (Jones & Bartlett Learning, 1999)
226. S.C. Singh, H. Zeng, C. Guo et al., Nanomaterials: processing and characterization with lasers. (John Wiley & Sons, 2012)
227. K. Ramesh, Nanomaterials : Mechanics and Mechanisms. (Dordrecht: Springer, 2009)
228. L. Reimer, Meas. Sci. Technol. **11**, 1826 (2000)
229. C. Zweben, JOM **50**, 47 (1998)
230. S. Sharma, Atomic and Nuclear Physics. (Pearson Education India, 2008)
231. R. Jalilian, Pulse Laser Assisted Growth of Nanowires and Nano-heterojunctions and Their Properties. (ProQuest, 2008)
232. J. Henzie, M. Grünwald, A. Widmer-Cooper et al., Nat. Mater. **11**, 131 (2012)
233. X. Ye, Y. Gao, J. Chen et al., Nano Lett. **13**, 2163 (2013)
234. G. Haugstad, Atomic force microscopy : understanding basic modes and advanced applications. (John Wiley & Sons, 2012)
235. K.S. Breuer, Microscale diagnostic techniques. (Springer, 2005)
236. R.W. Kelsall, I.W. Hamley, M. Geoghegan, Nanoscale science and technology. (Wiley Online Library, 2005)
237. G. Binnig, C.F. Quate, C. Gerber, Phys. Rev. Lett. **56**, 930 (1986)
238. X.C. Tong, Advanced materials for thermal management of electronic packaging. (Springer, 2011)

239. Y. Waseda, E. Matsubara, K. Shinoda, X-ray diffraction crystallography: introduction, examples and solved problems. (Springer, 2011)
240. M. Rai, N. Duran, G. Southam, Metal nanoparticles in microbiology. (Springer, 2011)
241. R. Saravanan, M.P. Rani, Metal and Alloy Bonding: An Experimental Analysis. (Springer, 2012)
242. D. Seo, J.C. Park, H. Song, J. Am. Chem. Soc. **128**, 14863 (2006)
243. X. Zou, S. Hovmöller, P. Oleynikov, Electron Crystallography : electron microscopy and electron diffraction. (Oxford University Press, 2011)
244. P.J. Goodhew, J. Humphreys, R. Beanland, Electron microscopy and analysis. (CRC Press, 2000)
245. D.L. Schodek, P. Ferreira, M.F. Ashby, Nanomaterials, nanotechnologies and design : an introduction for engineers and architects. (Butterworth-Heinemann, 2009)
246. P.D. Brown, Transmission Electron Microscopy-A Textbook for Materials Science. (Cambridge Univ Press, 1999)
247. W. Niu, L. Zhang, G. Xu, Acs Nano **4**, 1987 (2010)
248. T. Liu, P. Jiang, Q. You et al., CrystEngComm **15**, 2350 (2013)
249. S. Hofmann, Auger-and X-ray Photoelectron Spectroscopy in Materials Science: A User-oriented Guide. (Springer, 2012)
250. K. Nogi, M. Hosokawa, M. Naito et al., Nanoparticle technology handbook. (Elsevier, 2012)
251. J.P. Davim, Surface integrity in machining. (Springer, 2010)
252. A. Dimoulas, E. Gusev, P.C. Mcintyre et al., Advanced gate stacks for high-mobility semiconductors. (Springer, 2008)
253. M.L. Personick, M.R. Langille, J. Zhang et al., Nano Lett. **11**, 3394 (2011)
254. D. Bell, A. Garratt-Reed, Energy dispersive X-ray analysis in the electron microscope. (Garland Science, 2003)
255. C.R. Brundle, C.A. Evans, S. Wilson, Encyclopedia of materials characterization: surfaces, interfaces, thin films. (Gulf Professional Publishing, 1992)
256. G. Agostini, C. Lamberti, Characterization of semiconductor heterostructures and nanostructures. (Elsevier, 2011)
257. W. Zhang, H.Y.J. Goh, S. Firdoz et al., Chem. Eur. J **19**, 12732 (2013)
258. G.L. Moore, Introduction to inductively coupled plasma atomic emission spectrometry. (Elsevier, 2012)
259. H.E. Taylor, Inductively coupled plasma-mass spectrometry : practices and techniques. (Academic Press, 2001)
260. J.R. Dean, Practical inductively coupled plasma spectroscopy. (John Wiley & Sons, 2005)
261. M.J. Brisson, A.A. Ekechukwu, Beryllium : environmental analysis and monitoring. (Royal Society of Chemistry, 2009)
262. P. Szefer, J.O. Nriagu, Mineral components in foods. (CRC Press, 2006)
263. R.A. Scott, C.M. Lukehart, Applications of physical methods to inorganic and bioinorganic chemistry. (John Wiley & Sons, 2013)
264. H.-C. Peng, S. Xie, J. Park et al., J. Am. Chem. Soc. **135**, 3780 (2013)
265. S.E. Lohse, C.J. Murphy, Chem. Mater. **25**, 1250 (2013)
266. K.R. Brown, L.A. Lyon, A.P. Fox et al., Chem. Mater. **12**, 314 (1999)
267. K.R. Brown, M.J. Natan, Langmuir **14**, 726 (1998)
268. K.R. Brown, D.G. Walter, M.J. Natan, Chem. Mater. **12**, 306 (1999)
269. N.R. Jana, L. Gearheart, C.J. Murphy, Chem . Commun. 617 (2001)
270. N.R. Jana, L. Gearheart, C.J. Murphy, J. Phys. Chem. B **105**, 4065 (2001)
271. C.J. Johnson, E. Dujardin, S.A. Davis et al., J. Mater. Chem. **12**, 1765 (2002)
272. A. Gole, C.J. Murphy, Chem. Mater. **16**, 3633 (2004)
273. N.R. Jana, Chem. Commun. 1950 (2003)
274. B. Nikoobakht, M.A. El-Sayed, Chem. Mater. **15**, 1957 (2003)
275. T.K. Sau, C.J. Murphy, Langmuir **20**, 6414 (2004)
276. H.-Y. Wu, H.-C. Chu, T.-J. Kuo et al., Chem. Mater. **17**, 6447 (2005)

277. H.-Y. Wu, W.-L. Huang, M.H. Huang, Cryst. Growth Des. **7**, 831 (2007)
278. L. Gou, C.J. Murphy, Chem. Mater. **17**, 3668 (2005)
279. M. Liu, P. Guyot-Sionnest, J. Phys. Chem. B **109**, 22192 (2005)
280. X. Kou, S. Zhang, C.-K. Tsung et al., J. Phys. Chem. B **110**, 16377 (2006)
281. X. Kou, S. Zhang, C.K. Tsung et al., Chem. Eur. J **13**, 2929 (2007)
282. X. Ye, C. Zheng, J. Chen et al., Nano Lett. **13**, 765 (2013)
283. T.K. Sau, C.J. Murphy, J. Am. Chem. Soc. **126**, 8648 (2004)
284. W. Niu, S. Zheng, D. Wang et al., J. Am. Chem. Soc. **131**, 697 (2008)
285. H.-L. Wu, C.-H. Kuo, M.H. Huang, Langmuir **26**, 12307 (2010)
286. M. Eguchi, D. Mitsui, H.-L. Wu et al., Langmuir **28**, 9021 (2012)
287. Y. Yu, Q. Zhang, X. Lu et al., J. Phys. Chem. C **114**, 11119 (2010)
288. M.R. Langille, M.L. Personick, J. Zhang et al., J. Am. Chem. Soc. **134**, 14542 (2012)
289. M.L. Personick, C.A. Mirkin, J. Am. Chem. Soc. **135**, 18238 (2013)
290. T. Ming, W. Feng, Q. Tang et al., J. Am. Chem. Soc. **131**, 16350 (2009)
291. J. Li, L. Wang, L. Liu et al., Chem. Commun. **46**, 5109 (2010)
292. M.L. Personick, M.R. Langille, J. Zhang et al., J. Am. Chem. Soc. **133**, 6170 (2011)
293. F. Lu, Y. Zhang, L. Zhang et al., J. Am. Chem. Soc. **133**, 18074 (2011)
294. M.L. Personick, M.R. Langille, J. Wu et al., J. Am. Chem. Soc. **135**, 3800 (2013)
295. J. Zhang, M.R. Langille, M.L. Personick et al., J. Am. Chem. Soc. **132**, 14012 (2010)
296. Y. Yu, Q. Zhang, J. Xie et al., Nat. Commun. **4**, 1454 (2013)
297. W. Niu, Z.-Y. Li, L. Shi et al., Cryst. Growth Des. **8**, 4440 (2008)
298. B.T. Sneed, C.-H. Kuo, C.N. Brodsky et al., J. Am. Chem. Soc. **134**, 18417 (2012)
299. F.-R. Fan, D.-Y. Liu, Y.-F. Wu et al., J. Am. Chem. Soc. **130**, 6949 (2008)
300. Y.-C. Tsao, S. Rej, C.-Y. Chiu et al., J. Am. Chem. Soc. (2013)
301. C.-W. Yang, K. Chanda, P.-H. Lin et al., J. Am. Chem. Soc. **133**, 19993 (2011)
302. C.-L. Lu, K.S. Prasad, H.-L. Wu et al., J. Am. Chem. Soc. **132**, 14546 (2010)
303. Y. Yu, Q. Zhang, B. Liu et al., J. Am. Chem. Soc. **132**, 18258 (2010)
304. S.E. Habas, H. Lee, V. Radmilovic et al., Nat. Mater. **6**, 692 (2007)
305. Y.-H. Chen, H.-H. Hung, M.H. Huang, J. Am. Chem. Soc. **131**, 9114 (2009)
306. P.J. Chung, L.M. Lyu, M.H. Huang, Chem. Eur. J **17**, 9746 (2011)
307. B. Wiley, Y. Sun, B. Mayers et al., Chem. Eur. J **11**, 454 (2005)
308. D. Seo, C.I. Yoo, I.S. Chung et al., J. Phys. Chem. C **112**, 2469 (2008)
309. A. Tao, P. Sinsermsuksakul, P. Yang, Angew. Chem. Int. Ed. **45**, 4597 (2006)
310. Y. Sun, Y. Xia, Science **298**, 2176 (2002)
311. Y. Zheng, J. Zeng, A. Ruditskiy et al., Chem. Mater. (2013)
312. B. Wiley, T. Herricks, Y. Sun et al., Nano Lett. **4**, 1733 (2004)
313. S.H. Im, Y.T. Lee, B. Wiley et al., Angew. Chem. Int. Ed. **44**, 2154 (2005)
314. Y. Sun, B. Gates, B. Mayers et al., Nano Lett. **2**, 165 (2002)
315. B.J. Wiley, Y. Xiong, Z.-Y. Li et al., Nano Lett. **6**, 765 (2006)
316. Y. Xiong, J. Chen, B. Wiley et al., J. Am. Chem. Soc. **127**, 7332 (2005)
317. Y. Xiong, H. Cai, B.J. Wiley et al., J. Am. Chem. Soc. **129**, 3665 (2007)
318. Y. Xiong, J. Chen, B. Wiley et al., Nano Lett. **5**, 1237 (2005)
319. Y.J. Xiong, J.M. Mclellan, Y.D. Yin et al., Angew. Chem. Int. Ed. **46**, 790 (2007)
320. F. Kim, S. Connor, H. Song et al., Angew. Chem. **116**, 3759 (2004)
321. D. Seo, C.I. Yoo, J.C. Park et al., Angew. Chem. **120**, 775 (2008)
322. H. Song, F. Kim, S. Connor et al., J. Phys. Chem. B **109**, 188 (2005)
323. C. Li, K.L. Shuford, M. Chen et al., ACS nano **2**, 1760 (2008)
324. T.T. Tran, X. Lu, J. Phys. Chem. C **115**, 3638 (2011)
325. X.L. Li, Y. Yang, G.J. Zhou et al., Nanoscale **5**, 4976 (2013)
326. Y. Yang, W. Wang, X. Li et al., Chem. Mater. **25**, 34 (2012)
327. C. Li, L. Sun, Y. Sun et al., Chem. Mater. **25**, 2580 (2013)
328. J.Y. Chen, B. Wiley, Z.Y. Li et al., Adv. Mater. **17**, 2255 (2005)
329. W. Niu, W. Zhang, S. Firdoz et al., J. Am. Chem. Soc. **136**, 3010 (2014)
330. I. Pastoriza-Santos, L.M. Liz-Marzán, Nano Lett. **2**, 903 (2002)

331. Y. Gao, P. Jiang, L. Song et al., J. Cryst. Growth **289**, 376 (2006)
332. G.H. Jeong, M. Kim, Y.W. Lee et al., J. Am. Chem. Soc. **131**, 1672 (2009)
333. X. Huang, Z. Zhao, J. Fan et al., J. Am. Chem. Soc. **133**, 4718 (2011)
334. Y. Dai, X. Mu, Y. Tan et al., J. Am. Chem. Soc. **134**, 7073 (2012)
335. I. Pastoriza-Santos, L.M. Liz-Marzán, Adv. Funct. Mater. **19**, 679 (2009)
336. I. Pastoriza-Santos, L.M. Liz-Marzán, Langmuir **15**, 948 (1999)
337. Y. Chen, X. Gu, C.-G. Nie et al., Chem. Commun. 4181 (2005)
338. D. Wang, Y. Liu, T. You, CrystEngComm **12**, 4028 (2010)
339. M. Chen, B. Wu, J. Yang et al., Adv. Mater. **24**, 862 (2012)
340. C. Cui, L. Gan, M. Neumann et al., J. Am. Chem. Soc. **136**, 4813 (2014)
341. S. Mourdikoudis, L.M. Liz-MarzáN, Chem. Mater. **25**, 1465 (2013)
342. S. Peng, J.M. Mcmahon, G.C. Schatz et al., Proc. Natl. Acad. Sci. U.S.A. **107**, 14530 (2010)
343. S. Sun, C. Murray, D. Weller et al., Science **287**, 1989 (2000)
344. V. Mazumder, M. Chi, K.L. More et al., J. Am. Chem. Soc. **132**, 7848 (2010)
345. C. Wang, D. Van Der Vliet, K.L. More et al., Nano Lett. **11**, 919 (2010)
346. V. Mazumder, S. Sun, J. Am. Chem. Soc. **131**, 4588 (2009)
347. S. Zhang, Ö. Metin, D. Su et al., Angew. Chem. Int. Ed. **52**, 3681 (2013)
348. C. Wang, W. Tian, Y. Ding et al., J. Am. Chem. Soc. **132**, 6524 (2010)
349. S. Zhang, S. Guo, H. Zhu et al., J. Am. Chem. Soc. **134**, 5060 (2012)
350. V. Mazumder, M. Chi, M.N. Mankin et al., Nano Lett. **12**, 1102 (2012)
351. Y. Yu, W. Yang, X. Sun et al., Nano Lett. **14**, 2778 (2014)
352. C.B. Murray, S.H. Sun, H. Doyle et al., MRS Bull. **26**, 985 (2001)
353. Z. Huo, C.-K. Tsung, W. Huang et al., Nano Lett. **8**, 2041 (2008)
354. C. Wang, Y. Hou, J. Kim et al., Angew. Chem. Int. Ed. **46**, 6333 (2007)
355. Z. Niu, Q. Peng, M. Gong et al., Angew. Chem. **123**, 6439 (2011)
356. X. Lu, M.S. Yavuz, H.-Y. Tuan et al., J. Am. Chem. Soc. **130**, 8900 (2008)
357. C. Wang, Y. Hu, C.M. Lieber et al., J. Am. Chem. Soc. **130**, 8902 (2008)
358. Z. Li, J. Tao, X. Lu et al., Nano Lett. **8**, 3052 (2008)
359. H. Feng, Y. Yang, Y. You et al., Chem. Commun. 1984 (2009)
360. N. Pazos-Pérez, D. Baranov, S. Irsen et al., Langmuir **24**, 9855 (2008)
361. S. Guo, D. Li, H. Zhu et al., Angew. Chem. Int. Ed. **52**, 3465 (2013)
362. H. Zhu, S. Zhang, S. Guo et al., J. Am. Chem. Soc. **135**, 7130 (2013)
363. S. Guo, S. Zhang, X. Sun et al., J. Am. Chem. Soc. **133**, 15354 (2011)
364. M. Chen, T. Pica, Y.-B. Jiang et al., J. Am. Chem. Soc. **129**, 6348 (2007)
365. S. Peng, Y. Sun, Chem. Mater. **22**, 6272 (2010)
366. L. Polavarapu, L.M. Liz-Marzán, Nanoscale **5**, 4355 (2013)
367. Y. Ma, W. Li, J. Zeng et al., J. Mater. Chem. **20**, 3586 (2010)
368. J. Zhang, H. Yang, J. Fang et al., Nano Lett. **10**, 638 (2010)
369. J. Zhang, H. Yang, K. Yang et al., Adv. Funct. Mater. **20**, 3727 (2010)
370. H. Yang, J. Zhang, K. Sun et al., Angew. Chem. **122**, 7000 (2010)
371. C. Wang, H. Daimon, T. Onodera et al., Angew. Chem. Int. Ed. **47**, 3588 (2008)
372. R. Loukrakpam, P. Chang, J. Luo et al., Chem. Commun. **46**, 7184 (2010)
373. C. Wang, H. Daimon, Y. Lee et al., J. Am. Chem. Soc. **129**, 6974 (2007)
374. J. Zhang, J. Fang, J. Am. Chem. Soc. **131**, 18543 (2009)
375. Y. Kang, M. Li, Y. Cai et al., J. Am. Chem. Soc. **135**, 2741 (2013)
376. J. Wu, L. Qi, H. You et al., J. Am. Chem. Soc. **134**, 11880 (2012)
377. J. Wu, A. Gross, H. Yang, Nano Lett. **11**, 798 (2011)
378. J. Watt, N. Young, S. Haigh et al., Adv. Mater. **21**, 2288 (2009)
379. J. Watt, S. Cheong, M.F. Toney et al., ACS nano **4**, 396 (2009)
380. J. Ren, R.D. Tilley, J. Am. Chem. Soc. **129**, 3287 (2007)
381. J. Zhang, M.R. Langille, C.A. Mirkin, J. Am. Chem. Soc. **132**, 12502 (2010)
382. M.L. Personick, M.R. Langille, J. Zhang et al., Small **9**, 1947 (2013)
383. J. Zhou, J. An, B. Tang et al., Langmuir **24**, 10407 (2008)
384. B. Pietrobon, V. Kitaev, Chem. Mater. **20**, 5186 (2008)

385. J. Zhang, M.R. Langille, C.A. Mirkin, Nano Lett. **11**, 2495 (2011)
386. R. Keunen, N. Cathcart, V. Kitaev, Nanoscale **6**, 8045 (2014)
387. C. Xue, J.E. Millstone, S. Li et al., Angew. Chem. **119**, 8588 (2007)
388. M.R. Langille, J. Zhang, C.A. Mirkin, Angew. Chem. Int. Ed. **50**, 3543 (2011)
389. C. Xue, G.S. Metraux, J.E. Millstone et al., J. Am. Chem. Soc. **130**, 8337 (2008)
390. P.L. Redmond, X. Wu, L. Brus, J. Phys. Chem. C **111**, 8942 (2007)
391. P.L. Redmond, L.E. Brus, J. Phys. Chem. C **111**, 14849 (2007)
392. C. Xue, C.A. Mirkin, Angew. Chem. **119**, 2082 (2007)
393. J. Zhang, S. Li, J. Wu et al., Angew. Chem. **121**, 7927 (2009)
394. M.R. Langille, J. Zhang, M.L. Personick et al., Science **337**, 954 (2012)
395. J. Xiao, S. Liu, N. Tian et al., J. Am. Chem. Soc. **135**, 18754 (2013)
396. L. Wei, Z.-Y. Zhou, S.-P. Chen et al., Chem. Commun. **49**, 11152 (2013)
397. L. Wei, Y.-J. Fan, N. Tian et al., J. Phys. Chem. C **116**, 2040 (2011)
398. N. Tian, Z.-Y. Zhou, S.-G. Sun, Chem . Commun. 1502 (2009)
399. Y.-X. Chen, S.-P. Chen, Z.-Y. Zhou et al., J. Am. Chem. Soc. **131**, 10860 (2009)
400. L. Zhang, W. Niu, G. Xu, Nano Today **7**, 586 (2012)
401. Z.Y. Jiang, Q. Kuang, Z.X. Xie et al., Adv. Funct. Mater. **20**, 3634 (2010)
402. Z.W. Quan, Y.X. Wang, J.Y. Fang, Acc. Chem. Res. **46**, 191 (2013)
403. Y. Li, Y. Jiang, M. Chen et al., Chem. Commun. **48**, 9531 (2012)
404. N.F. Yu, N. Tian, Z.Y. Zhou et al., Angew. Chem. Int. Ed. **126**, 5197 (2014)
405. Y.-J. Deng, N. Tian, Z.-Y. Zhou et al., Chem. Sci. **3**, 1157 (2012)
406. N. Tian, Z.-Y. Zhou, N.-F. Yu et al., J. Am. Chem. Soc. **132**, 7580 (2010)

Chapter 2
Controlled Synthesis: Nucleation and Growth in Solution

Pedro H. C. Camargo, Thenner S. Rodrigues, Anderson G. M. da Silva and Jiale Wang

Abstract The controlled synthesis of metallic nanomaterials in solution is central to realize many applications that arise from their fascinating properties. As properties in metal nanomaterials are strongly dependent upon size, shape, composition, structure (solid versus hollow interiors), and surface functionality, controlled synthesis is a powerful approach to tailor and optimize properties as well as to establish how they are dependent on the several physical and chemical parameters that define a nanostructure. In this context, this chapter focuses on the fundamentals of the controlled synthesis of metal nanomaterials in solution phase in terms of the available theoretical framework. Specifically, it starts by introducing the mechanisms employed for the stabilization of nanomaterials during solution-phase synthesis (Sect. 2.2). The basics of nucleation and growth in solution will be discussed in Sect. 2.3. After that, the shape-controlled synthesis of Ag nanomaterials will be employed as proof-of-concept example of how thermodynamic versus kinetic considerations, oxidative etching, and surface capping can be employed to effectively maneuver the shape of a metal nanocrystal in solution (Sect. 2.4). Finally, some of the current challenges and outlook regarding the controlled synthesis of metal-based nanomaterials will be presented (Sect. 2.5).

2.1 Motivation for Controlled Synthesis of Metallic Nanomaterials

The motivation for the controlled synthesis of metallic nanomaterials comes from the fact that their properties are strongly dependent upon size, shape, composition, structure (solid versus hollow interiors), and surface functionality [1]. Thus, the precise control over these parameters opens up the possibility of designing nanomaterials with desired or optimized performances for a given application. Taking catalysis as an example, in addition to small sizes (that are imperative to achieve high-surface areas), the control over the shape enables the exposure of specific surface facets that

P. H. C. Camargo (✉) · T. S. Rodrigues · A. G. M. da Silva · J. Wang
Departamento de Química Fundamental, Instituto de Química, Universidade de São Paulo, Av. Prof. Lineu Prestes, 748, São Paulo 05508-000, Brazil
e-mail: camargo@iq.usp.br

© Springer International Publishing Switzerland 2015
Y. Xiong, X. Lu (eds.), *Metallic Nanostructures,* DOI 10.1007/978-3-319-11304-3_2

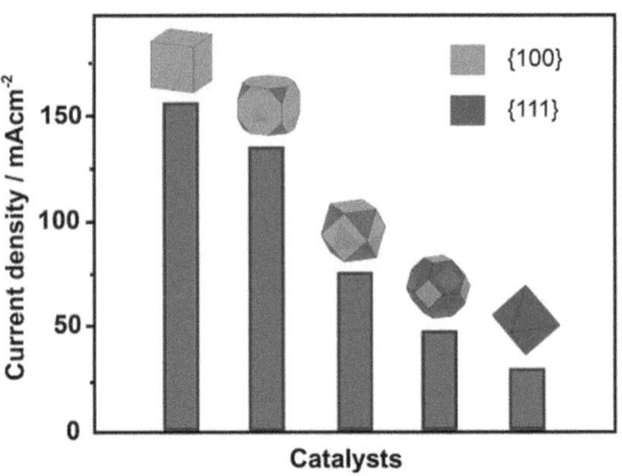

Fig. 2.1 Maximum current densities for formic acid oxidation employing Pd nanocrystals displaying controlled shapes as electrocatalysts. The current densities (electrocatalytic activities) were strongly dependent on the relative exposure of {100} and {111} surface facets, decreasing as the shapes transitioned from cubes to octahedrons [3]. Copyright 2012 American Chemical Society

can be more catalytically active or selective for a reaction of interest [2]. The shape-dependent catalytic activity of Pd nanocrystals towards the electroxidation of formic acid is shown in Fig. 2.1 and represents a notable example on how the control over shape can be employed for optimizing catalytic performance. As {100} surface facets are more active relative to {111} for the electroxidation of formic acid, a gradual increase in the current densities was observed as the shape was changed from cubes to truncated cubes, cuboctrahedrons, truncated octahedrons, and finally octahedrons, as these shapes gradually enable an increased exposure of {111} relative to {100} surface facets [3]. In addition to catalytic activity, optical, magnetic, and electronic properties have also shown shape/size-dependency [4].

It is noteworthy that, in order to reach its full potential, the controlled synthesis of metallic nanomaterials must meet several requirements that include monodispersity in terms of size, shape, structure, and composition. Among different methods for the controlled synthesis of metallic nanomaterials, solution-phase approaches offer many advantages that include the facile stabilization by the addition of proper capping agents, easy extraction/separation from the reaction mixture (by centrifugation, for example), straightforward surface modification/functionalization, potential for large-scale production, and versatility regarding the several experimental parameters that can be controlled during the synthesis (temperature, nature of precursors, stabilizers, solvent, reducing agents, their molar ratios, concentrations, etc.) [5]. Surprisingly, substantial developments on the solution-phase-controlled synthesis of metallic nanomaterials with a variety of shapes and sizes were made only recently. Figure 2.2 illustrates a variety of shapes that have been realized by the controlled synthesis of metal nanostructures [1]. Despite this progress, a detailed atomic understanding of the mechanisms of nucleation and growth stages during nanoparticle formation is still not completely understood, and most fundamentals are currently borrowed from classical colloidal theories [6].

In this chapter, rather than describing a variety of protocols for the controlled synthesis of different classes of metal nanomaterials, we will focus on the main fundamentals and concepts behind the controlled synthesis of metal nanomaterials in solution phase. Specifically, we will start by introducing the mechanisms employed

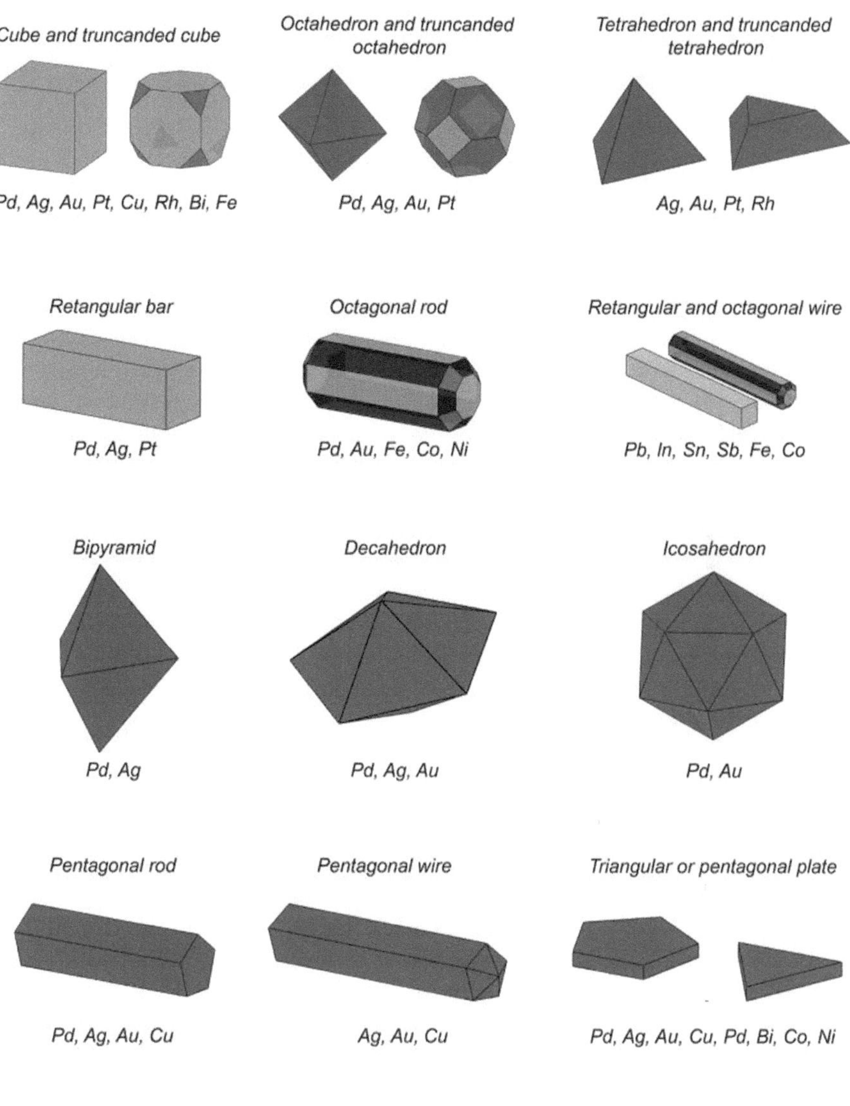

Fig. 2.2 Different shapes that have been enabled for a range of metals nanocrystals by solution-phase synthesis

for the stabilization of nanomaterials in the context of solution-phase synthesis (Sect. 2.2). The fundamentals of nucleation and growth in solution will be discussed in Sect. 2.3. After that, the shape-controlled synthesis of Ag nanomaterials will be employed as proof-of-concept example of how thermodynamic versus kinetic considerations, oxidative etching, and surface capping can be employed to effectively maneuver the shape of a metal nanocrystal in solution (Sect. 2.4). Finally, we will present some of the current challenges and outlook regarding the controlled synthesis of metal-based nanomaterials (Sect. 2.5).

2.2 Stabilization in Solution-Phase Synthesis

The main feature that sets nanomaterials apart from their micro- and macro-counterparts is their large surface to volume ratios, which is responsible for many unique or improved properties that nanomaterials possess [7]. A large surface to volume ratio implies that nanomaterials display high-surface energies (as surface energy increases with surface area), which makes them thermodynamically unstable or metastable [8]. Consequently, the synthesis of nanomaterials in solution requires one to overcome their large surface energies and prevent them from agglomerating. Agglomeration is thermodynamically driven by the reduction in surface area and thus surface energy. In fact, the reduction of surface energy is also the driving force for other processes in solution-phase syntheses such as surface restructuring, formation of facets, and Ostwald ripening [5]. In the following section, we will discuss two main approaches that are commonly employed to stabilize metallic nanomaterials against agglomeration: the electrostatic and steric (or polymeric) stabilization. Interestingly, as will be discussed in Sect. 3 and 4, the steric stabilization can also be put to work for favoring the formation of nanomaterials with monodisperse sizes and controlled shapes.

2.2.1 Electrostatic Stabilization

The concept behind the electrostatic stabilization is illustrated in Fig. 2.3 and it is based on the repulsion of electrical charges having the same sign present in the electrical double layers between neighboring particles [9]. The electrostatic stabilization can also be understood in terms of osmotic flow, in which solvent from the suspension flows into the region where the double layers overlap until the distance between the nanoparticles equals to or become larger than the sum of their individual double layers. As the overlap of double layers lead to a substantial increase

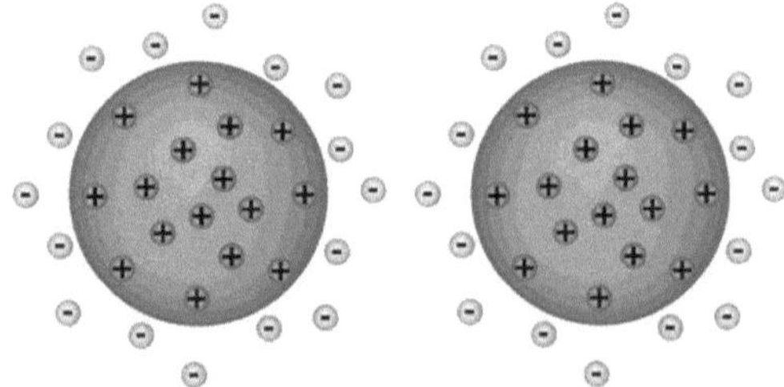

Fig. 2.3 Electrostatic stabilization between two approaching particles: the electrostatic repulsion of like-charges drive particles apart as their double layers overlap

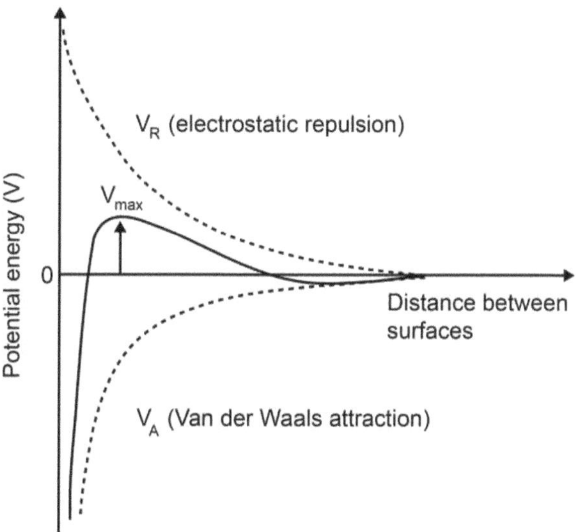

Fig. 2.4 Overall potential energy (V, *solid line*) expressed as a combination of the attractive Van der Waals and repulsive electrostatic potentials (V_A and V_R, respectively, *dashed lines*) [9]. Copyright © 2002, John Wiley and Sons

in the concentration of ions in the inter-particle region, the osmotic flow works to restore the equilibrium concentration.

The formation of a double layer arises from the adsorption of charged species at the surface, dissociation of surface-charged species, and/or accumulation or depletion of electrons at the surface following the formation of a solid surface in a polar solvent or electrolyte. An equal number of counter-ions with the opposite charge will then surround the system to produce a charge-neutral double layer, as depicted in Fig. 2.3. In addition to the contribution from columbic forces, Brownian motion and entropy also contribute to the location of species in solution, resulting in an inhomogeneous distribution of surface ions and counter ions [10]. The DLVO theory, which was named after Derjaguin, Landau, Verwey and Overbeek, successfully describes how the electrostatic stabilization between neighboring nanoparticles in suspension works [11]. The DLVO theory assumes that the interaction between two electrically charged approaching particles in suspension consists of a combination of Van der Waals attraction and electrostatic repulsion arising from the neighboring double layers [12, 13]. Figure 2.4 shows a plot of the overall potential (V, solid line) illustrated as a sum of Van der Waals attraction (V_A, dashed line) and electrostatic repulsion (V_R, dashed line) as a function of the distance between the approaching particles. When the two particles are far away from each other (distance is large), both V_A and V_R tend to zero. As the particles start to approach each other, the electrostatic repulsion is stronger than the Van der Waals attraction ($V_R > V_A$), leading to an increase in the overall potential (V) with the decrease in distance until a maximum is reached at V_{MAX}, which corresponds to the repulsive energy barrier. As the distance is further decreased, the Van der Waals attraction surpasses the electrostatic repulsion ($V_A > V_R$), and the overall potential becomes strongly dominated by V_A, which leads to the agglomeration of the particles. If $V_{MAX} > \sim 10\ kT$ (k is the

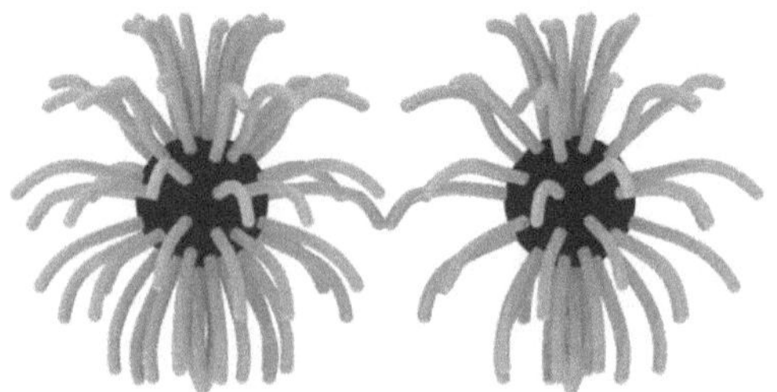

Fig. 2.5 Steric stabilization between two approaching particles containing anchored polymers at their surface. Under both low- and high-surface coverage conditions, the polymer layers repel one another as they approach and/or interact

Boltzmann constant and T the temperature), collision from Brownian motion will not overcome V_{MAX} and agglomeration will not take place. On the other hand, if $V_{MAX} < \sim 10\ kT$, agglomeration takes place and the particle suspension would not be stable.

It is important to note that these considerations indicate that electrostatic stabilization mechanism only applies to dilute systems, is dependent on the concentration of ions in suspension, agglomerated particles cannot be re-dispersed, and comprises a kinetic stabilization method. As an example, the widely employed synthesis of Au nanoparticles by the reduction of $AuCl^-_{4(aq)}$ by trisodium citrate pioneered by Turchevich [14] and refined by Frens [15] is based on electrostatic stabilization, in which citrate anions (negative charge) bind to the surface of the Au NPs during the synthesis.

2.2.2 Steric (or Polymeric) Stabilization

The steric (also called polymeric) stabilization consists in preventing neighboring particles of getting close to each other in the range of attractive forces where Van der Waals attraction would lead to agglomeration by the presence of polymers at the particles surface [16]. In general, a polymer can interact with a solid surface by one-end binding (anchored polymer) or via weak interactions from random points along its backbone (adsorbing polymer) [9]. When two particles covered with polymer approach each other, the polymer layers from each particle will interact only when the distance between particles become lower than the sum of the thicknesses of the neighboring polymer layers. Figure 2.5 depicts two approaching nanoparticles presenting an anchored polymer at their surface. As the distance between the particles becomes lower than the sum of the thicknesses of the neighboring polymer layers, they can overlap or become compressed resulting in strong repulsion between the neighboring nanoparticles. When the polymer is at good solvent conditions and neighboring adsorbed layers can overlap (under low-surface coverage), stabilization occurs as a result of the unfavorable mixing of the adsorbed layers, as an increase in entropy accompanied by an increase in the Gibbs free energy results from the interpenetration of polymers (assuming the enthalpy variations are negligible due to the interpenetration). At higher coverage (and/or under poor solvent conditions), the approaching adsorbed layers become compressed at the surface as the nanoparticles approach each other, and therefore, cannot interpenetrate. This process leads to an increase in the Gibbs free energy, driving the nanoparticles apart. Hence, steric stabilization emerges from the volume restriction effect (exclusion by space) due to the decrease in the possible configurations in the region between the particles and from the osmotic effect as the concentration of polymeric molecules in the region between the two particles becomes high [11].

It is important to mention that the steric stabilization is not so well understood in comparison to the electrostatic stabilization. Nevertheless, it is widely employed in the solution-phase synthesis of nanomaterials, offering many advantages when compared with electrostatic stabilization. For instance, steric stabilization is thermodynamic (agglomerated particles can be redispersed), it is not restricted to dilute systems, and is not sensitive to the presence of electrolytes, which is highly desirable for several applications. Moreover, it contributes to the production of monodisperse nanoparticles as the polymer layer adsorbed at the nanoparticle surface can also work as a diffusion barrier to the addition of growth species, favoring the diffusion-limited growth (Sect. 3.2) [5]. Polymers employed for steric stabilization can also play other roles in solution-phase syntheses such as serving as reducing agents, helping in manipulating precursor reduction rates, and interacting with distinct nanocrystals surface facets with different binding affinities, which is central in the context of shape-controlled synthesis [1, 17].

2.3 Fundamentals of Nucleation and Growth

The controlled synthesis of metal nanomaterials requires (and refers to) nanomaterials with monodisperse sizes and shapes. Therefore, all nanostructures produced from a reaction mixture must have sizes and shapes as similar as possible (narrow size and shape distribution). The controlled synthesis of metal nanomaterials from a homogeneous solution comprises two main stages: nucleation and growth. In this section, we will discuss the theory behind nucleation and growth stages in order to understand how these steps can be employed to obtain nanomaterials with monodisperse sizes.

2.3.1 Homogeneous Nucleation

Nucleation refers to the extremely localized building of a distinct thermodynamic phase and represents the first stages during a crystallization process. More specifically, it can be defined as the process by which building blocks (metal atoms in the synthesis of metal nanomaterials) arrange themselves according to their crystalline structure to form a site upon which additional building blocks can deposit over and undergo subsequent growth [18]. In this context, nuclei correspond to infinitesimal clusters consisting of very few atoms of the growth species.

Lamer et al. pioneered the study on the synthesis of uniform colloids and the related nucleation and growth mechanisms [6]. These studies date back from the 1940s and are still the basis of our understanding on the synthesis of uniform nano-materials (including metallic, polymeric, semiconducting, and oxide systems) [19]. From these studies, the concept of "burst nucleation" has been established to be crucial for the synthesis of monodispersed particles. Burst nucleation refers to the formation of a large number of nuclei in a short period of time, followed by growth without additional nucleation. This allows nuclei to have similar growth histories and, therefore, yield nanoparticles with same sizes. This concept is normally re-ferred to as the separation of nucleation and growth stages during the synthesis [19]. For instance, if nucleation and growth could occur simultaneously, the nuclei growth histories would be significantly different and heterogeneous size distribu-tions would be obtained.

The supersaturation of growth species comprises the first requirement for the occurrence of homogeneous nucleation. In the synthesis of metal nanoparticles, su-persaturation may be achieved via the in situ generation of the solute growth species (metal atoms) through the conversion of soluble precursors into metal atoms by re-duction or decomposition reactions. Alternatively, supersaturation could be induced by the decrease in the temperature of an equilibrium mixture containing the growth species. Although nucleation can take place in liquid, gas, and solid phases, we will center our discussion on the liquid phase, as the fundamentals are essentially the same in all cases.

The reduction of the overall Gibbs free energy to generate a new solid phase from a supersaturated solution containing the growth species as the solute repre-sents the driving force for homogenous nucleation. The change in Gibbs free energy per unit volume of the solid phase (ΔG_v) is dependent on the concentration of the solute and can be expressed as follows:

$$\Delta G_v = -kT / \Omega \ln(C / C_0) \tag{2.1}$$

where T is the temperature, Ω the atomic volume, C the concentration of the solute, and C_0 the solubility or equilibrium concentration [20]. When $C > C_0$, ΔG_v becomes negative and nucleation is spontaneously favored. However, this decrease in the Gibbs free energy is accompanied by an increase in the surface energy per unit area (γ) of the newly formed solid phase. Thus, the overall Gibbs free energy change (ΔG) for the formation of spherical nuclei with radius r from a supersaturated solu-tion corresponds to:

$$\Delta G = (4/3)\pi r^3 \Delta G_v + 4\pi r^2 \gamma \tag{2.2}$$

where the first term corresponds to the change of volume free energy (volume term, $\Delta \mu_v$) and the second to the change in surface free energy (surface term, $\Delta \mu_s$). Figure 2.6 displays a plot of ΔG as a function of r (solid line) [20]. The surface and volume terms are also shown as dashed lines. As γ is always positive and ΔG_v nega-tive as long as the solution is supersaturated, the plot of ΔG as a function of r has a

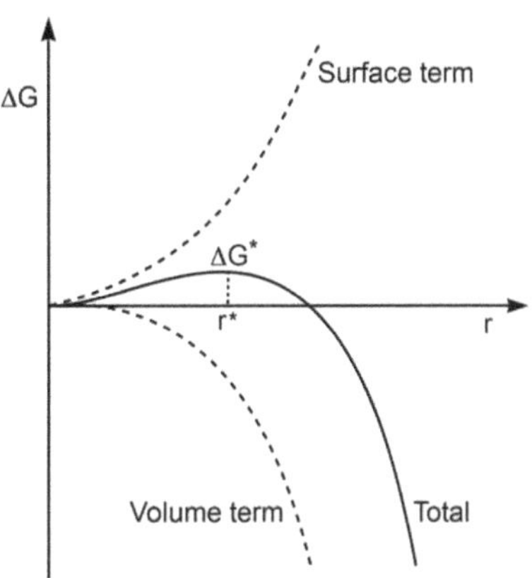

Fig. 2.6 Change in the Gibbs free energy (G, *solid line*) as a function of the nucleus radius (*r*) as a sum of the volume and surface free energies [20]. Copyright © 2001, Elsevier

maximum that corresponds to the critical radius r*. The critical radius (r*) can be defined as the minimum radius that a nucleus must possess in order to spontaneously grow from a supersaturated solution, or the minimum size required for the formation of a stable nucleus by precipitation from a supersaturated solution. When r<r*, the surface term is higher than the volume term and any nucleus smaller than r* will dissolve into the solution. In other words, the increase in the surface energy due to the nucleus formation is higher than the stabilization due to the formation of a new solid phase from the supersaturated solution, and nucleation is not favored. On the other hand, when r>r*, the volume term exceeds the contribution from the surface term so that the created nucleus is stable and undergoes subsequent growth (the stabilization due to the formation of the new solid phase from the supersaturated solution becomes higher than the destabilization due to the increase in surface energy). Setting dΔG/dr allows us to find r* and ΔG* according to the following equations:

$$r^* = -2\gamma/\Delta G_v \tag{2.3}$$

$$\Delta G^* = 16\pi\gamma/3(\Delta G_v)^2 \tag{2.4}$$

where ΔG* corresponds to the critical Gibbs free energy, i.e., the minimum energy barrier that must be reached for nucleation to take place. These equations illustrate that r* can be decreased by lowering the γ of the new solid phase. Also, increasing ΔG_v will lead to a decrease in both r* and ΔG* (ΔG_v increases with the supersaturation, Eq. 1). Surface energy, on the other hand, can be manipulated by changes in the solvent, temperature, and the use of "impurities" in the reaction mixture such as capping agents [19]. As an example, Fig. 2.7 shows how r* and ΔG* for a spherical nucleus can vary as a function of temperature. It can be observed as an increase in both r* and ΔG* with temperature. This is because supersaturation (and thus ΔG_v) decreases as the temperature is increased (Eq. 1) [5].

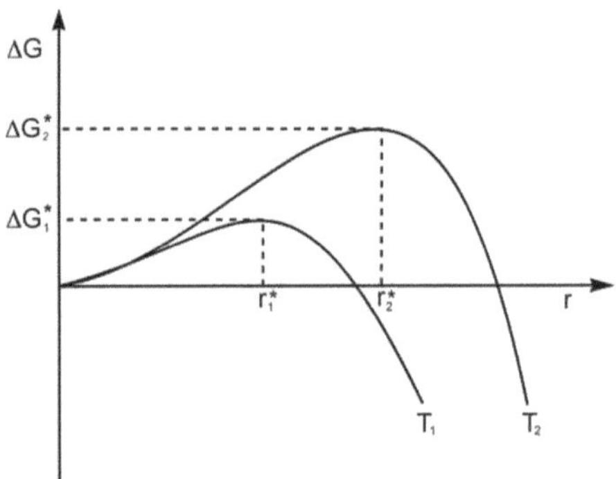

Fig. 2.7 Change in the Gibbs free energy (G) as a function of the nucleus radius (r) for two different temperatures, where $T_1 < T_2$ [20]. Copyright © 2001, Elsevier

The nucleation rate (N_R, per unit volume and unit time) is defined as the increase in the number of particles as a function of time and can be written as a function of ΔG^* in the Arrhenius form:

$$dN/dt = N_R = \exp(-\Delta G^*/kT) \tag{2.5}$$

N_R is also dependent on the number of growth species per unit volume (number of nucleation centers, which is determined by the initial concentration C_0) and the jump frequency of growth species from one site to another. Thus, N_R can be written as:

$$N_R = \{(C_0 kT / (3\pi\eta d^3)\}\exp(-\Delta G^*/kT) \tag{2.6}$$

where η corresponds to the viscosity of the solution and d to the diameter of growth species. This equation shows that high nucleation rates are promoted by high initial concentration of growth species (high supersaturation), low viscosity, and low ΔG^*. Moreover, in order to start the accumulation of nuclei for subsequent growth, N_R must be high enough to surpass the re-dissolution of nuclei. It is important to mention, however, that the thermodynamic model discussed above has some limitations when the synthesis of nanocrystals is regarded, as both γ and ΔG_v should not be constant and vary with size [19].

Owing to their sub-nanometer sizes, there is actually little information and knowledge regarding the true identity of nuclei during the synthesis of metal nanomaterials due to the lack of experimental tools capable of precisely capturing and characterizing them. According to theoretical investigations and evidence from mass spectrometry as well as adsorption and emission spectroscopies, it has been proposed that the crystal embryos seem to be closely related to their corresponding metal clusters [1, 21].

Figure 2.8 depicts the Lamer plot, which shows the change in the atomic concentration of the solute (growth species) during the nucleation and growth processes as a function of time. This plot is also very useful for illustrating the concept of burst nucleation. In the synthesis of metal nanocrystals, the concentration of solute (metal atoms) is increased by the decomposition or reduction of the corresponding

Fig. 2.8 Variations in the atomic concentration of growth species in solution as a function of time during **a** the generation of atoms, **b** nucleation, and **c** growth stages [1]. Copyright © 2009 WILEY-VCH Verlag GmbH & Co. KGaA, Weinheim

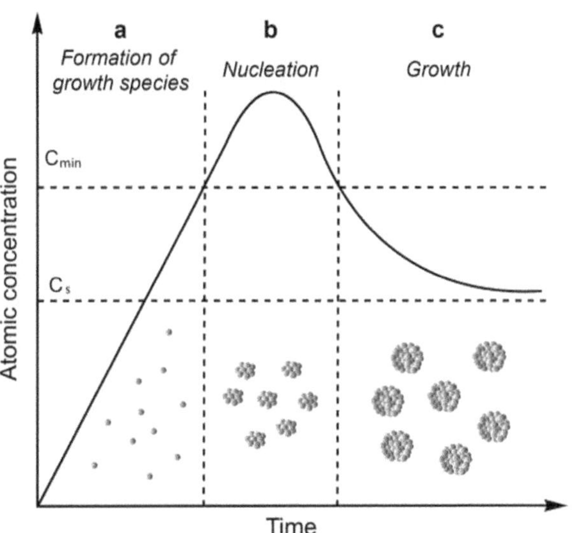

soluble precursors. In the first stage, although the concentration of metal atoms increases, no nucleation/precipitation occurs even after the concentration of solute is above the equilibrium concentration (C_s). At this stage, although the solution is supersaturated, the energy barrier for nucleation (ΔG^*) is high due to the contribution from surface energy. Nucleation takes place only when the supersaturation reaches a specific value above the solubility (C_{min} = minimum solute concentration required for precipitation), which corresponds to the point where ΔG^* is reached for the formation of nuclei. At this point, atoms start to aggregate into nuclei and homogeneous nucleation begins. As nucleation takes place, the concentration or supersaturation of metal atoms in solution starts to decrease, leading to a decrease in ΔG_v. When the concentration is below C_{min} (where ΔG^* is reached), nucleation stops and growth starts. Growth then occurs under a continuous supply of metal atoms (solute) as long as their concentration is above the equilibrium solubility (C_s). In addition to growth via atomic addition and Ostwald ripening, the nuclei can also merge into larger objects via coalescence and oriented attachment [22]. The stabilization mechanisms discussed in Sect. 2 are essential to avoid extensive agglomeration that would hamper the production of nanoscaled materials. According to the Lamer plot, the concentration of the growth species should be increased abruptly to a very high supersaturation level followed by quickly bringing it below C_{min} in order to achieve burst nucleation [6].

2.3.2 Growth

In addition to the "burst nucleation" concept and the subsequent separation of nucleation and growth during the synthesis, the size distribution of nanoparticles in solution-phase synthesis is dependent on the growth stage. More specifically, the

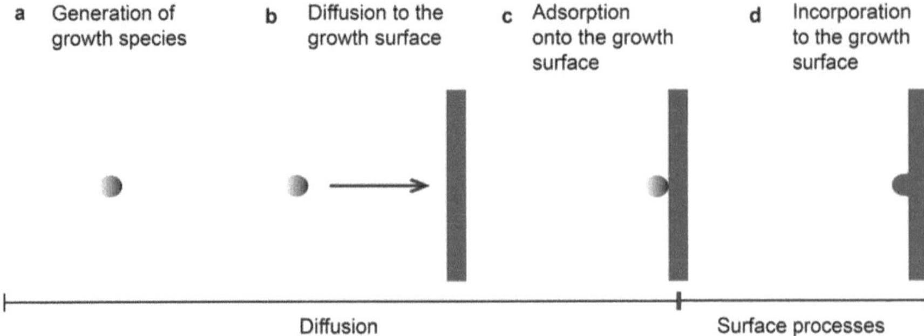

Fig. 2.9 Schematics of the main steps that occur during growth: generation of growth species (**a**), followed by their diffusion (**b**), adsorption (**c**), and irreversible incorporation (**d**) to the growth surface. Steps (**a**–**c**) comprise diffusion while (**d**) is regarded as growth itself

size distribution of the initial nuclei may increase or decrease depending on the kinetics of growth [23]. The growth of nuclei can be subdivided in four major steps as illustrated in Fig. 2.9: generation of growth species (Fig. 2.9a), their diffusion (Fig. 2.9b), adsorption (Fig. 2.9c), and irreversible incorporation to the growth surface (Fig. 2.9d). Steps (a–c) can be further grouped and regarded as diffusion (supplying the growth species to the growth surface), while (d) corresponds to growth itself (by surface processes). Therefore, growth can be controlled either by diffusion or surface processes, and each of these regimes lead to differences in the final size distribution.

First, let us consider the case in which the growth is controlled by diffusion, i.e., controlled by the diffusion of growth species from the bulk of the solution to the growth surface. This situation is favored when diffusion is the slowest step during growth. In this condition, the growth rate (dr/dt) of a spherical nucleus of radius r depends on the flux of growth species to its surface (J) according to the equation:

$$\mathrm{dr/dt} = \mathrm{JV_m}/4\pi\mathrm{r}^2 \tag{2.7}$$

where $\mathrm{V_m}$ is the molar volume of the nuclei. If the distance among nuclei is large enough, their growth can be considered independent from one another, and thus the diffusion layer formed at the surface of each growing particle will be undisturbed. The Fick's law of diffusion expresses the flux of growing species (J) through the diffusion layer of each growing particle as:

$$\mathrm{J} = 4\pi\mathrm{x}^2\mathrm{D}(\mathrm{dC/dx}) \tag{2.8}$$

where D is the diffusion coefficient, C the concentration, and x the distance from the center of the growing particle. Assuming that J is constant with respect to x and integrating C(x) from x to x+δ, we get:

$$\mathrm{J} = 4\pi\mathrm{Dr}(\mathrm{r}+\delta)/\delta\left[\mathrm{C}(\mathrm{r}+\delta)-\mathrm{C_s}\right] \tag{2.9}$$

where C_s is the concentration at the surface of the growing particle. At large δ values ($\delta > r$), Eq. 9 becomes:

$$J = 4\pi Dr(C_{bulk} - C_s) \qquad (2.10)$$

where C_{bulk} is the concentration of the bulk solution [19]. Hence, from Eq. 7 and 10 we get the growth rate as:

$$dr/dt = V_m D\ (C_{bulk} - C_s)/r \qquad (2.11)$$

indicating that the growth rate of a particle is inversely proportional to its radius. Assuming the initial radius of the nuclei to be r_0 and the change in C_{bulk} to be negligible, this equation can be solved as:

$$r^2 = k_D t + r_0^2 \qquad (2.12)$$

where $k_D = 2DV_m(C - C_s)$. Thus, for two particles having an initial radius difference δr_0, the radius difference δr as the particles grow decreases with time according to Eq. 13 and 14, demonstrating that the growth limited by diffusion induces the formation of nanoparticles with monodisperse sizes [5].

$$\delta r = r_0 \delta r_0 / r \qquad (2.13)$$

$$\delta r = r_0 \delta r_0 / (k_D t + r_0^2)^{1/2} \qquad (2.14)$$

From the experimental point of view, growth limited by diffusion can be favored by: (i) keeping the concentration of growth species very low, so that the distance for diffusion from the solution to the growth surface would be large; (ii) increasing the viscosity of the solvent; (iii) controlling the supply of growth species to the growth surface; and (iv) introducing a diffusion barrier at the particle's surface. Interestingly, in the controlled synthesis of metallic nanomaterials, many polymers employed to stabilize the nanoparticles during the synthesis (by steric stabilization) can also serve as a diffusion barrier that aids to promote diffusion-limited growth.

Now let us turn our attention to the opposite scenario in which the diffusion of growth species from the solution to the surface is fast, and the growth rate becomes controlled by surface processes [24]. The surface processes in which the adsorbed growth species becomes irreversibly incorporated at the surface (growth itself, Fig. 2.9d) can follow a layer-by-layer or polynuclear pathway. In the layer-by-layer pathway, the growth species have enough time to diffuse on the surface so that a second layer if formed only after a complete layer is produced. In this condition, the growth rate is proportional to the surface area and can be written as:

$$dr/dt = k_m r^2 \qquad (2.15)$$

where k_m is a proportionality constant. Solving this equation we have:

$$1/r = 1/r_0 - k_m t \qquad (2.16)$$

Thus, the radius difference between two growing particles increases with size (r) and time (t) according to the following expressions, showing that this mechanism does not promote the formation of monodisperse sizes:

$$\delta r = r^2 \delta r_0 / r_0^2 \qquad (2.17)$$

$$\delta r = \delta r_0 / (1 - k_m t r_0)^2 \qquad (2.18)$$

In the polynuclear mechanism, the concentration of growth species at the surface becomes so high that the formation of a second growing layer is favored before first growth layer is complete. In this case, growth species do not have enough time to diffuse on the surface and the growth rate becomes independent of both particle size and time. Particles then grow linearly with time and the radius difference between distinct growing particles remains constant. Although the absolute radius difference is constant, relative radius difference becomes smaller with the increase in size and time, also favoring the formation of monodisperse nanoparticles (to a lesser extent relative to growth controlled by diffusion). Therefore, the above discussion indicates that growth limited by diffusion represents the best growth condition for achieving monodispersed nanoparticles.

In addition to the classical crystal growth, which can be roughly described by diffusion and incorporation of metal atoms to the growth surface as well as the diffusion of metal atoms from more soluble to the less soluble crystals (Ostwald ripening), the so-called non-classical mechanisms also play an important role during the growth of metal nanostructures [25, 26]. In fact, recent advances regarding in situ transmission electron microscopy techniques and time-resolved small-angle X-ray scattering measurements have allowed the direct observation of solution-phase growth in metal nanoparticles and demonstrated that non-classical mechanisms such as aggregation, coalescence, and oriented attachment affect the growth route and final morphology of metal nanoparticles [22, 27–29]. Specifically, these non-classical mechanisms refer to growth in which primary crystallites assemble into secondary crystals, and can be understood as follows: first, the reversible formation of loose assemblies of primary particles is observed, which are free to rotate and rearrange with respect to one another through Brownian motion. These primary particles can either dissociate or irreversibly attach to each other along some preferential crystallographic orientations to produce a new secondary crystal. In this process, the formation of unique architectures, the incorporation of stacking faults, twin defects, and edge dislocations can be observed [26, 30, 31]. It is important to note that an aggregate corresponds to an irreversible assembly, and thus oriented attachment refers to an irreversible assembly of oriented primary particles. Although the detailed assembly process and the driving forces for the oriented attachment of primary particles along certain directions are still unclear, reduction in the overall surface energy by eliminating some high-energy facets represents one plausible explanation [26].

2.3.3 *Growth by Heterogeneous Nucleation (or Seeded Growth)*

Heterogeneous nucleation refers to the formation of the new solid phase (nucleation) onto the surface of another material [32]. Heterogeneous nucleation possess lower ΔG^* (critical Gibbs free energy barrier for nucleation), and thus are more thermodynamically favored relative to homogeneous nucleation [18]. While in homogeneous nucleation the nucleus is approximated by a sphere whose surface area is $4\pi r^2$, the nuclei in heterogeneous nucleation have a surface area lower than $4\pi r^2$ due to the introduction of the interface between the nucleus and the surface, which leads to a decrease in the surface free energy (recall that the increase in ΔG is due to the surface free energy, Eq. 2). The synthesis of metal nanomaterials by heterogeneous nucleation (seeded growth) in solution employing nanoparticles obtained in a separate/previous synthetic step as seeds is very helpful, as it enables one to effectively separate nucleation and growth processes as well as to apply a much wider range of growth conditions, milder reducing agents, lower temperatures, and different solvents. This approach has been extensively employed for the shape-controlled synthesis of a variety of metal nanocrystals, including bimetallic (core-shell) systems where the composition of seeds differs from the growth species [32–34]. For example, if the seed particles possess well-defined shapes and are monodisperse, nanocrystal growth via heterogeneous nucleation can not only produce monodisperse nanocrystal sizes but can also influence (and in some cases template) the shape of the resulting nanocrystal after the final growth step [35].

2.4 Manipulating Nucleation and Growth for Shape-Control

2.4.1 *Growth Mechanisms*

We have mentioned in the previous sections that the solution-phase synthesis of metal nanomaterials is comprised of two main stages: nucleation and growth. In the shape-controlled synthesis of metal nanomaterials, these stages can be further subdivided into (i) nucleation; (ii) growth of nuclei into seeds; (iii) and evolution of the seeds into the final nanocrystal. These steps are schematically illustrated in Fig. 2.10 [1]. As nuclei correspond to infinitesimal clusters consisting of very few atoms formed as the new solid phase in solution, structure fluctuations may occur in the nuclei with the incorporation of defects in order to minimize surface energy until they evolve into seeds. The seeds can then be defined as species that are larger than the nuclei that are locked into a specific structure, in which structure

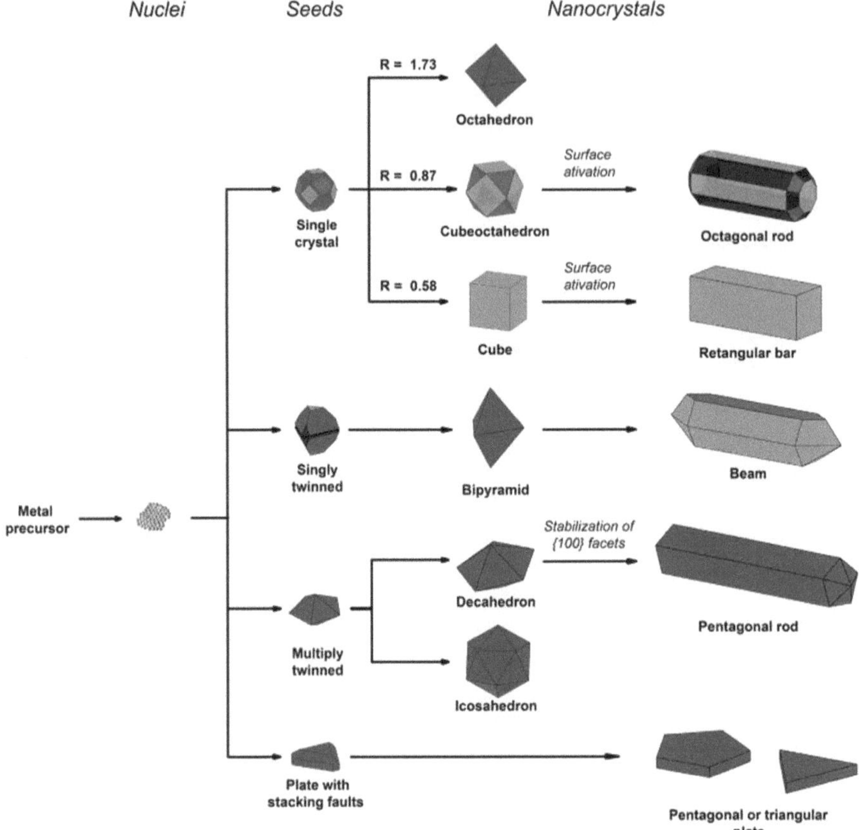

Fig. 2.10 Main stages of controlled synthesis of metal nanomaterials (with face-centered cubic (fcc) structure): (i) nucleation; (ii) growth of nuclei into seeds; (iii) and evolution of the seeds into the final nanocrystal. The shape assumed by the nanocrystal depends on the structure of its corresponding seed and the relative binding affinities of capping agents to its distinct surface facets. R, for example, refers to the ratio between the growth rates along the {100} and {111} directions [38]. Copyright © 2007 WILEY-VCH Verlag GmbH & Co. KGaA, Weinheim

fluctuation does not occur as they become energetically more costly as the size increases. Interestingly, for several systems (including Ag, Au, and Pd), it has been established that the final shape assumed by the nanocrystal is determined by both the structure of its corresponding seed and the relative binding affinities of capping agents to its distinct surface facets [36–38].

Figure 2.10 illustrates how the structure of the seeds play an important role in determining the final shape of a metal nanocrystal in solution-phase synthesis [38]. For example, a single-crystalline seed (truncated octahedron) can evolve into an octahedron, cuboctahedron, or cube depending on the relative growth rate of {111} and {100} surface facets (expressed as R in Fig. 2.10). Considering their structure, seeds often assume single-crystalline, single-twinned, and/or multiple-twinned

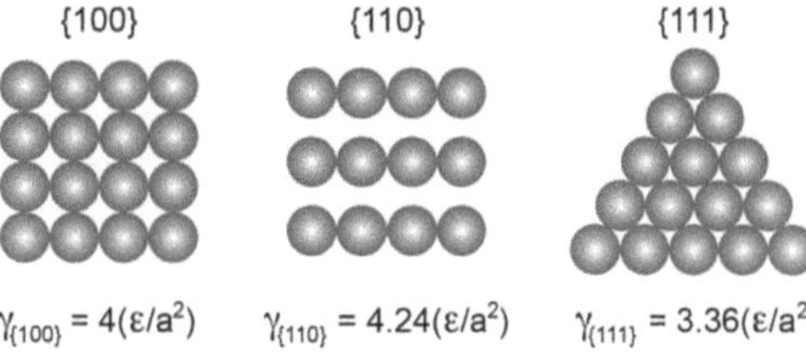

Fig. 2.11 Atomic arrangements and surface energy values for {100}, {110}, and {111} surface facets (from *left* to *right*, respectively)

structures. Thus, the key for the synthesis of metal nanocrystals with controlled shapes relies on maneuvering nucleation and growth processes so that only one kind of seed, with the proper structure, is present in the reaction mixture. The structure of seeds during solution-phase synthesis (single crystalline, single-twinned, or multiple-twinned) are dependent upon many parameters that include thermodynamic and kinetic considerations, while the relative population distribution of seeds can be further influenced by oxidative etching processes. Moreover, surface capping plays an important role in controlling the shape as the final nanocrystal evolves from its corresponding seed during nanocrystal growth.

When the synthesis in under thermodynamic control, the population of seeds displaying different structures will be determined by the statistical thermodynamics of their free energies. The most stable (lowest energy) seeds will be favored, and surface energy considerations are crucial in understanding and predicting the structure of the produced seeds. Surface energy (γ) can be defined as the energy required for creating a unit area of "new" surface, or as the excess free energy per unit area for a particular crystallographic face according to Eq. 19 and 20:

$$\gamma = (\delta G / \delta A)_{n,T,P} \tag{2.19}$$

$$\gamma = (1/2)N_b \varepsilon \rho \tag{2.20}$$

where G is the free energy, A the surface area, N_b the number of bonds that need to the broken to produce the new surface, ε the bond strength, and ρ the density of surface atoms. [39] When we consider an amorphous solid or liquid droplet, these equations show that simply decreasing the surface area can lower the total surface energy, and the perfectly symmetric sphere is the resulting shape. However, noble metals assume a face-centered cubic (fcc) crystal structure, in which the distinct surface facets possess different atomic arrangements and surface energy values as illustrated in Fig. 2.11. Stable morphologies that enable the minimization of free energy can be achieved by the exposure of low-index crystal planes that exhibit closest atomic packing (for fcc metals, $\gamma\{111\}<\gamma\{100\}<\gamma\{110\}$). Although these surface energy values suggest that single-crystal seeds should assume octahedral or tetrahedral shapes (which enables the exposure of {111} facets), they have higher surface areas relative to a cube of the same size. The formation of single-crystal seeds can then be considered in the context of Wulff's theorem, which represents a method to determine the equilibrium shape of a droplet or crystal having a fixed volume [40]. Energy minimization arguments are used to show that certain crystal planes are preferred over others, giving the crystal its final shape. According to the Wulff construction, single-crystal seeds are expected to exist as truncated octahedrons enclosed by a mix of {111} and {100} facets. This shape, which presents a nearly spherical profile, displays the smallest surface area and leads to the minimization of the total surface energy. Single- and multiple-twinned seeds have also been observed in the synthesis of metal nanocrystals under thermodynamic control. Here, a twin defect refers to single atomic layer in the form of a (111) mirror plane. The surface of a single-twinned seed tends to be enclosed by a mix of {111} and {100} facets to lower the surface free energy, while a multiple-twinned seed is enclosed by {111} facets [41]. As the strain energy caused by the twin defects greatly increases as the seeds grow in size, multiple-twinned seeds are stable only at smaller sizes [42].

Kinetic control relies on the manipulation over the structure/population of seeds containing different numbers of twin defects by decreasing the reduction or decomposition rate of a precursor that is responsible for providing metal atoms (growth species). In fact, when the decomposition or reduction becomes considerably low, atoms tend to form nuclei and seeds through random hexagonal close packing together with the inclusion of stacking faults, and the resulting seeds typically takes a shape that deviates (having higher surface energies) from those favored by thermodynamics [43]. In practice, structures that deviate from thermodynamic considerations can be achieved by a set of special conditions that enable an extremely low concentration of metal atoms in solution so the nuclei will not be able to grow into polyhedral structures. These include substantially slowing down precursor decomposition or reduction rates, coupling the reduction to an oxidation process, and/or by taking advantage of Ostwald ripening (process by which relatively large structures grow at the expense of smaller ones) [1].

In addition to thermodynamic and kinetic control, which dictate the identity and structure of the produced seeds, the relative population of single-crystal and twinned seeds can be influenced by oxidative etching processes, in which metal atoms can be oxidized back to the ionic form by O_2 from air (present in the synthesis) and a ligand such as Cl^- and Br^- [44]. The defect zones in twinned seeds have higher energy relative to single-crystal regions, making them more susceptible to undergo oxidation. Thus, multiply-twinned seeds are more susceptible to oxidation than single-twinned and single-crystal seeds, respectively [45]. Consequently, if properly controlled, oxidative etching can enable the selective removal of a specific kind of seed from the reaction mixture as well as to control the number of seeds [46, 47].

When we consider the evolution of the seeds into the nanocrystal, capping agents that facilitate the interaction of distinct surface facets with different binding affinities can be put to work to control the shape of the produced nanocrystal. Upon interaction with the surface, capping agents can change the order of free energies from distinct facets, changing their relative growth rates. Therefore, the use of a capping agent can be considered as a thermodynamic approach. For example, a capping agent that binds more strongly to a specific facet can slow down its growth relative to the other facets, allowing for the control over its relative surface area in the final nanocrystal.

This is illustrated by the possibility of producing octahedron, cuboctahedron, or cube nanocrystals from a single-crystal seed as shown in Fig. 2.10. If a capping agent that binds more strongly to the {111} relative to the {100} facet from the single-crystal seed is employed, the growth along {111} is significantly slowed down relative to the {100} direction producing octahedrons (contain only {111} surface facets). In the opposite scenario, cubes would be obtained. Intermediary conditions would lead to the formation of cubeoctahedrons (enclosed by a mix of {111} and {100} surface facets). It is important to emphasize that although many successful examples have been reported on the use of capping agents to control the shape of metal nanocrystals, the exact mechanisms are still not completely understood due to the lack of experimental techniques to precisely probe the molecular structure and resulting surface energy of different facets upon the binding of a capping agent.

Now that we have discussed the fundamentals of nucleation and growth, as well as the general concepts involved in synthesis of metal nanomaterials with controlled shapes, let us discuss the controlled synthesis of Ag nanostructures as an example on how thermodynamic and kinetic control, oxidative etching, and surface capping can be employed to give birth to a myriad of nanocrystals with monodispersed sizes and well-defined shapes.

2.4.2 Controlled Synthesis of Silver Nanomaterials

In the last decade, many methods have been developed for the synthesis of Ag nanocrystals displaying uniform size distribution and variety of shapes that include cubes, bypiramides, wires, octahedrons, triangular plates, bars, among others [1, 17, 37, 48]. Representative scanning electron microscopy images of some of these shapes are as illustrated in Fig. 2.12. Among the various solution-phase methods for obtaining Ag nanocrystals, we will focus our discussion on the polyol reduction. Polyol reduction is probably the best-established protocol for generating Ag nanomaterials with controllable shapes [37]. This approach is illustrated in Fig. 2.13 and it is based on the utilization of ethylene glycol as both a solvent and source of reducing agent (glycoaldehyde), high temperatures ($\sim 140\text{--}160\,^\circ$C), Ag^+ as the Ag precursor, and polyvinylpyrrolidone (PVP) as the capping agent. As the reduction of Ag^+ to Ag starts, the concentration Ag^0 in solution starts to increase. As discussed in Sect. 3.1, nucleation takes place only when the supersaturation with Ag reaches a value above the solubility where ΔG^* is satisfied. As the nuclei are formed, they can undergo structure fluctuations at small sizes due to the input of thermal energy so that defects can be incorporated to reduce their surface energy [49]. For example, the formation of a twin defect can be thermodynamically favored at small sizes as they result in the exposure of {111} (lowest energy) facets. As nuclei grow, these structural fluctuations start to become energetically more costly. The birth of the seeds is reached when these structure fluctuations stop (thermal energy is not enough to enable structure fluctuation). In the polyol synthesis of Ag nanomaterials, a mixture of multiple-twinned, single-twinned, and single-crystal seeds is produced in the reaction mixture as shown in Fig. 2.13, which can evolve into wires, bypiramids, and cubes as shown in Fig. 2.13a–c, respectively [50].

Fig. 2.12 Scanning electron micrographs of Ag wires (**a**), cubes (**b**), truncated octahedrons (**c**), octahedrons (**d**), bypiramids (**e**), and plates (**f**) [36, 47, 48, 55]. **a** Reproduced by permission of the PCCP Owner Societies, (**c–d**) Copyright © 2006 Wiley-VCH Verlag GmbH & Co. KGaA, Weinheim, (**e**) Copyright 2006 American Chemical Society, (**f**) Copyright © 2006 Wiley-VCH Verlag GmbH & Co. KGaA, Weinheim

The multiple-twinned (decahedral) seeds are the most abundant ones (relative to single crystal and single-twinned seeds) in the reaction mixture as introduction of twin planes is thermodynamically favored at small sizes [51]. Therefore, the removal of the other seeds is not required to achieve the synthesis of Ag nanowires. As their twin defects are more reactive relative to single-crystal regions, Ag atoms (growth species) preferentially add to the twin defects of the multiple-twinned seeds during growth, leading to the elongation of the seeds producing small pentagonal rods. As PVP is employed as a surface capping agent and has a higher binding affinity with the {100} side facets relative to the {111} end facets, growth along the side facets is hindered, leading anisotropic growth into nanowires tens of micrometers long (Fig. 2.12a and 2.13a) [47, 52]. Although the other seeds (single crystal and single-twinned) may also grow in the same reaction mixture, their smaller sizes relative to nanowires facilitate their separation by centrifugation.

The removal of twinned seeds is required to achieve the synthesis of Ag nanocubes (Fig. 2.12b and 2.13c). This selective removal of twinned seeds can be accomplished by oxidative etching upon the addition of Cl$^-$ ions in the reaction mixture [53]. In this case, the O$_2$/Cl$^-$ pair becomes an efficient etchant of twinned seeds owing to their higher susceptibility to oxidation relative to the single-crystal seeds [54]. As PVP binds more strongly {100} relative to the {111} facets in the single-crystal seeds that are left behind, cubes are produced. Interestingly, it has been observed that under a continued supply of Ag atoms by extending the polyol reaction, the Ag nanocrystal shape can evolve from a cube to a cuboctahedron and then to an octahedron, with an increasing ratio of {111} to {100} facets (Fig. 2.12c and d,

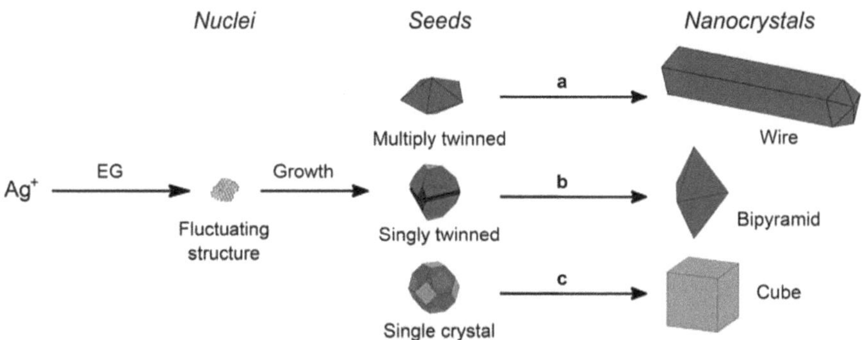

Fig. 2.13 Polyol approach for the synthesis of Ag nanocrystals displaying controlled shapes. A mixture of multiple-twinned, single-twinned, and single-crystal seeds is produced in the reaction mixture, which can grow into wires (**a**), bypiramids (**b**), and cubes (**c**) in the presence of polyvinyl-pyrrolidone (PVP) as the capping agent [50]. Copyright 2006 American Chemical Society

respectively) [36]. In this case, rather than a layer by layer growth, Ag selectively deposits over the {100} cube facets, slowing down the growth rate along the {111} direction and enabling its increase exposure at the surface of the final nanocrystal [36].

In order to selectively produce single-twinned seeds, the degree of oxidative etching must be moderated so that only the multiple-twinned seeds (that are the most reactive) are removed from the reaction mixture by oxidation. This can be performed by replacing Cl^- with the milder etchant Br^- and has enabled the synthesis of right bypiramids (Fig. 2.12e and 2.13b) [48].

All these shapes discussed so far (wires, cubes, octahedrons, and bypiramids) are thermodynamically favored. In order to obtain thermodynamically unfavorable shapes, such as triangular nanoplates (Fig. 2.12f), different strategies must be employed, such as the substantial reduction on the supply of Ag atoms by greatly reducing the precursor reduction rate [1]. In this case, Ag atoms may form plate-like seeds through random hexagonal close packing with the inclusion of stacking faults, that can subsequently grow to yield Ag triangular plates [43]. The use of the hydroxyl end groups of PVP as an extremely mild reducing agent has been employed as an effective approach to slow down the Ag^+ reduction rate and yield the formation of Ag triangular plates [55].

It is important to mention that these Ag nanocrystals could be further employed as seeds for overgrowth by heterogeneous nucleation (seeded growth) in order to obtain nanomaterials with monodisperse sizes. Moreover, the shape can be further manipulated by the proper choice of capping agents. In a typical example, it has been demonstrated that Ag cuboctahedrons (single-crystal structures enclosed by a mix of {111} and {100} surface facets) could be employed as seeds for Ag heterogeneous nucleation and growth, as shown in Fig. 2.14a [35]. In this seeded-growth process, the utilization of sodium citrate (Na_3CA) as capping agent led to the production of Ag octahedrons enclosed by {111} facets (Fig. 2.14b), while the utilization of PVP led to the formation of cubes/bars enclosed by {100} facets (Fig. 2.14c).

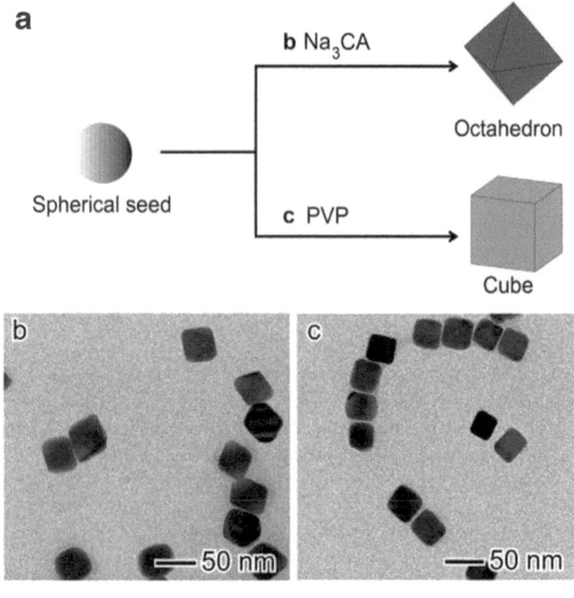

Fig. 2.14 **a** Scheme for the formation of Ag octahedrons or cubes by overgrowth using Ag cuboctahedrons as seeds and Na₃CA or PVP as capping agents. **b** and **c** Show transmission electron micrographs for the produced octahedrons and cubes, respectively [35]. Copyright 2010 American Chemical Society

It has been established that, in the case of Ag, citrate binds more strongly to {111} relative to {100} facets.[56] Therefore, the utilization of citrate as capping agent allows for the reduction on the growth rate along the [111] direction and thus leading to the exposure of {111} facets in the final nanocrystal (formation of octahedrons). Conversely, PVP binds more strongly to {100} than {111} facets of Ag, reducing the growth rate along the [100] direction and resulting in the exposure of {100} facets in the final nanocrystal (formation of cubes and bars) [52]. Other Ag nanocrystals, such as Ag triangular plates, can also be employed as seeds for Ag overgrowth using this principle as depicted in Fig. 2.15a [57]. Ag triangular plates are covered by {111} faces in both top and bottom surfaces, while their sides contain twin planes and stacking faults. When Na₃CA is employed as capping agent, growth along the side facets (lateral growth) occurs preferentially relative to growth along the top and bottom surfaces (Fig. 2.15b and c). Even though the side surface of the plates also contains {111} planes, the stacking faults and twin defects make them more reactive towards Ag nucleation relative to the top and bottom surfaces in the presence of Na₃CA. On the other hand, vertical growth is induced in the presence of PVP (Fig. 2.15d and e). As PVP can bind more strongly to {100} facets, the growth along the top and bottom {111} surfaces becomes faster relative to lateral growth (that also occurs).

Seeded growth has been widely applied for producing a variety of controlled bimetallic nanomaterials with a myriad of architectures by the utilization of Ag, Au, Pd, and Pt nanocrystals of different shapes as seeds for the overgrowth of Ag, Au, Pd, and Pt, for example [32, 34]. In addition to the utilization of pre-formed nanocrystals as seeds for heterogeneous nucleation and growth, they can also be employed as chemical templates for the synthesis of controlled nanomaterials. One example is the utilization of Ag nanocrystals as sacrificial templates in a galvanic replacement

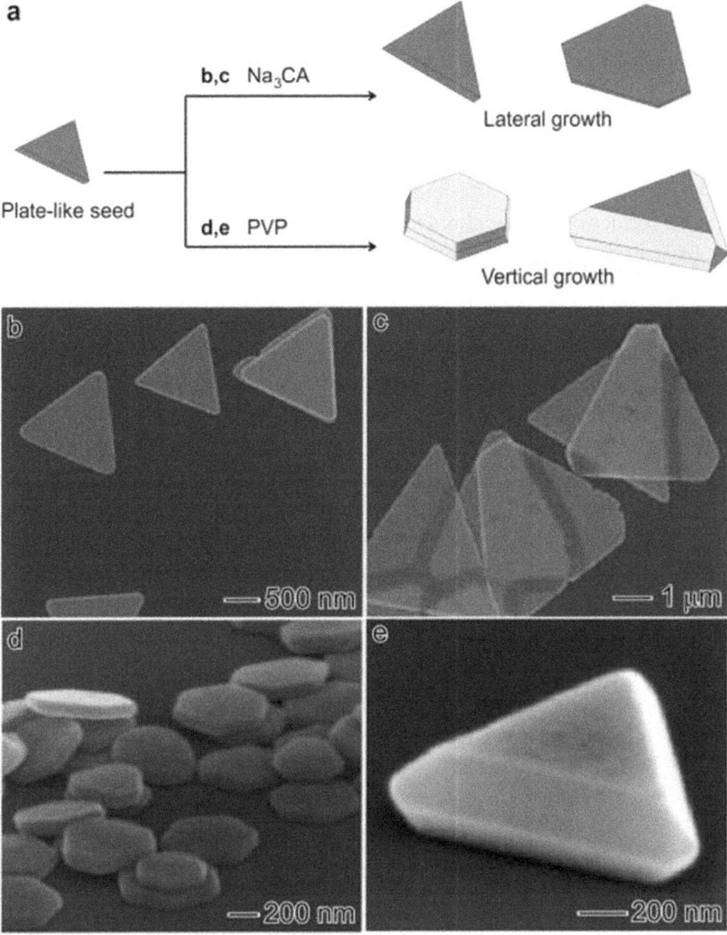

Fig. 2.15 **a** Scheme for the lateral or vertical overgrowth of Ag employing Ag triangular plates as seeds and Na$_3$CA or PVP as capping agents. **b–c** and **d–e** Show scanning electron micrographs for the produced plates after lateral and vertical growth, respectively [57]. Copyright © 2011 Wiley-VCH Verlag GmbH & Co. KGaA, Weinheim

reaction with ions $AuCl_4^-$, $PdCl_4^{2-}$, and $PtCl_6^{2-}$ to produce Ag-Au, Ag-Pd, and Ag-Pt nanocrystals, respectively, having bimetallic compositions, ultrathin walls, hollow interiors, and different morphologies (boxes, cages, tubes, frames, shells, etc.) [58].

2.5   Final Remarks

This chapter described the fundamentals of the controlled synthesis of metal nano-materials in solution phase in terms of the available theoretical framework. Our discussion encompassed the mechanisms for the stabilization of nanostructures in solution, the theory behind nucleation and growth stages, and the role that thermodynamic and kinetic considerations, oxidative etching, and surface capping can play in the synthesis of metal nanomaterials with controlled shapes using Ag as a representative example. It is obvious that solution-phase synthesis represents a powerful approach to produce metal nanomaterials with monodisperse and controlled sizes, shapes, architectures, and composition. Solution-phase methods enable the control over several experimental parameters (temperature, choice of precursors, stabilizers, solvent, reducing agents, their molar ratios, concentrations, among others) and have, at least in theory, the potential for large-scale fabrication. However, progress on the truly controlled synthesis of metal nanomaterials has only been achieved in the last decade, and the synthesis of many metal nanocrystals in a variety of shapes has been realized. Despite these attractive features, several challenges (both fundamental and practical) still remain and need to be addressed so that solution-phase synthesis can reach its full potential. These include: (i) a better understanding on the atomistic details and mechanisms behind the nucleation and growth processes is required; (ii) the precise roles played by solvent molecules, capping agents, and trace impurities must be unraveled; (iii) the scope of controlled synthesis needs to be increased, i.e., the development of protocols for the controlled synthesis of several metals still need to be developed; (iv) protocols must be reproducible; (v) it is desirable that solution-phase protocols are facile, environmentally friendly, and carried out in water as the solvent; and (vi) the controlled synthesis of metal nanomaterials in large scale must be realized in order to allow their transition from the lab to industry. It is safe to say that addressing these shortcomings will allow not only the synthesis variety of metal nanocrystals with controlled shapes/sizes by design, but also enable them to have a more profound impact in various fields of knowledge.

References

1. Y. N. Xia, Y. J. Xiong, B. Lim, and S. E. Skrabalak, Angew. Chem. Int. Ed. **48**, 60 (2009).
2. I. Lee, F. Delbecq, R. Morales, M. A. Albiter, and F. Zaera, Nat. Mater. **8**, 132 (2009).
3. H. Zhang, M. Jin, Y. Xiong, B. Lim, and Y. Xia, Acc. Chem. Res. **46**, 1783 (2012).
4. H. Goesmann and C. Feldmann, Angew. Chem. Int. Ed. **49**, 1362 (2010).
5. G. Cao and Y. Wang, *Nanostructures and Nanomaterials Synthesis, Properties, and Applications*, 2nd ed. (World Scientific Publishing Company, 2011).
6. V. K. LaMer and R. H. Dinegar, J. Am. Chem. Soc. **72**, 4847 (1950).
7. E. Roduner, Chem. Soc. Rev. **35**, 583 (2006).
8. A. W. Adamson and A. P. Gast, *Physical Chemistry of Surfaces*, 6th ed. (John Wiley & Sons, Ltd., New York, 1997).
9. K. Holmberg, *Handbook of Applied Surface and Colloid Chemistry* (John Wiley & Sons Ltd, West Sussex, 2002).
10. P. C. Hiemenz and R. Rajagopalan, *Principles of Colloid and Surface Chemistry*, 3rd ed. (Marcel Dekker, Inc, 1997).
11. J. Goodwin, *Colloids and Interfaces with Surfactants and Polymers*, 2nd ed. (John Wiley & Sons Ltd, Chichester, 2009).
12. B. V Derjaguin and L. Landau, Acta Physicochim. URSS 633 (1941).

13. E. J. W. Verwey, J. T. J. Overbeek, and K. van Ness, *Theory of the Stability of Lyophobic Colloids. The Interaction of Particles Having an Electric Double Layer* (Elsevier Inc., New Tork-Amsterdan, 1948).
14. J. Turkevich, P. C. Stevenson, and J. Hillier, Discuss. Faraday Soc. **11**, 55 (1951).
15. G. Frens, Nat. Phys. Sci. **241**, 20 (1973).
16. D. Napper, J. Colloid Interface Sci. **58**, 390 (1977).
17. A. R. Tao, S. Habas, and P. Yang, Small **4**, 310 (2008).
18. R. P. Sear, J. Phys. Condens. Matter **19**, 33101 (2007).
19. J. Park, J. Joo, S. G. Kwon, Y. Jang, and T. Hyeon, Angew. Chem. Int. Ed. **46**, 4630 (2007).
20. J. W. Mullin, *Crystallization* (Elsevier, 2001), pp. 181–215.
21. Y. Wu, D. Wang, and Y. Li, Chem. Soc. Rev. **43**, 2112 (2014).
22. J. M. Yuk, J. Park, P. Ercius, K. Kim, D. J. Hellebusch, M. F. Crommie, J. Y. Lee, A. Zettl, and A. P. Alivisatos, Science **336**, 61 (2012).
23. H. Reiss, J. Chem. Phys. **19**, 482 (1951).
24. R. Williams, P. N. Yocom, and F. S. Stofko, J. Colloid Interface Sci. **106**, 388 (1985).
25. Q. Zhang, S.-J. Liu, and S.-H. Yu, J. Mater. Chem. **19**, 191 (2009).
26. X. Xue, R. L. Penn, E. R. Leite, F. Huang, and Z. Lin, CrystEngComm **16**, 1419 (2014).
27. T. J. Woehl, C. Park, J. E. Evans, I. Arslan, W. D. Ristenpart, and N. D. Browning, (2014).
28. M. Harada, N. Tamura, and M. Takenaka, J. Phys. Chem. C **115**, 14081 (2011).
29. J. Polte, R. Kraehnert, M. Radtke, U. Reinholz, H. Riesemeier, A. F. Thünemann, and F. Emmerling, J. Phys. Conf. Ser. **247**, 012051 (2010).
30. R. L. Penn and J. A. Soltis, CrystEngComm **16**, 1409 (2014).
31. W. Lv, W. He, X. Wang, Y. Niu, H. Cao, J. H. Dickerson, and Z. Wang, Nanoscale **6**, 2531 (2014).
32. F.-R. Fan, D.-Y. Liu, Y.-F. Wu, S. Duan, Z.-X. Xie, Z.-Y. Jiang, and Z.-Q. Tian, J. Am. Chem. Soc. **130**, 6949 (2008).
33. C. J. DeSantis, R. G. Weiner, A. Radmilovic, M. M. Bower, and S. E. Skrabalak, J. Phys. Chem. Lett. **4**, 3072 (2013).
34. S. E. Habas, H. Lee, V. Radmilovic, G. A. Somorjai, and P. Yang, Nat. Mater. **6**, 692 (2007).
35. J. Zeng, Y. Zheng, M. Rycenga, J. Tao, Z.-Y. Li, Q. Zhang, Y. Zhu, and Y. Xia, J. Am. Chem. Soc. **132**, 8552 (2010).
36. A. Tao, P. Sinsermsuksakul, and P. Yang, Angew. Chem. Int. Ed. **45**, 4597 (2006).
37. B. Wiley, Y. G. Sun, and Y. N. Xia, Acc. Chem. Res. **40**, 1067 (2007).
38. Y. Xiong and Y. Xia, Adv. Mater. **19**, 3385 (2007).
39. J.-M. Zhang, F. Ma, and K.-W. Xu, Appl. Surf. Sci. **229**, 34 (2004).
40. A. Pimpinelli and J. Villain, *Physics of Crystal Growth* (Cambridge University Press, Cambridge, 1998).
41. P. M. Ajayan and L. D. Marks, Phys. Rev. Lett. **60**, 585 (1988).
42. C. L. Cleveland and U. Landman, J. Chem. Phys. **94**, 7376 (1991).
43. V. Germain, J. Li, D. Ingert, Z. L. Wang, and M. P. Pileni, J. Phys. Chem. B **107**, 8717 (2003).
44. Y. Zheng, J. Zeng, A. Ruditskiy, M. Liu, and Y. Xia, Chem. Mater. **26**, 22 (2013).
45. H. Hofmeister, S. A. Nepijko, D. N. Ievlev, W. Schulze, and G. Ertl, J. Cryst. Growth **234**, 773 (2002).
46. B. Li, R. Long, X. Zhong, Y. Bai, Z. Zhu, X. Zhang, M. Zhi, J. He, C. Wang, Z. Y. Li, and Y. Xiong, Small **8**, 1710 (2012).
47. C. C. S. de Oliveira, R. A. Ando, and P. H. C. Camargo, Phys. Chem. Chem. Phys. **15**, 1887 (2013).
48. B. J. Wiley, Y. Xiong, Z.-Y. Li, Y. Yin, and Y. Xia, Nano Lett. **6**, 765 (2006).
49. C. M. Cobley, S. E. Skrabalak, D. J. Campbell, and Y. N. Xia, Plasmonics **4**, 171 (2009).
50. B. J. Wiley, S. H. Im, Z. Y. Li, J. McLellan, A. Siekkinen, and Y. N. Xia, J. Phys. Chem. B **110**, 15666 (2006).
51. C. Lofton and W. Sigmund, Adv. Funct. Mater. **15**, 1197 (2005).
52. Y. G. Sun, B. Mayers, T. Herricks, and Y. N. Xia, Nano Lett. **3**, 955 (2003).

53. Y. G. Sun and Y. N. Xia, Adv. Mater. **14**, 833 (2002).
54. B. Wiley, T. Herricks, Y. Sun, and Y. Xia, Nano Lett. **4**, 1733 (2004).
55. I. Washio, Y. Xiong, Y. Yin, and Y. Xia, Adv. Mater. **18**, 1745 (2006).
56. D. S. Kilin, O. V. Prezhdo, and Y. Xia, Chem. Phys. Lett. **458**, 113 (2008).
57. J. Zeng, X. Xia, M. Rycenga, P. Henneghan, Q. Li, and Y. Xia, Angew. Chem. Int. Ed. **50**, 244 (2011).
58. X. Xia, Y. Wang, A. Ruditskiy, and Y. Xia, Adv. Mater. **25**, 6313 (2013).

Chapter 3
Bimetallic Nanocrystals: Growth Models and Controlled Synthesis

Zhenni Wang and Mingshang Jin

Abstract In recent years, improvements in the fundamental understanding of synergistic effect between two distinct metal elements in bimetallic nanocrystals have started to revolutionize the development of alternative energy systems for clean energy production, storage, and conversion. By using an interdisciplinary approach that combines both experimental and computational methods, the scientists are possible to rational-design and predict the properties that can enhance the performance of bimetallic nanocrystals for required applications. Although the choice of metal elements is an important aspect of bimetallic nanocrystals, their structures (e.g., core-shell, dendrite, and alloy) are a concern for nearly all bimetallic nanocrystals' applications. This chapter will mainly describe the influence of bimetallic structures for the applications and their related synthetic strategies, aiming to provide guidance to rationally design and synthesize bimetallic nanocrystals with desired properties.

3.1 Introduction

Bimetallic nanocrystals composed of two distinct metal elements usually possess novel physical and chemical properties distinct from monometallic nanocrystals due to their complex components and structures. That is, when two metals with complementary properties are properly combined, enhanced performance relative to monometallic component can occur. This enhancement is known as synergy, which is resulted from the electronic and/or ensemble effects of two metals in bimetallic nanocrystals. Due to this synergistic effect, bimetallic nanocrystals, therefore, can display not only a combination of the properties associated with two different metals, but sometimes even new properties and capabilities. This synergistic effect is particularly interesting, which provides possibilities to the findings of new materials for energy and environment-related applications. We anticipate that this synergistic effect could be affected by the component and structure of bimetallic nanocrystals,

M. Jin (✉) · Z. Wang
Center for Materials Chemistry, Frontier Institute of Science and Technology,
Xi'an Jiaotong University, Shaanxi, China
e-mail: jinm@mail.xjtu.edu.cn

© Springer International Publishing Switzerland 2015
Y. Xiong, X. Lu (eds.), *Metallic Nanostructures*, DOI 10.1007/978-3-319-11304-3_3

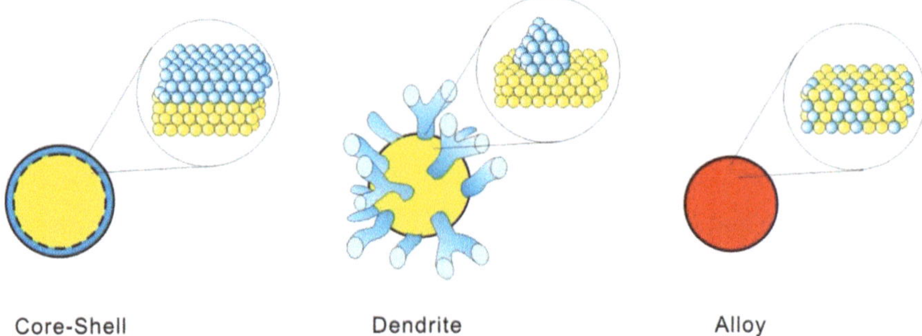

Fig. 3.1 Three types of bimetallic nanocrystal structures

which are two main parameters of bimetallic nanocrystals. Studies of these two parameters should therefore yield information, which complements the extensive literature describing nanosize effect with extended, monometallic nanocrystals. Recently, bimetallic nanocrystals formed by noble and non-noble metals used in catalysis closely resemble practical heterogeneous catalysts for fuel cells, organic reactions, and catalytic convertors, so important new scientific and technical progress may be gained regarding the synergistic effect on activity and specificity.

In this chapter, we will emphasize the important role of bimetallic nanocrystal structure for their physical and chemical properties, including optical and catalytic properties of bimetallic nanocrystals, and how can we rationally design the structure of bimetallic nanocrystals for the desired properties. Section 3.1 presents a brief introduction of three types of bimetallic nanocrystals. The principle for bimetallic design is concluded in Sect. 3.2. Methods for characterizing bimetallic nanocrystals are discussed in Sect. 3.3. The relationship between the structure and property is introduced in Sect. 3.4. Finally, Sect. 3.5 deals with four traditional synthetic strategies for synthesizing bimetallic nanocrystals with designed structures.

As a modification to monometallic nanocrystals, bimetallic nanocrystals are formed by two metal elements, and they can be classified according to the types of atoms of which they are composed and the nature of the structure. Many reviews have mainly categorized bimetallic nanocrystals into three types of structures [1–3]. Examples (see Fig. 3.1) include core-shell nanocrystals, dendrite nanocrystals, and alloy nanocrystals.

Core-shell nanocrystals consist of a shell of one type of atom (B, blue-colored atoms) surrounding a core of another (A, yellow-colored atoms), and therefore A is fully capsulated inside B and forms $A_{core}B_{shell}$ core-shell nanocrystals. The core-shell structure can provide bimetallic nanocrystals with varied optical properties and unusual stabilities. That is, the dielectric constant of the medium surrounding a metal nanocrystal can be tailored by encasing the metal core inside the second metal shell, thereby tuning the optical properties of metal nanocrystals; typical examples will be given in Sect. 3.4, as well as others. In addition, a noble metal shell can lend stability to non-noble or active metal cores from oxidation. Yin et al. [4] prepared a

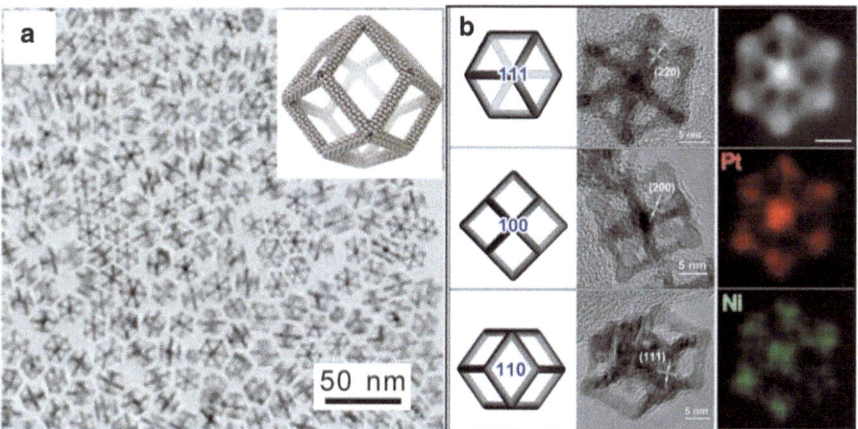

Fig. 3.2 HRTEM and energy-dispersive X-ray spectroscopy elemental mapping characterizations of Pt$_3$Ni nanoframes. (Reproduced from Chen et al. [5])

thin Au shell on Ag nanoplates and observed a significant enhancement of stability of Ag nanoplates. More examples of core-shell nanocrystals will be discussed in Sect. 3.4 and 3.5.

Dendrite nanocrystals consist of A and B subcrystals, with a mixed A–B interface. These are examples of the so-called "branched nanocrystals," as they have branches of B on A cores (see Fig. 3.1). Dendrite nanocrystals have great potential as catalysts, sensing materials, and building blocks for nanoscale devices due to their large surface area and high densities of atomic steps and kinks. Various strategies have recently been developed for the solution-phase synthesis of bimetallic dendrite nanocrystals. In Sect. 3.2, the procedures and mechanisms, as well as examples underlying the formation of dendrite nanocrystals, will be presented in parallel with other types of bimetallic nanocrystals.

Alloy nanocrystals may either be ordered (pseudocrystalline) or random (solid solution). The intermixed pattern is found in many systems, with the ordered structure of Pt$_3$Ni being important for their fuel cell applications. Based on the phase diagrams of bimetallic materials, it is possible to optimize the experimental parameters (e.g., pressure, temperature) for the preparation of desired alloy nanocrystals. However, some metals, like Cu and Ag, are immiscible in the bulk thus cannot form alloy bulk materials, while some others, like Pt and Cu/Co/Ni, can form ordered alloys. Figure 3.2a shows the transmission electron microscopy (TEM) images of Pt$_3$Ni nanoframe along different directions, with high-resolution TEM (HRTEM) and energy-dispersive X-ray spectroscopy (EDX) mapping results shown in Fig. 3.2b [5]. However, when the size of materials is down to a critical value, such as Fe and Ag, which are immiscible in the bulk, it can form alloys [6]. There is considerable interest in mapping out for diverse bimetallic systems in the nanoscale. On the other hand, alloy nanocrystals, for example, alloying noble metals with cheaper metals, such as Fe, Co, and Ni, can reduce the cost of catalysts. A number of reported alloy nanocrystals were found to exhibit much better catalytic behavior than those of monometallic ones due to the influence of different neighboring atoms, named synergistic effect. For example, a clear synergistic effect has been observed in the catalytic oxygen reduction reactions (ORR) by Pt_3Ni mentioned in Fig. 3.2, for which a factor of 36 enhancement in mass activity and a factor of 22 enhancement in specific activity relative to state-of-the-art platinum-carbon catalysts.

There are also some other types of bimetallic nanocrystals, such as multishell nanocrystals, which present onion-like alternating concentric -A-B-A- shells. Recently, three-shell Pd–Au nanocrystals have been produced experimentally and characterized by Wang et al. [7]. As this multishell bimetallic nanocrystal is more likely to be an extension of the core-shell nanocrystals, we classify this structure to be core-shell nanocrystals.

3.2 Influence Factors of Bimetallic Nanocrystal Structures

When two metal elements are mixed together, three possible structures can be expected as aforelisted, including core-shell, dendrite, and alloy nanocrystals. Different structures may result in significant different properties, thus their potential applications. Whether we can finely control the structure of bimetallic nanocrystals during their growth is the most important issue for rational design of bimetallic nanomaterials.

As early as 1986, Bauer and van der Merwe have proposed three growth modes for the overgrowth of the second metal on a substance [8]. That is, the layered growth (Frank-van der Merwe mode), the island growth (Volmer–Weber mode), and the intermediate type of growth (Stranski–Krastanow mode; see Fig. 3.3). These growth modes have been recently adopted to control the growth of bimetallic nanocrystals. Fan et al. [9] propose three rules for the epitaxial layered growth of heterogeneous core-shell nanocrystals, which is (i) the lattice constants of two metals should be comparable with the lattice mismatch smaller than approximately 5 %; the shell metal with smaller atom radius is easier to epitaxially grow on the core, as they could uniformly release the lattice strain resulting from the lattice mismatch; (ii) the electronegativity of the shell metal is lower than the core metal to avoid the displacement reaction and to easily wet the surface of the core; and (iii) the bond

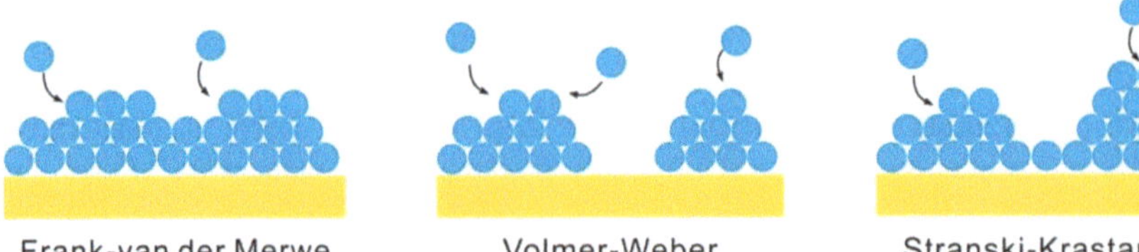

Fig. 3.3 Schematic illustrations of three typical growth modes

energy between metal atoms of the shell should be smaller than that between the shell atoms and substrate atoms to ensure the growth in the Frank-van der Merwe mode, with the experimental observations for the layered growth mode of four types of noble metals.

These three fundamental rules play key role and provide the guidance to the growth of bimetallic nanocrystals with rationally designed structures. As for lattice mismatch, which generally exists between a second metal (B) and the substance (A), it can induce positive strain energy (γ_{strain}) in the prepared bimetallic $A_{core}B_{shell}$ nanocrystals. Along with the increasing of lattice mismatch between core and shell metals, γ_{strain} increases rapidly, thus influencing the overall excess energy of bimetallic nanocrystals ($\Delta\gamma$). $\Delta\gamma$ is defined as the following equation [10, 11]:

$$\Delta\gamma = \gamma_B + \gamma_i + \gamma_{strain} - \gamma_A \tag{3.1}$$

γ_i Is the interfacial energy of bimetallic $A_{core}B_{shell}$ nanocrystals due to the formation of interfaces between metals A and B during the nucleation and growth. γ_A And γ_B are the surface energy of metals A and B, respectively. In case of a relatively small lattice mismatch between metals A and B, the total energy of γ_B, γ_i, and γ_{strain} would be smaller than the surface energy of the substrate metal (γ_A, Eq. 3.1), resulting in a negative $\Delta\gamma$ during growth and thus layer-by-layer growth of B atoms on A substrate. Contrarily, $\Delta\gamma$ is positive in case of a large lattice mismatch so that metal B atoms grow on those high-energy sites of substrate metal and form islands to minimize strain energy and reduce $\Delta\gamma$. This is the so-called island growth mode. When the lattice mismatch is relatively large but $\Delta\gamma$ is still negative at the initial stage, the layer-by-layer growth is preferable initially. However, the strain energy is increased with the number of layers of B atoms increased. When the number reaches a critical value, $\Delta\gamma$ becomes positive and the growth mode changes from layer-by-layer growth to island growth on the wetting layers to release strain energy. In addition to the lattice mismatch, the metal bond energy is also a key factor during the preparation of bimetallic nanocrystals. The interaction among atoms in the deposited overlayer should be smaller than that between the substrate and overlayer. Moreover, the proper interaction (electron transfer) of the overlayer and substrate is another key factor when the crystal growth occurs in solutions, which is determined by electronegativity of two kinds of metal atoms. For instance, electrochemical underpotential deposition (UPD) in electrolyte solution is a good example [12]. The foreign metal atoms with lower electronegativity tend to wet the heterogeneous metal surface and form the two-dimensional UPD overlayer. It also prevents the galvanic replacement reaction, that is, the deposition of less active metal (with higher electronegativity) on the metal core. Based on this achievement on the preparation of core-shell nanocrystals, Pt@Pd, Au@Pd, and Au@Ag core-shell nanocrystals have been synthesized and well explained by Frank-van der Merwe growth modes.

Besides the aforementioned three rules for layered overgrowth for the preparation of core-shell nanocrystals, study of bimetallic nanocrystals has been restricted by many factors, including (1) relative strengths of A–A, B–B, and A–B bonds; (2) surface energies of bulk elements A and B; (3) relative atomic sizes; (4) charge transfer; (5) strength of binding to substrates or surface ligands (surfactants); and (6) specific electronic/magnetic effects [13]. The structure of bimetallic nanocrystals depends critically on the balance of the factors outlined earlier in the text, as well as on the preparation method and experimental conditions, which may give rise to kinetic, rather than thermodynamic, products.

3.3 Characterizations of Bimetallic Nanocrystals

Similar to the characterization of monometallic nanocrystals, a variety of experimental techniques can be used to characterize and measure the properties of bimetallic nanocrystals, including X-ray diffraction (XRD), TEM, X-ray photoelectron spectroscopy (XPS), EDX, and scanning electron microscopy (SEM). However, the analyses of bimetallic nanocrystals are more complicated than those of monometallic nanocrystals due to the different elements they are composed of and complex structures. Therefore, accurate characterization of bimetallic nanocrystals presents a formidable challenge and typically requires the combination of several types of characterization techniques. Due to the important role of XRD, TEM, high-angle annular dark field scanning TEM (HAADF-STEM), and EDX in characterizing bimetallic nanocrystals, we will give an introduction to these technologies in this section.

3.3.1 XRD Analysis

As one of the most widely used techniques, XRD can give the information on bimetallic structure, crystallinity, lattice spacing, crystal size, and qualitative chemical composition information. Due to its strong ability in determining the composition of bimetallic nanocrystals, XRD is particularly effective to differentiate bimetallic alloy nanocrystals with core-shell/dendrite bimetallic ones. From the peak positions of XRD pattern, the lattice parameters a can be calculated. Lipson [14] has discussed in detail how to figure out the composition of binary materials with different structures through XRD. Formation of alloy nanocrystals, as well as the dissolution of B in A, will cause a shift in diffraction peak positions due to the changes of the lattice constants. In comparison, XRD patterns of core-shell or dendrite nanocrystals will be a combination of the two individual metal phases. By far, comparison of the XRD patterns of bimetallic nanocrystals has provided researchers with correction information of structures. For example, Au–Cu, Au–Pd, Pt–Ni, and Pt–Co alloy nanocrystals have been indentified based on calculations of a values, which are further compared with their standard a constants [15–19]. The size of bimetallic nanocrystals can also be calculated based on XRD patterns. Denton and Ashcroft [20] developed Vegard's law in 1991, which is an approximate empirical rule that holds that a linear relation exists, at constant temperature, between

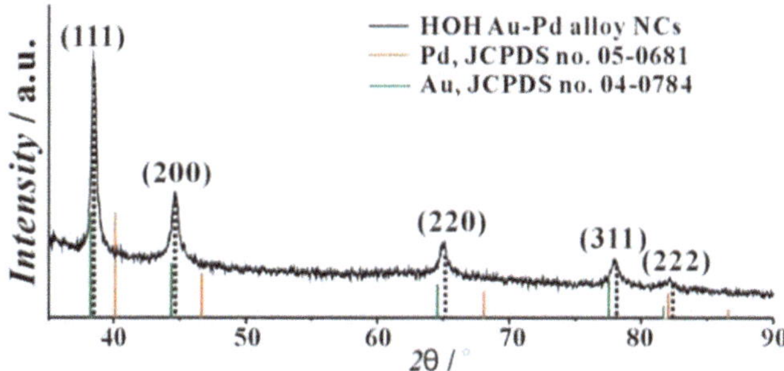

Fig. 3.4 XRD patterns of Au–Pd alloy nanocrystals. (Reproduced from Zhang et al. [21])

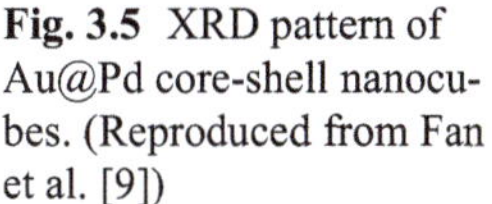

Fig. 3.5 XRD pattern of Au@Pd core-shell nanocubes. (Reproduced from Fan et al. [9])

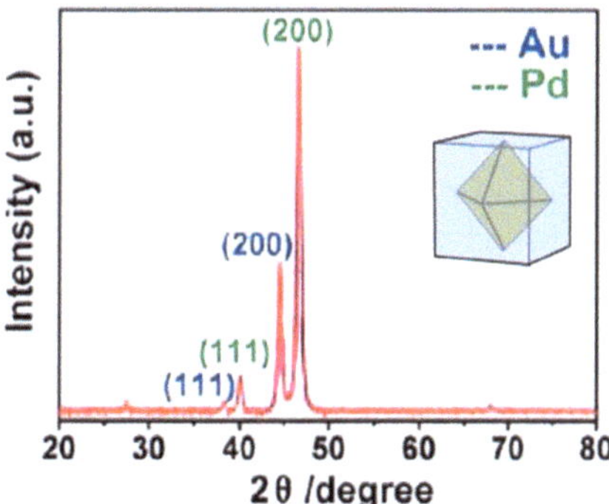

the crystal lattice constant of an alloy and the concentrations of the constituent elements. Figure 3.4 presents the XRD patterns of Au–Pd alloy, Au, and Pd, respectively [21]. As can be seen, the peaks of Au–Pd alloy locate besides the Au and Pd metals. The composition of alloy nanocrystals, thus, can be calculated by using Vegard's law. On the other hand, XRD patterns of core-shell/dendrite nanocrystals will exhibit two different sets of peaks, indicating the existence of two separate phases, as shown in Fig. 3.5.

In some cases, for example, core-shell cubes, plates, or octahedrons, a preferential orientation effect (parallel to the substrate) will occur during the XRD characterizations. This effect will result in an enhancement of the diffraction intensity of a particular facet that is parallel to the substrate. Based on the preferential orientation of core-shell nanocrystals, it is possible to fix out the orientation relationship between the core and shell metals via XRD analysis. Figure 3.5 shows XRD patterns of Au@Pd core-shell nanocubes [9]. Clearly, the preferential orientation of the sample can be viewed, as the {200} diffraction of the sample is significantly enhanced compared with other facets, which implies the {100} facets preferential orientation of cubes due to their regular shapes. The same preferential orientation direction of core and shell metals indicates that the growth of the shell is affected and induced by the core metals.

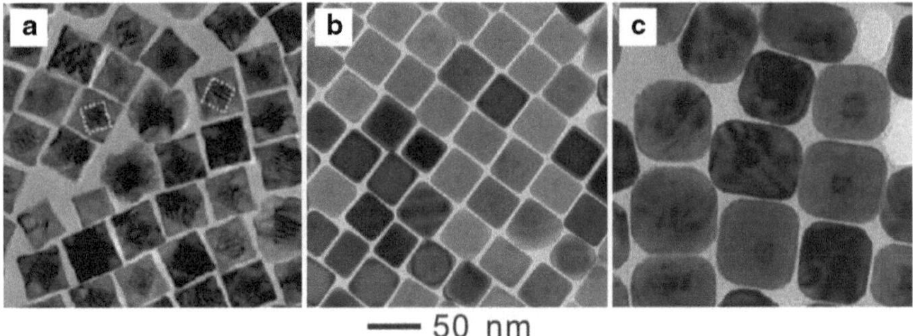

Fig. 3.6 TEM images of Au@Pd, Au@Ag, and Pd@Cu nanocubes. (Reproduced from Fang et al. [9], Ma et al. [22, and Jin et al. [23])

3.3.2 TEM Observations

While XRD mainly provide the composition information of bimetallic nanocrystals, TEM and its higher-resolution version HRTEM has traditionally been the main tools used to obtain the nanocrystal shape, size, and structure information. Especially for core-shell and dendrite bimetallic nanocrystals, TEM and HRTEM are able to show different contrasts of the separate phases of two metals. That is, the metal atoms with different atomic numbers possess different contrasts under TEM and HRTEM characterizations, such as Au@Pd, Pd@Cu, and Au@Ag core-shell nanocubes [9, 22, 23]. Figure 3.6 is the TEM images of the aforementioned core-shell nanocrystals, in which the dark and light contrast can be easily observed. However, some cases, like Pd and Pt, which possess similar contrast under these characterizations, are nearly impossible to differentiate only by TEM and HRTEM.

3.3.3 Elemental Information Analyzed by HAADF-STEM and EDX

Generally speaking, TEM and HRTEM is essentially a phase contrast technique with relatively little elemental information. However, in High-angle annular dark field scanning transmission electron microscopy (HAADF-STEM), STEM coupled with HAADF detector provides mainly incoherent images that have strong elemental dependence. Therefore, HAADF-STEM has recently been recognized as a powerful tool for imaging bimetallic nanocrystals. Xia et al. demonstrated that even the monolayer growth of Pt on Pd nanocubes can be visible via the HAADF-STEM characterization (see Fig. 3.7) [24]. Due to the elemental dependence of imaging contrast in the HAADF-STEM, the technique is particularly suited for study of bimetallic nanocrystals. Ideally, atomic resolved spectrum imaging of bimetallic nanocrystals can also be achieved by coupling TEM or STEM with EDX. They are going to increasingly provide complementary and elemental-specific information toward characterization of bimetallic nanocrystals.

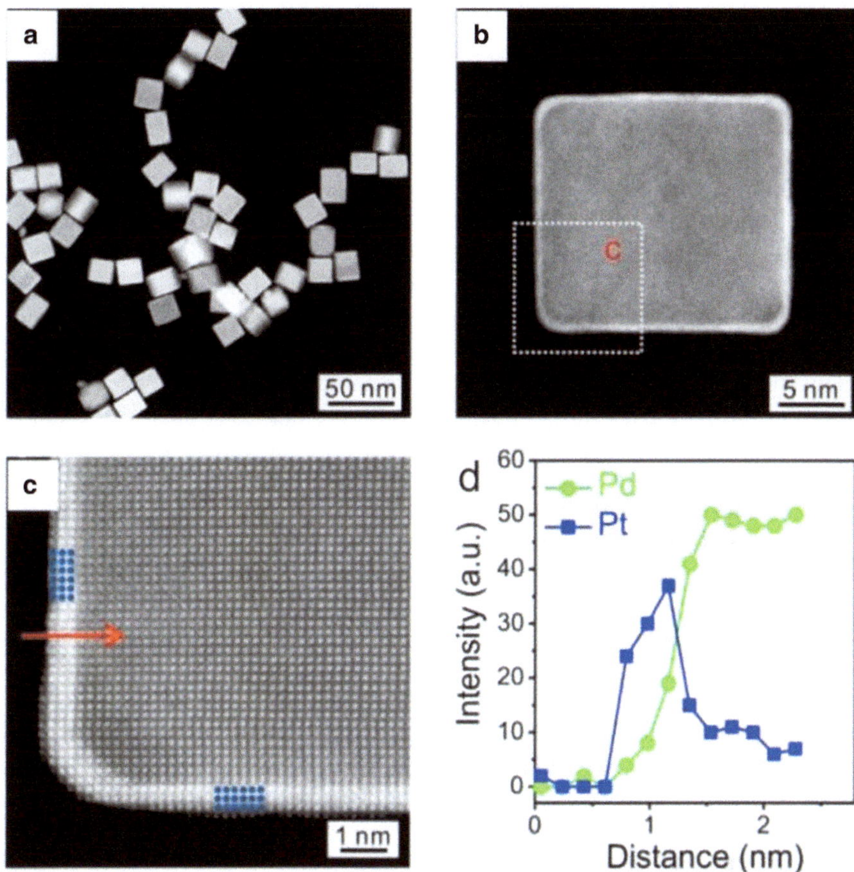

Fig. 3.7 HAADF-STEM images showing the numbers of Pt atomic layers. (Reproduced from Xie et al. [24])

3.4   Properties of Bimetallic Nanocrystals with Different Structures

Commonly, the properties of bimetallic nanocrystals are mainly dependent on their structures with the same components. Therefore, there are numerous scientific reasons to investigate the relationship between structure and their corresponding properties. The strong influence of the bimetallic nanocrystals structure is general, affecting the magnetic response, optical properties, and chemical activities. The implication of these observations is that structure can be utilized to optimize the physical properties of nanocrystals. Even a monolayer shell of a second metal on the core of another metal can markedly influence the properties of a nanocrystal, such as Fermi level, thus resulting in the different optical, catalytic, and magnetic properties of a nanocrystal.

3.4.1 Optical Properties of Au@Ag Core-Shell and Alloy Nanocrystals

Shape-controlled synthesis of noble metal, such as Au and Ag, has attracted a great deal of attention in recent years because surface plasmon resonance (SPR) excitation within the Au and Ag nanostructures can greatly enhance the local electric field. The frequency and intensity of SPR is highly sensitive to their compositions, structures, size, morphologies, and embedding mediums. In general, the field enhancement can be significantly improved by the nanostructures with sharp corners. However, the composite metallic nanostructures with multiple elements are of significant interests for their improving catalytic and optical properties. Especially, bimetallic Au and Ag nanocrystals are particularly attractive due to broader range of Plasmon tunability as compared with the individual unit of Ag and Au nanocrystal. By forming Au and Ag into core-shell structures, the LSPR can be controlled by not only varying the size and shape of the core metals but also the shell thickness.

Figure 3.8 is the extinction spectra of Au@Ag cube, tetrahedra, octahedra, and dodecahedra [25]. The ratio of OD (outside diameter) to ID (inside diameter) is varied for each polyhedron to test the influence of the core size. For Au@Ag cubes, when the core size is 0 nm (Ag nanocubes), two strong LSPR peaks can be observed at 467 and 520 nm, representing the dipolar and quadrupolar modes, respectively. When the core size is increased to 16 nm, both peaks are found to red-shift to 478 and 535 nm, respectively, which should be attributed to the differences in the dielectric constants of the two metals. Meanwhile, as Ag exhibits much stronger optical resonance than Au, the intensities of all the adsorption peaks are decreased along with the increasing of the Au core size. However, in case of Au@Ag tetrahedra, the peaks are shifted to 460, 562, 639, and 757 nm, respectively, compared with the Ag tetrahedra when the size of the Au core is 6 nm. In comparison, Ag octahedra can exhibit two LSPR peaks at 493 and 562 nm, respectively. If the size of the Au core is 10 nm, the peak located at 562 nm blue-shifted, while the other red-shifted, resulting in peaks relocation at 522 and 562 nm, respectively. When the core size is further increased to 14 nm, these two peaks shift closer and finally combine into one peak at 575 nm. In addition, the intensity of the peak positioned around 500 nm is reduced as the size of the Au core increases. In another case, Ag dodecahedra can exhibit one resonance peak at 433 nm as shown in Fig. 3.8d. However, there is a significant decrease in extinction efficiency, along with a red-shifting of the extinction maximum of Au@Ag dodecahedra when the size of the Au core increases from 0 to 16 nm.

On the other hand, the shell thickness of core-shell nanocrystals can also affect the optical properties of bimetallic nanocrystals. Xia et al. recently synthesized Au@Ag core-shell nanocubes with the shell thickness tunable from 1.2 to 4.5 nm, and studied their corresponding optical properties [22]. These Au@Ag core-shell nanocubes can exhibit two characteristic peaks located at 510 and 410 nm, representative of the Au core and the Ag shell, respectively. Further tailoring the shell thickness will result in the peak intensity decrease of the Au cores (See Fig. 3.9).

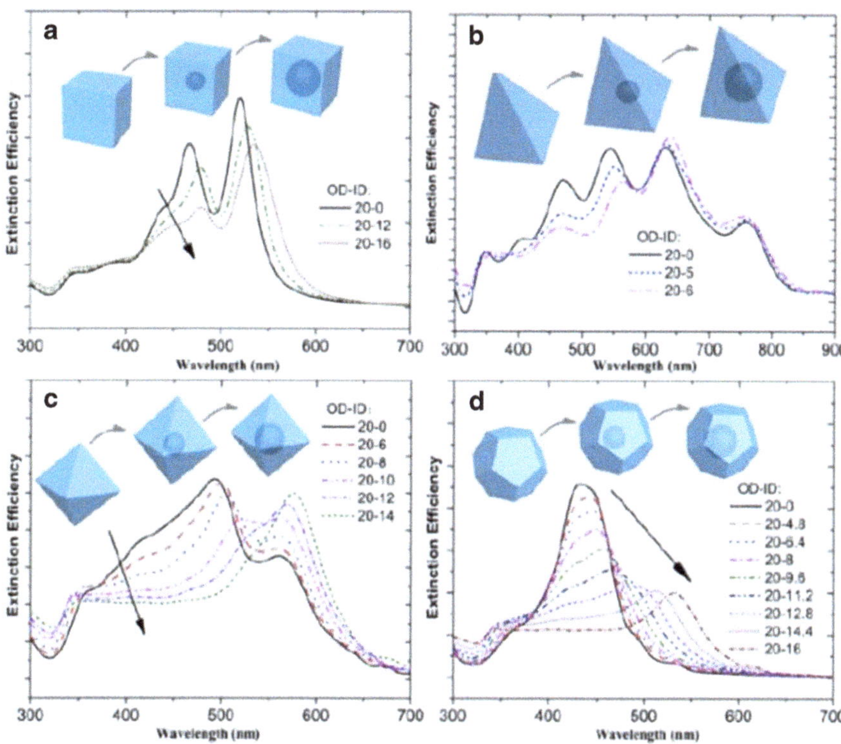

Fig. 3.8 Extinction spectra of Au@Ag polyhedra, the outside diameter is equal to 20 nm; **a** cubes, the core diameter varies from 0 to 16 nm; **b** tetrahedra, the core diameter varies from 0 to 6 nm; **c** octahedra, the core diameter varies from 0 to 14 nm; **d** dodecahedra, the core diameter varies from 0 to 16 nm. (Reproduced from Zhang et al. [25])

Fig. 3.9 Ultraviolet-visible extinction spectra of Au@ Ag core-shell nanocubes with different thicknesses for the Ag shells. (Reproduced from Ma et al. [22])

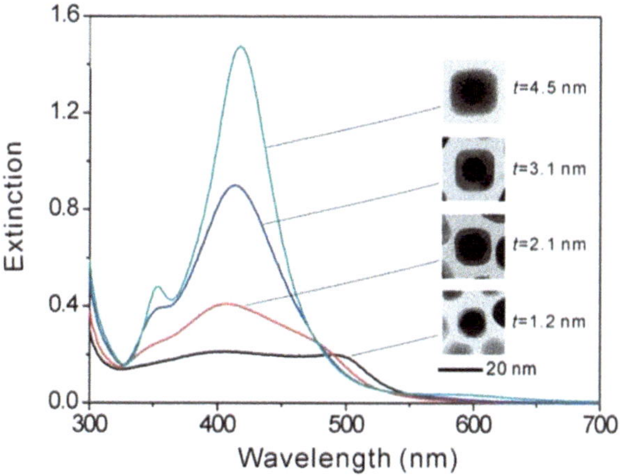

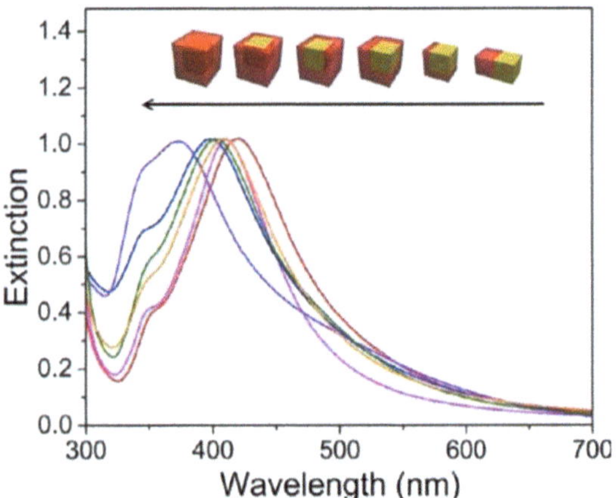

Fig. 3.10 Ultraviolet-visible extinction spectra of the different types of Pd–Ag bimetallic nanocrystals. (Reproduced from Zhu et al. [26])

The shell material can also help to impart novel, desired properties on the nanocrystals. Pd nanocrystals, which are known to show weak optical absorbance property, can serve as seeds for the overgrowth of different Ag structures (the material that can exhibit excellent optical property), thus enabling the bimetallic nanocrystals with enhanced optical absorbance properties. Figure 3.10 shows the extinction spectra of the as-obtained Pd–Ag bimetallic nanocrystals with Ag growing on different numbers of the faces of a cubic Pd seed [26]. When Ag only grows on one of the six faces of a Pd seed, the major LSPR peak was located at 421 nm. However, when Ag was deposited on two, three, four, and five faces, the LSPR peaks can increasingly blue-shift to the region between 415 and 400 nm along with an increase in the number of faces being covered by Ag.

On the other hand, alloy nanocrystals can show different optical responses, such as Au–Ag alloy nanoboxes. Xia et al. [27] reported that Au–Ag nanocages with alloy walls can be prepared via galvanic replacement reaction between Ag nanocubes and $HAuCl_4$. By controlling the molar ratio of Ag to $HAuCl_4$, the bimetallic nanostructures can be changed from solid cubes to hollow boxes and porous cages. Through this way, the extinction peaks of bimetallic nanocrystals can be continuously tuned from 400 nm to near-infrared (1200 nm). The discrete-dipole approximation (DDA) calculations suggest that peak broadening is likely due to the small variations in the wall thickness. Figure 3.11 compares the calculated extinction cross-section (C_{ext}), C_{abs}, and C_{sca} ($C_{ext} = C_{abs} + C_{sca}$) for Au nanoboxes with 60- and 40-nm edge length, respectively, with a wall thickness of 5 nm for both samples. Similar to solid Au nanoparticles, light absorption dominates the extinction spectra for Au nanoboxes of relatively small sizes (<30 nm), and light scattering increases for nanoboxes of larger dimensions (>60 nm). Figure 3.11c shows the cross-sections calculated for nanoboxes whose edge length and wall thickness are 36 and 3 nm, respectively. Results indicate that the extinction peak can red-shift from 710 to 820 nm as the wall thickness decreases from 5 to 3 nm. Figure 3.11d shows the calculated spectra of nanoboxes similar to that of Fig. 3.11b, except that all eight corners are replaced with eight holes. Obviously, an increase in the magnitude of scattering and adsorption cross-sections

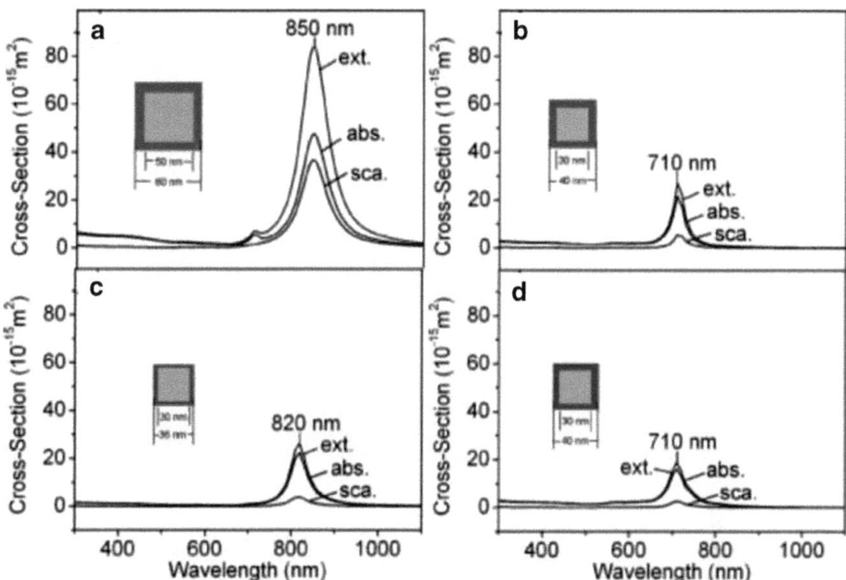

Fig. 3.11 Extinction spectra calculated using the discrete-dipole-approximation method for Au nanocages having four different sets of geometric parameters: **a** an Au nanobox 50 nm in inner edge length and 5 nm in wall thickness; **b** an Au nanobox 30 nm in inner edge length and 5 nm in wall thickness; **c** the same as in (b), except that the wall thickness is 3 nm; and **d** the same as in (b), except that the eight corners are decorated with triangular holes 5 nm in edge length. (Reproduced from Chen et al. 2012) [27]

can be observed. In contrast to the wall thickness, the position of the extinction peak does not show any significant dependence on the porosity of the nanoboxes.

3.4.2   Catalytic Properties of Bimetallic Nanocrystals with Different structures

Not only the optical property but also the catalytic activity and/or even selectivity of bimetallic nanocrystals can be enhanced by the rational design of components and structures. Catalysis by bimetallic nanocrystals is therefore a very popular area of research in recent years. Another important driving force for research into catalysis by bimetallic nanocrystals is the cost/rarity of the metals typically used in catalysis. It is clearly desirable to use cheaper metals, such as Fe, Co, and Ni, to replace expensive metals, such as Pt and Au, while achieving bimetallic nanostructures that possess much better catalytic properties than that of the monometallic catalyst. Generally, a catalytic process can be divided into three steps: the adsorption of reactants, reaction on the surface, and desorption of products. The relationship between catalytic activity and adsorption energy shows a volcano shape. Two factors have an influence on adsorption energy, one is the atomic arrangement of the

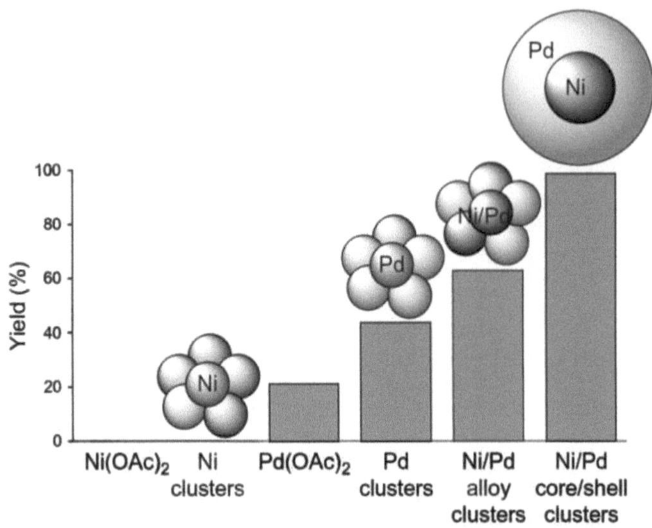

Fig. 3.12 Comparison of the catalytic activity for six different systems in the Hiyama cross-coupling of iodotoluene and trimethoxyphenylsilane. (Reproduced from Durán Pachón et al. [28])

surface atoms, and other one is electronic structure, especially the d-band center position of the surface atoms. The d-band center position can be directly controlled by preparing alloy catalysts and changing the shape of bimetallic nanocrystals. Bimetallic nanocrystals with different shapes expose various facets, and thus the coordination number of surface atoms is different. For instance, core-shell bimetallic nanocrystals are very promising for the design of new catalysts. Figure 3.12 shows the comparison of the catalytic activity for six different systems in the Hiyama cross-coupling of iodotoluene and trimethoxyphenylsilane. As can be seen, Ni–Pd in the form of core-shell structure exhibited the highest catalytic activity than monometallic Pd nanocrystals and mixed Ni–Pd nanocrystals [28]. That is, attribute to the core-shell structure results in more Pd atoms on the surface, and thus more accessible catalytic sites per mole of palladium.

Dendrite bimetallic nanocrystals also provide a way to enhance the catalytic performance of bimetallic catalysts. A number of researchers have successfully deposited Pt branches/atomic layers onto fine particles of other cheaper metals. In this way, the total surface area available for catalysis, for a given mass of Pt catalyst, is increased, leading to greater catalytic activity. As been reported by Xia and other groups, dendrite bimetallic nanocrystals consisting of a dense array of Pt branches on a core of Pd nanocrystal can be synthesized by reduction of K_2PtCl_4 with L-ascorbic acid (AA) in the presence of uniform Pd nanocrystal seeds in an aqueous solution [29]. The Pt branches supported on faceted Pd nanocrystals exhibited relatively large surface areas and particularly active facets toward formic acid oxidation and the ORR; in the ORR, the Pd–Pt nanodendrites were 2.5 times more active on the basis of equivalent Pt mass for the ORR than the state-of-the-art Pt/C

Fig. 3.13 Comparisons of electrocatalytic properties of the Pd–Pt nanodendrites, Pt/C catalysts (E-TEK), and Pt black. (Reproduced from Lim et al. [29])

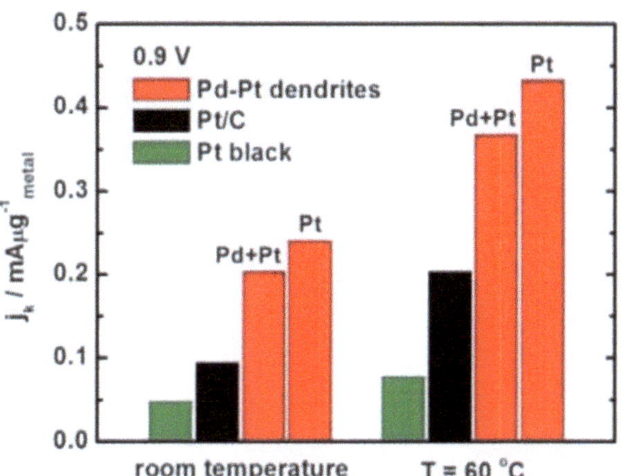

catalyst and five times more active than the first-generation support-less Pt-black catalyst (Fig. 3.13).

The catalytic activity of metals can also be modified and fine-tuned by alloying, which can alter the d-band center position. Schmid stated that the mutual influence of different neighboring atoms in alloy nanocrystals can lead to catalytic behavior better than that of monometallic nanocrystals. For example, a clear enhancement has been observed in the ORR by bimetallic Pt_3Ni alloy nanocrystals. Markovic and coworkers demonstrate the improved oxygen reduction activity on Pt_3Ni (111), as the Pt_3Ni (111) surface has an unusual electronic structure and arrangement of surface atoms, which can greatly increase the number of active sites for oxygen adsorption; thus, the Pt_3Ni (111) surface is 10-fold more active for the ORR than the corresponding Pt (111) surface and 90-fold more active than the current Pt/C catalysts for polymer electrolyte membrane fuel cells (PEMFCS; Fig. 3.14) [30].

Based on these facts, bimetallic nanocrystals have raised more and more significant concern from worldwide researchers in recent years. Generally speaking, synthesis of high-quality bimetallic nanocrystals with controllable size, morphology, composition, and structure is the most important key for tuning their properties, thus their future applications.

3.5   Synthetic Approaches of Bimetallic Nanocrystals

Bimetallic nanocrystals can be generated in a variety of media: in the gas phase, in solution, supported on a substrate, embedded in a matrix, or even in a bacterial cell. Among all these media, growth of nanocrystals in solution have been proven to be the most effective and controllable, especially those bimetallic nanocrystals with well-defined facets. This method can be induced by thermal, photo, or reducing

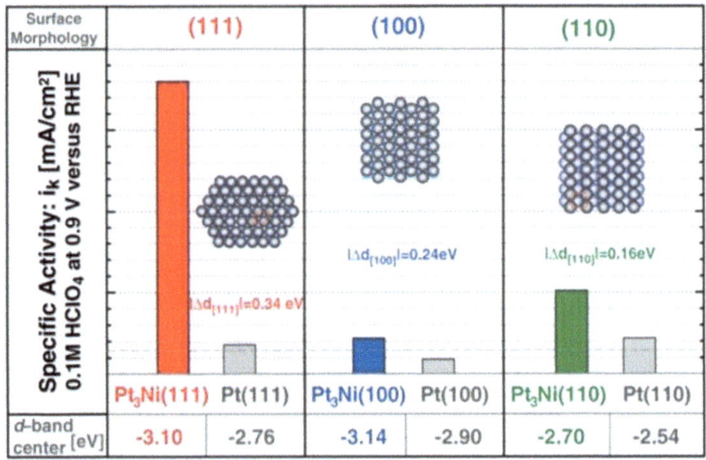

Fig. 3.14 Comparison of electrocatalytic properties of different facets of Pt$_3$Ni and Pt nanocrystals. (Reproduced from Stamenkovic et al. [30])

agent. In a typical solution-based synthetic procedure, one must select a proper chemical reaction and corresponding reactants to generate the target products; then, a rational solution system including solvents and surfactants to carry out the selected reaction should be designed; finally, the nucleation and growth process of nanocrystals must be precisely controlled by adjusting the thermodynamic and kinetic parameters of the reaction. In this section, we will summarize and discuss the strategies used in the solution phase for tailoring the structure of bimetallic nanocrystals, including thermal decomposition, coreduction, galvanic replacement reaction, and seed-mediated growth.

3.5.1 Thermal Decomposition

Decomposition of a thermally labile, oil soluble, metal-organic precursor is the oldest approach for non-aqueous synthesis of metal nanocrystals. This approach requires the presence of a surfactant-like stabilizer (e.g., a block copolymer), a high-boiling-point solvent (usually between 200 and 300 °C), and an available thermally unstable metal-organic precursor. Theoretically, this method is typically only used for base metal nanocrystals synthesis like Co, Fe, and Ni, via the decomposition of $Co_2(CO)_8$, $Fe(CO)_5$, and $Ni(CO)_4$.

Simultaneous decomposition of two metal precursors can result in the formation of bimetallic nanocrystals. Hyeon et al. synthesized Pd–Ni bimetallic nanocrystals with a core-shell structure from the thermal decomposition of Pd-TOP and Ni-TOP complexes [31]. As the Ni-TOP complex is more readily decomposed thermally compared with the Pd-TOP complex, Ni nanocrystals will nucleate first, followed by the deposition of Pd shell, resulting in the formation of Ni-rich core/Pd-rich shell bimetallic nanocrystals. The different decomposition rates of two metals can be finely tuned by the addition of proper catalysts. For example, $Co_2(CO)_8$ can decompose at 150 °C in tetrahydronaphthalene solvent, while $Fe(CO)_5$ decomposes at temperatures as high as 200 °C. If the experiment is carried out at 200 °C, separate phases of monometallic Co and Fe are formed due to the faster decomposition of $Co_2(CO)_8$ at 200 °C. However, by the addition of aluminum trialkyl as a catalyst for the thermolysis of $Fe(CO)_5$, the decomposition temperature of $Fe(CO)_5$ can be tuned to 150 °C, and thus, the Fe–Co alloy nanocrystals can be obtained [32]. Commonly, controlling the decomposition rates of two different organometallic precursors is relatively hard to achieve, although there may be some efficient catalysts that can be used. Therefore, many researchers have further modified "simultaneous decomposition" method by using bimetallic precursors as single-source molecular precursor. The use of a bimetallic precursor can avoid the problems associated with the different reaction kinetics when using more than one precursor. As a typical example, Fe–Pt alloy nanoparticles could be prepared by using $Pt_3Fe_3(CO)_{15}$ as single-source precursor [33]. Other bimetallic single-source precursors like $[FeCo_3(CO)_{12}]^-$, $[Fe_3Pt_3(CO)_{15}]^{2-}$, $[FeNi_5(CO)_{13}]^{2-}$, and $[Fe_4Pt(CO)_{16}]^{2-}$ could also be used to synthesize $FeCo_3$, FePt, $FeNi_4$, and Fe_4Pt alloys, respectively, with their compositions determined by the molar ratio of metals of bimetallic precursors [34].

In some cases, bimetallic nanocrystals can also be synthesized in solution through combining decomposition method with high-temperature reduction of a second metal precursor. A certain organometallic precursors, like $Pt(acac)_2$, can be reduced by alcohol in coordinating solvents like diol and triol. Sun et al. [35, 36] have successfully refined this "poly-ol" method to produce monodispersed iron–platinum (FePt) nanocrystals. FePt nanocyrstals with controlled size and composition were obtained via reduction of $Pt(acac)_2$ by a long-chain 1,2-hexadecanediol and decomposition of $Fe(CO)_5$ in the presence of OA and OAm (Fig. 3.15a). The composition of FePt bimetallic nanocrystals can be easily adjusted by controlling the molar ratio of iron carbonyl to the platinum salt. A seeded growth can be further applied to achieve an optimal size or shape. Sun and coworkers [37] further extended this method to one-step synthesis of FePt nanocrystals with controlled sizes and shapes via thermal decomposition of $Fe(CO)_5$ and reduction of $Pt(acac)_2$ in the absence of 1,2-alkanediol. It is believed that 1,2-alkanediol can lead to the facile reduction of $Pt(acac)_2$ to Pt, resulting in fast nucleation of FePt, thus the formation of smaller-sized bimetallic nanocrystals (see Fig. 3.15b). Without the addition of a reducing agent, the nucleation rate may slow down, allowing more metal precursors to be deposited around the nuclei, resulting in a larger size (see Fig. 3.15c). The morphology of bimetallic nanocrystals can be tailored by tuning the reaction kinetic parameters. FePt nanocubes (Fig. 3.15d) could be prepared by controlling the addition of the surfactants (OA and OAm) and Fe/Pt ratio [38]. During the preparation of FePt nanocubes, OA must be added at the initial stage of the reaction to generate cubic Pt-rich nuclei due to its weak tendency to bind to Pt. In the following stage, Fe-rich species would prefer to deposit on the {100} facet, leading to the formation of cubic nanocrystals. Therefore, empirically chosen surfactants are typically used

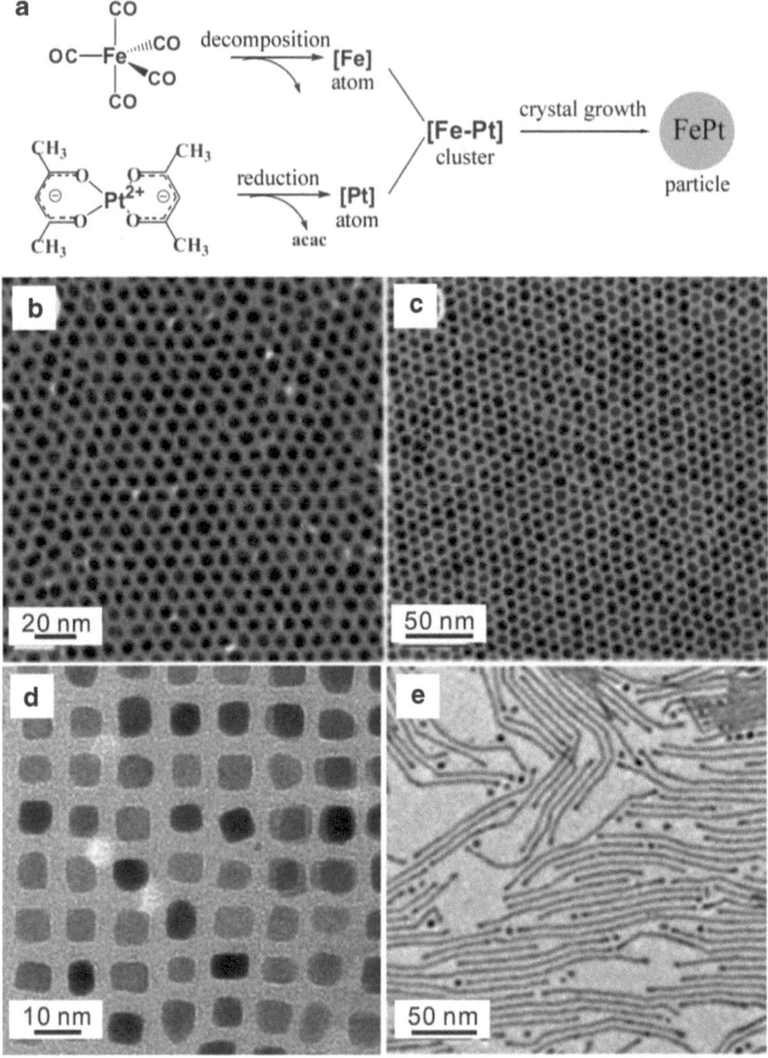

Fig. 3.15 TEM images of FePt alloy nanocrystals. (Reproduced from Sun et al. [35, 36], Chen et al. [37, 38], and Wang et al. [39])

to stabilize the growth of bimetallic nanocrystals. Sometimes, more than one stabilizer may be employed, a common combination being a long-chain alkyl primary amine and a long-alkyl chain acid. Some empirical variation of solvent is necessary to achieve an optimal growth path of bimetallic nanocrystals. In the aforementioned reaction, when only OAm was used as both surfactant and solvent, FePt nanowires (Fig. 3.15e) with a length of more than 200 nm could be obtained [39].

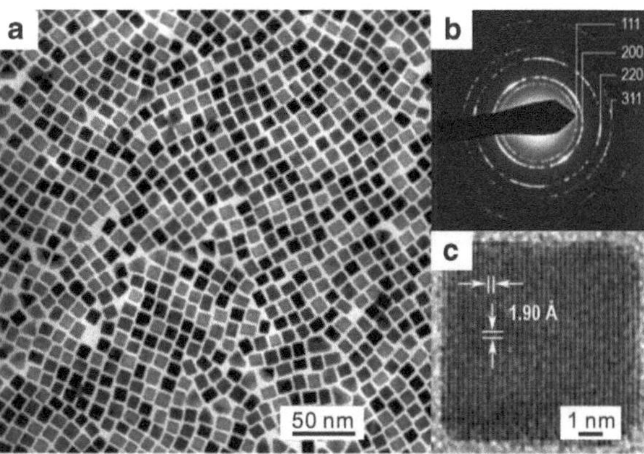

Fig. 3.16 **a** Low-magnification TEM image of the overall morphology of Pt–Cu nanocubes. **b** TEM selected-area electron diffraction pattern of the Pt–Cu nanocubes. **c** High-resolution TEM image of a selected Pt–Cu nanocube. (Reproduced from Xu et al. [42])

3.5.2 Coreduction

Coreduction is a straightforward method for the synthesis of alloy nanocrystals. The key aspect of this method is to manipulate the redox potential and chemical behavior of two distinct metals. Normally, two types of metals would nucleate separately due to their different redox potential, which is known as sequential reduction with respect to coreduction process, resulting in the formation of core-shell/dendrite bimetallic nanocrystals. However, the selection of a proper reducing agent and reaction system can modify the redox potentials of the two metals, and thus their reduction rates, aiming to reduce two metals simultaneously (coreduction). Taking the preparation of Ni–Co alloy for example, by choosing hydrazine as the reducing agent and $NiSO_4$ and $CoSO_4$ as precursors, Ni–Co alloy can be synthesized [40].

Compared with other synthetic strategies, coreduction synthesis technique has been described extensively in several papers and patents for the preparation of high-quality alloys with tunable size, composition, and controlled morphology. The success of this method relies on the reaction parameters, including proper surfactants, reducing agents, solvent, and others like temperature. Schaak and coworkers reported the synthesis of Cu–Pt alloy nanorods with tunable lengths and aspect ratios by reducing $Pt(acac)_2$ and $Cu(acac)_2$ with 1,2-hexadecanediol in a mixed solvent containing OA, OAm, and 1-octadecence [41]. Control of the morphology and length of alloy nanorods can be achieved by varying the OAm/OA ratio, as well as reduction rate, solvent, and temperature. Generally speaking, the surfactants and solvents can influence the growth rates of different crystal facets by changing the energy of different facets, and thus can affect the size and morphology of desired alloy nanocrystals. Taking Br^- ions for a simple example, due to the specific absorption property of Br^- ions toward {100} facets of Pt–Cu alloy nanocrystals, Pt–Cu nanocubes can be prepared by simultaneous reduction process of $Pt(acac)_2$ and $Cu(acac)_2$ with 1,2-tetradecanediol (TDD) in ODE solvent, which contains tetraoctylammonium bromide (TOAB) as an effective surfactant (see Fig. 3.16) [42].

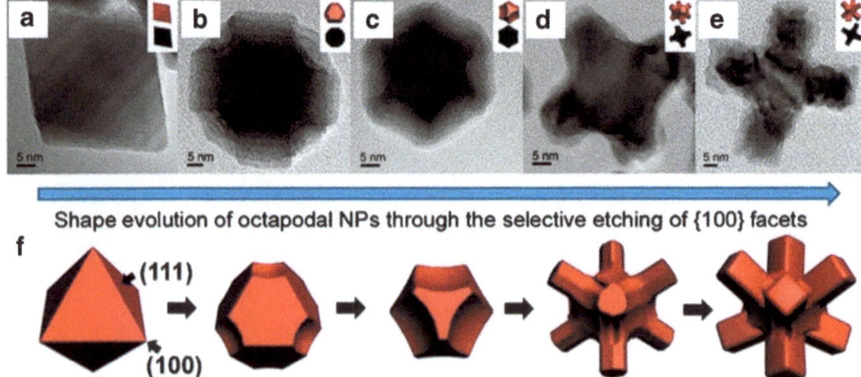

Fig. 3.17 TEM images of nanostructures collected at different reaction times: **a** 10 min, **b** 15 min, **c** 20 min, **d** 2 h, and **e** 4 h. *Black diagrams* in insets are the projection of the *red*-colored three-dimensional structures under the electron beam. **f** Scheme for the shape evolution of nanoparticles by the selective etching process. (Reproduced from Hong et al. [43])

Fig. 3.18 **a** Schematic illustration of shape-selective synthesis of Pt–Pd alloy tetrahedrons and nanocubes. **b** TEM image of Pt–Pd alloy tetrahedrons, and **c** TEM image of Pt–Pt alloy nanocubes. (Reproduced from Yin et al. [44])

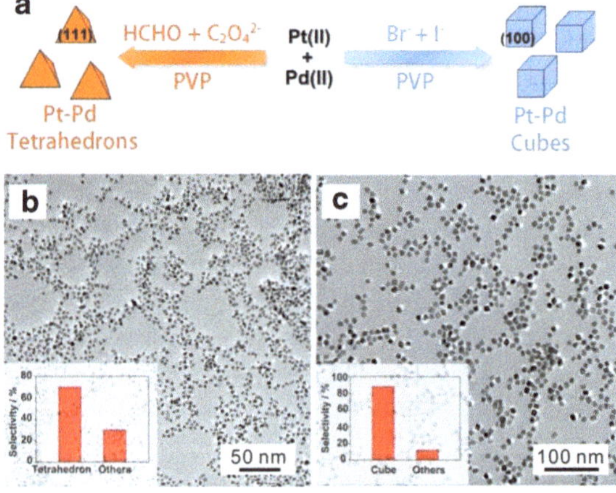

Not only the growth of alloy nanocrystals would be induced by specific capping agent, selective etching of alloy nanocrystals may also be affected by the selected surfactants. Han and coworkers recently reported the successful synthesis of bimetallic octapodal alloy Au–Pd nanoparticles via coreduction of $NaAuBr_4$ and K_2PdCl_4 by AA in the presence of cetyltrimethylammonium chloride (CTAC) [43]. In this synthesis, the unprecedented octapodal shape is attributed to the selective etching of {100} facets by Br^- ions generated via the reduction of $AuBr_4^-$ (see Fig. 3.17). Small molecules, like $C_2O_4^{2-}$ species and Br^- or I^- ions, can also act as surfactants for other bimetallic systems. Yan and coworkers have successfully prepared Pt–Pd alloy tetrahedrons and cubes in the presence of $C_2O_4^{2-}$ and Br^-/I^- ions, respectively (see Fig. 3.18) [44]. The selective adsorption of $C_2O_4^{2-}$ species on the {111} facets

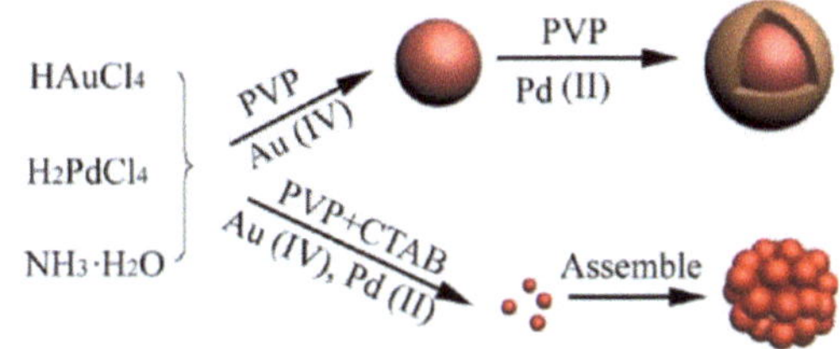

Fig. 3.19 Formation process of Au–Pd core-shell and alloy bimetallic nanostructures. (Reproduced from Kuai et al. [45])

and Br⁻/I⁻ ions toward {100} facets were found to be a key factor in directing the formation of Pd–Pt alloy tetrahedrons.

The coordination effect of surfactants toward specific metallic ions further enables them with the ability to tune the reduction rate. For example, Geng and coworkers synthesized Au–Pd alloy and core-shell nanostructures via a simple one-pot hydrothermal coreduction route by using specific surfactants (CTAB) [45]. With the introduction of CTAB, uniform spherical polycrystalline alloy nanocrystals were obtained. Without CTAB, the products were core-shell nanostructures. Han and coworkers synthesized bimetallic core-shell Au–Pd nanoparticles with a well-defined octahedral shape in the presence of CTAC [46], which was acted as both a reducing agent and stabilizer. Due to the weak reduction power of CTAC, the reduction rate of metal precursors can be conducted at a relatively low speed. Due to the difference in reduction potentials of Au(III) and Pd(II) ($AuCl_4^-/Au$, $+1.002$ V vs. $PdCl_4^{2-}/Pd$, $+0.591$ V). Au (III) will be reduced first, followed by the reduction of Pd(II), thereby forming a core-shell nanocrystal (see Fig. 3.19).

Moreover, the type of reducing agents also plays a critical role in the synthesis of bimetallic nanocrystals. In a coreduction process, reducing agent can readily manipulate the reduction kinetics, thus influencing the crystallinity of seeds in the nucleation stage. Different structures of Pd–Pt alloys can be prepared by using PVP or ethylene glycol (EG) as reducing agents [47]. Smaller reduction rate due to the mild reducing ability of PVP can make coalesce of clusters formed in the nucleation stage much easier, resulting in the formation of twinned nuclei. Finally, star-shaped decahedrons and triangular nanoplates can be obtained (see Fig. 3.20a). In comparison, when a stronger reducing agent (EG) was used, the fast formation of metal atoms would lead to a rapid growth of nuclei, giving the formation of single-crystalline alloy with a truncated octahedral shape enclosed by a mix of {111} and {100} facets (see Fig. 3.20b). Other parameters, including capping agents and reaction conditions (concentration of reactants, temperature, time, etc.), can also be turned to control the formation of bimetallic nanocrystals with a specific shape, tunable size, and controlled composition.

For bimetallic nanocrystals, the composition-control synthesis is especially significant due to their composition-dependent physical and chemical properties. The composition of bimetallic alloys can be easily controlled by changing the molar ratio of two metal precursors during coreduction process. Sun and coworkers successfully tuned the composition of polyhedral Pd–Pt alloy nanocrystals by controlling the amount of metal precursors ($Pd_{88}Pt_{12}$ to $Pd_{34}Pt_{66}$) [48]. However, the compositions of the final products are not exactly consistent with the feeding

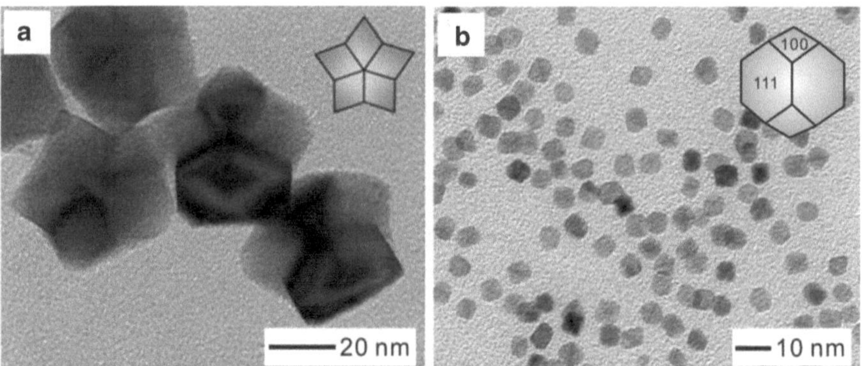

Fig. 3.20 a Transmission electron microscopy (TEM) image of star-shaped decahedron Pd–Pt alloy nanocrystals. **b** TEM image of truncated Pd–Pt alloy nanocrystals. (Reproduced from Lim et al. [47])

ratio in most cases due to the incomplete diffusion of the second metal into the first preformed metal, for example, during the synthesis of monodisperse Au–Ag alloy nanocrystals [49]. $Au_{0.82}Ag_{0.18}$ alloy nanoparticles could be obtained when 10:1 molar ratio of $AgNO_3$ and $HAuCl_4$ were reduced simultaneously at 120 °C for 0.5 h. Compared with Au, only a minor proportion of Ag precursor was reduced and alloyed with Au. The proportion of Ag in Au–Ag alloy nanocrystals could be increased if the reaction time was prolonged to 1 h ($Au_{0.60}Ag_{0.40}$ alloy nanoparticles). On the other hand, the proportion of Ag in Au–Ag alloy can be further increased by increasing the amount of silver precursor because the higher concentration of silver precursor in the reaction mixture can compensate its slower reduction. $Au_{0.52}Ag_{0.48}$ and $Au_{0.39}Ag_{0.61}$ alloy nanoparticles could be synthesized when 20:1 molar ratio of $AgNO_3$ and $HAuCl_4$ were coreduced at 120 °C for 1 and 2 h, respectively.

3.5.3　Galvanic Replacement Reaction

Compared with other synthetic strategies, galvanic replacement reaction has been extensively described in several papers for the preparation of high-quality Ag–M (M=Au, Pd, Pt) bimetallic nanocrystals in recent years. The redox potential difference is a key factor for galvanic replacement reaction. When the redox potential of a second metal is higher than that of the core metal, the galvanic replacement reaction occurs spontaneously between the second metal ions and the core metal. The precursor of the second metal is reduced by the core metal instead of the reducing agents. Cations of the second metal adsorbed on the surface of a sacrificial template can get electrons from the template and reduced metal atoms. Simultaneously, atoms on the surface of sacrificial templates would lose electrons and oxidize to metal ions. The deposition and dissolution sites on templates are strongly dependent on the surface capping agents. Facet-specific capping agents can be used to control the dissolution sites and deposition sites. Therefore, many complex

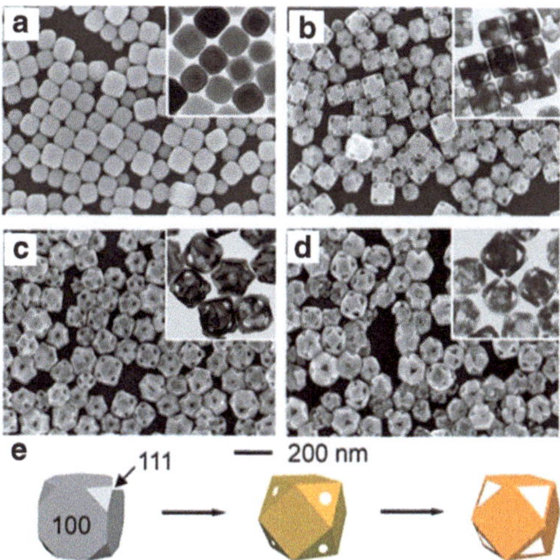

Fig. 3.21 SEM and TEM (inset) of **a** Ag nanocubes with rounded corners and **b–d** product after reaction with different amount of HAuCl$_4$ solution. **e** Illustration summarizing morphological changes. *Coloration* indicates conversion of an Ag nanocube into an Au/Ag nanocage then a predominately Au nanocage. (Reproduced from Skrabalak et al. [50] and Chen et al. [51])

bimetallic nanocrystals, such as concave nanocubes and multipod nanocrystals can be synthesized via this site-selected deposition and dissolution effect. In addition, as galvanic replacement reaction concerns the dissolution of the sacrificial metal and the growth of the deposition metal at the same time, this method can also be used to prepare bimetallic nanocrystals with hollow interiors. The final structure of hollow nanocrystals would prefer to take a shape similar to that of the sacrificial template. For example, Au/Ag alloy hollow nanostructures has been systematically explored via the galvanic replacement reaction between Ag nanomaterials and HAuCl$_4$ in boiling water by Xia and coworkers [50–52]. Figure 3.21 schematically illustrates the mechanism of this reaction and SEM images of the morphological transformation at different stages of galvanic replacement reaction.

Alloying and dealloying are two important processes involved in a galvanic replacement reaction [53], and both of them have significant impacts on the structure and morphology of the final product. Alloying occurred in the initial stage of a galvanic replacement reaction when a thin layer of the second metal is deposited on the template metal, which is the basis for retaining the morphology of the template. For example, the galvanic replacement between Ag cubes and NaPdCl$_4$ would result in the formation of smooth, single-crystal Pd–Pt alloy nanobox [54]. However, when Na$_2$PtCl$_4$ reacted with Ag nanocubes, polycrystalline nanobox with rough surfaces composed of Pt nanoparticles were formed (see Fig. 3.22). Differences in these two cases are the metal–metal bonding energies. Unlike Au and Pd, Pt does not readily undergo solid–solid interdiffusion over the entire surface of an Ag template to form a conformal Pt–Ag alloy. The much higher bonding energy of Pt–Pt relative to that of Pt–Ag not only would lead to the island growth pattern for the deposition of Pt, but also generated a relatively larger energy barrier to interdiffusion. Dealloying plays a key role in the later stages of a galvanic replacement reaction, especially in controlling the porosity of the wall of the hollow structures. Dealloying can be aroused by adding more second metal ions or by utilizing a conventional

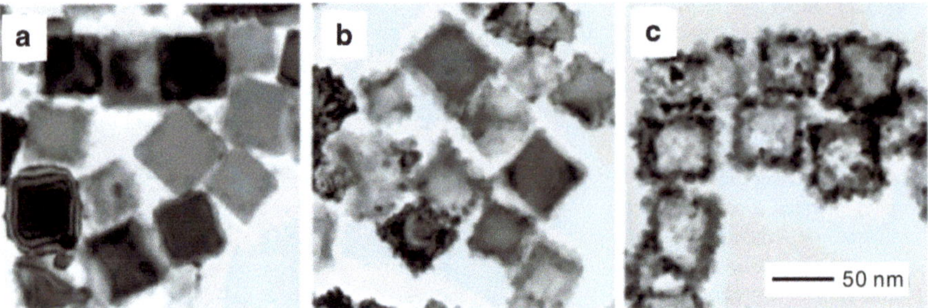

Fig. 3.22 TEM images of the Ag–Pt nanostructures formed with different amount of Na_2PtCl_4 aqueous solution. (Reproduced from Chen et al. [54])

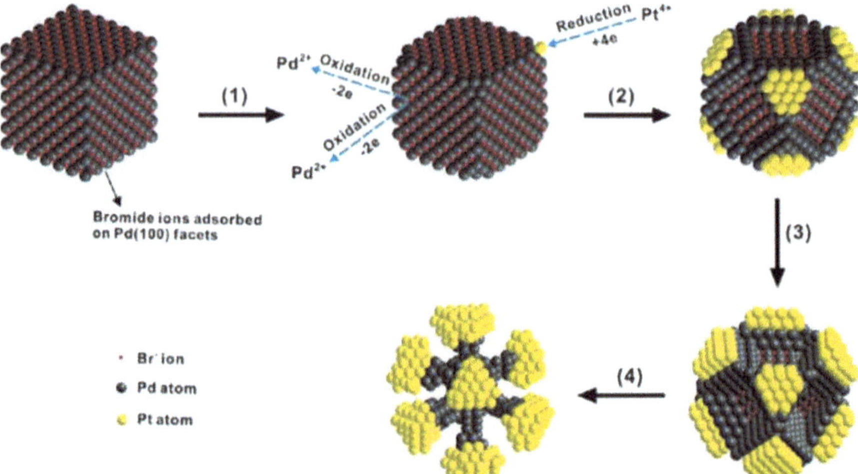

Fig. 3.23 Schematic illustration of a plausible mechanism for the formation of a Pd–Pt concave nanocube through the galvanic replacement between a Pd nanocube and $PtCl_6^{2-}$ ions. (Reproduced from Zhang et al. [58])

wet etchant [55–57]. Taking the galvanic replacement reaction between Ag nanocubes and $HAuCl_4$ aqueous solution for example, Au–Ag nanocages with pores on the surface can be obtained by extracting Ag atoms with $HAuCl_4$ from the Au–Ag alloyed walls.

In principle, galvanic replacement reaction should occur between any two metals with the favorable reduction potentials. Some specific species can change the reduction potentials of the metal/ion pairs and thus accelerate or inhibit the galvanic replacement. For example, the galvanic replacement reaction between Pd nanocube and H_2PtCl_6 could not occur or react slowly in the absence of Cl^-/Br^- ions [58]. With the addition of Cl^-/Br^- ions, the galvanic replacement between Pd nanocube and H_2PtCl_6 could occur at a relatively high speed, due to large solubility product constants (K_{sp}) of $[PdBr_4^{2-}]/[PdCl_4^{2-}]$ and $[PtBr_6^{2-}]/[PtCl_4^{2-}]$, and result in the formation of Pd–Pt concave nanocubes (see Fig. 3.23).

Fig. 3.24 TEM image of the Ag@Au core-shell nanoplates. (Reproduced from Gao et al. [4])

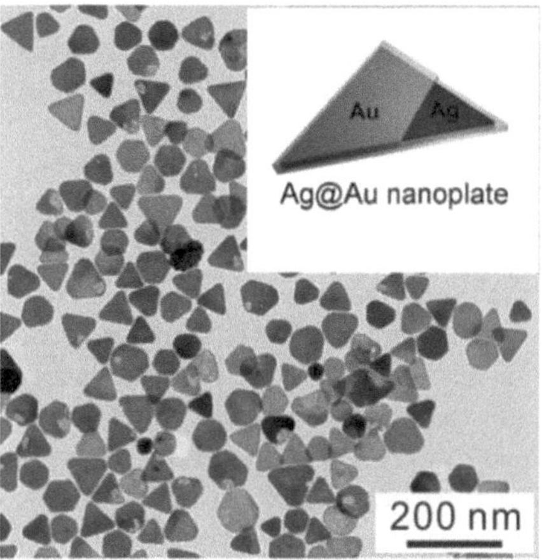

In some cases, it is necessary to prevent the galvanic replacement. For example, Yin and coworkers synthesized highly stable Ag@Au core-shell plates by deposition of a thin and uniform layer of Au on the surface of Ag nanoplates, to enhance the stability of Ag nanoplates [4]. The galvanic replacement between Ag nanoplates and $HAuCl_4$ would prevent the formation of Ag@Au core-shell nanoplates. To prevent this galvanic replacement reaction, I^- ions were introduced into the reaction system, as AuI_4^- complex would not be reduced by Ag nanoplates due to its high stability. As a result, bimetallic core-shell Ag–Au nanoplates were obtained (see Fig. 3.24). Another method for the prevention of galvanic replacement is the introduction of a strong antioxidant or reducing agent into the system. With the strong reducing agent, metal precursors will be rapidly reduced into metal atoms. For example, Ag–Au nanoplates were obtained by Xue and coworkers by introducing hydroxylamine and NaOH to prevent the galvanic replacement between $HAuCl_4$ and Ag nanoplates and generated Ag–Au core-shell nanoplates [59].

Overall, galvanic replacement offers a facile and versatile route to a variety of advanced multifunctional nanostructures often characterized by tightly controlled sized and shapes, hollow interiors, porous walls, and tunable elemental compositions. Ligand also plays an important role in the galvanic replacement, which can facilitate the oxidation of the core metal in many cases. Core-shell bimetallic nanocrystals can also be obtained when the galvanic reaction occurs slowly and evenly on the surface of the metal template.

3.5.4 Seed-Mediated Growth

Seed-mediated growth is the most powerful and widely used route to prepare core-shell and heterostructures bimetallic nanocrystals that are usually difficult to obtain via other routes. A typical seed-mediated process involves two steps: synthesis of seed nanocrystals with uniform and relatively small sizes, and heterogeneous nucleation and growth of another metal on the preformed seeds, leading to formation of bimetallic nanocrystals. It is particularly advantageous because it allows the thoughtful design of nanocrystals morphology through choosing seed nanocrystals and meticulously manipulating crystal growth conditions. When the second metal evenly deposits on the surface of the preformed seed, core-shell nanostructures will be obtained, while dendrite bimetallic nanocrystals are formed via the specific site deposition. Generally, during seed-mediated process, homogeneous nucleation should be prevented. As the free-energy barriers for nucleation on the preformed seeds are usually lower than that of a nucleation process, indicating heterogeneous nucleation is more thermodynamically favored [60]. However, if the reaction system can afford enough energy to overcome the energy barrier for a homogeneous nucleation process, it will still occur, leading to the generation of separate monometallic nanocrystals. Therefore, the reaction parameters must be rationally selected to meet the thermodynamic requirements for the synthesis of core/shell or dendrite bimetallic nanocrystals.

In a seed-mediated growth process, the structure, shape, and size of nanocrystal seeds have a considerable impact on the nucleation and growth of the secondary metal. Noble metal nanocrystals with face-centered cubic lattice can exist in several crystalline forms. The most common forms include single-crystal, singly twinned, or multiply twinned structures. Generally, single-crystal seeds, octahedrons, cuboctahedrons, or cubes will be formed depending on the relative growth rates between the <111> and <100> directions. Single-crystal seeds can evolve into single-crystal nanocrystals with various exposed crystal facets, such as cube, octahedrons, and polyhedrons, and if anisotropic growth is induced, the cuboctahedral and cubic seeds will grow into octagonal rods and rectangular bars. From single-twinned seeds bipyramids will be produced, but when anisotropic growth is induced, these seeds also can involve into nanobeams. Nevertheless, multiple-twinned seeds usually grow into decahedron and isosahedron, and pentagonal nanorods (or nanowires) can be produced, depending on whether the {100} planes on the side surface are stabilized. When there are multiple planar-twinned seeds with stacking fault, they form thin plates with a hexagonal cross-section due to the sixfold symmetry of an fcc system; as the growth is continued, the final products can also take a triangular shape by eliminating the {111} facets from the side surfaces [61, 62].

Seed-mediated growth is an effective route to prepare core-shell bimetallic nanocrystals. As indicated earlier in the text, there are three types of growth modes when a substance is deposited on a substrate in gas phase or vacuum; in the F-M mode growth, the metal bond energy is a key factor in addition to the lattice mismatch, and the interactions among atoms in the overlayer metal should be smaller than that between the substrate and overlayer. In addition to the lattice mismatch and bond dissociation energy, the electronegativity is also a key factor to the crystal growth in solution. Thus, three rules have been proposed for the epitaxial layered growth of heterogeneous core-shell nanocrystal. In the previous literatures, Au@Ag, Au@Pd, and Pt@Pd can well meet the aforementioned rules and have been prepared by Fan and other researchers [9, 63].

Many factors, including concentration of capping agent, reducing agent, and metal precursor ions as well as the reaction temperature, have been proved to have significant effect on the final structure of bimetallic nanocrystals. Xia and coworkers prepared Au@Pd core-shell nanocrystals with controlled size and morphology by manipulating the kinetics of seeded growth [64]. Along with the increase of reaction rate, core-shell nanocrystals evolved into octahedrons, concave octahedrons, rectangular bars, cubes, concave cubes, and dendrites. The reducing agent also plays an important role in a seed-mediated growth. For example, Pd–Au bimetallic nanocrystals with two different types of structures were obtained by choosing different reducing agents. AA and citric acid were used to reduce $HAuCl_4$ in the presence of cubic Pd seeds in the experiments, respectively. When AA was used to reduce $HAuCl_4$, conformal overgrowth of Au on the Pd nanocubes was observed and resulted in the formation of Pd@Au core-shell nanocrystals (see Fig. 3.25a) [65]. However, the reduction of $HAuCl_4$ by citric acid resulted in Pd–Au dimers consisting of Pd nanocubes and Au nanoparticles (see Fig. 3.25b).

Recently, Pd@Cu nanocrystal with a lattice mismatch of 7.1 % was also obtained via the reduction of $CuCl_2$ by glucose in the presence of Pd seeds and hexadecylamine as capping agent [23]. In this system, the formation of Pd@Cu core-shell nanocube follows a localized epitaxial growth mechanism; the Pd@Cu core-shell nanocrystals evolved into a cubic shape due to a strong affinity of hexadecylamine for the Cu {100} facet. In addition to the lattice mismatch, the bond dissociation energy also played a key role in the seed-mediated growth. For example, for the growth of Pt on Pd seeds, the island-on-wetting layer mode is preferred due to the following order in bond energy: E_{Pt-Pt} (307kJ/mol) > E_{Pt-Pd} (191kJ/mol) > E_{Pd-Pd} (136kJ/mol), resulting in the generation of Pd-Pt bimetallic nanodendrites structures. As reported by Xia and coworkers [30], the synthesis of bimetallic nanodendrite with Pt branches anchored to a Pd core. In this work, Pd truncated octahedrons with an average size of 9nm were used as seeds to direct the dendritic growth of Pt via reduction of K_2PtCl_4 by AA in an aqueous solution containing PVP as capping agent. High concentration of Pt atoms' fast reduction by AA was probably responsible for the branched growth of Pt. As shown in the figure, a large number of Pt branches had grown from each Pd core, leading to the formation of Pd@Pt nanodendrites; the Pd seeds play an important role in forming an open, dendritic structure by providing multiple nucleation sites for Pt that were spatially separated from each other, helping to avoid overlap and fusion between the Pt branches during the growth process. Such an open bimetallic architecture provides a new design strategy for generating advanced catalysts with enhanced activity by integrating several different functionalities in one structure. In another study, Pd-Pt nanodendrites with controlled Pt and Pd ratios were also generated by reducing Pd and Pt precursors in sequence with the use of AA as a reducing agent and Pluronic P123 (a triblock copolymer) as capping agent at room temperature [66].

Although core-shell and dendrites bimetallic nanocrystals are more favored via seed-mediated method, alloy and intermetallic NCs can also be prepared in some cases when the reaction system is elaborately designed. For example, CuAu and Cu_3Au nanocrystals can be synthesized by reducing $Cu(CH_3COO)_2 \cdot H_2O$ in the

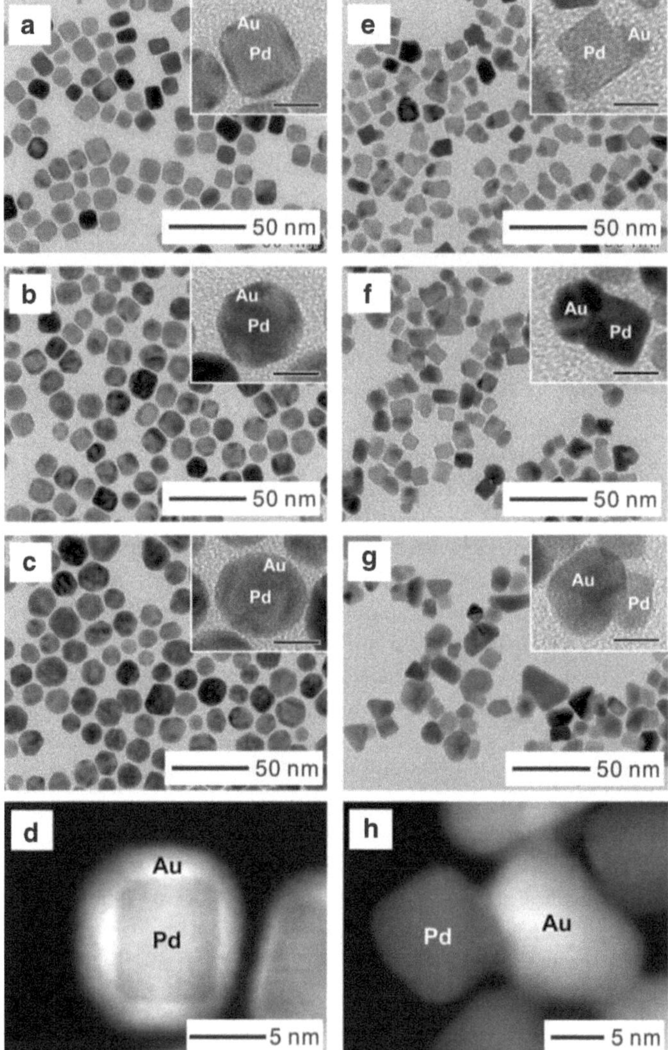

Fig. 3.25 TEM images of Pd–Au core-shell nanocrystals synthesized by reducing HAuCl₄ with L-ascorbic acid and citric acid, respectively. (Reproduced from Lim et al. [65])

presence of Au nanoparticles as seeds [67]. The molar ratio of Cu/Au precursors as well as the reaction temperature codetermines the composition of final products. Figure 3.26 illustrates this unique seed-mediated diffusion process for the synthesis of intermetallic nanocrystals. At high temperature, the capping molecules on the surface of Au seeds have a high probability to be desorbed, leaving active sites on the Au surface. Once the copper ions are reduced, it is very easy for the newly produced highly reactive Cu atoms or clusters to collide with the active sites of Au

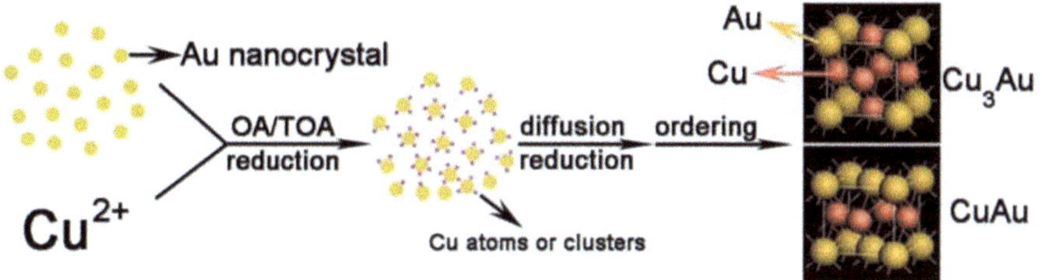

Fig. 3.26 Proposed mechanisms for the formation of the intermetallic CuAu nanocrystals. (Reproduced from Chen et al. [67])

particles and diffuse from the surface into the crystal lattice of Au seeds. The reduction and diffusion processes proceed continuously, leading to the formation of a Cu–Au solid solution. Finally, intermetallic nanocrystals were obtained.

3.6   Summary

In this chapter, we discussed three types of bimetallic nanocrystals, including core-shell, dendrite, and alloy nanocrystals. All three types of bimetallic nanocrystals have significant effect on their related properties, such as optical and catalytic activity. Four methods/techniques have been mostly adopted for the generation of bimetallic nanocrystals, including thermal decomposition, coreduction, galvanic replacement reaction, and seed-mediated growth. Role of reaction parameters (e.g., reaction temperature, surfactant, and reducing agents) has been discussed for these four methods. The advanced syntheses and characterization methods described in this chapter should allow optimization of the structure of bimetallic nanocrystals, thus enabling their application as new optical and catalytic materials.

Acknowledgment This work was performed at Xi'an Jiaotong University and supported by the "start-up fund," "the Fundamental Research Funds for the Central University," and Center for Materials Chemistry.

References

1. N. Toshima, T. Yonezawa, New J. Chem. **22**, 1179 (1998)
2. R. Ferrando, J. Jellinek, R. Johnston, Chem. Rev. **108**, 845 (2008)
3. D. S. Wang, Y. D. Li, Adv. Mater. **23**, 1044 (2011)
4. C. Gao, Z. Lu, Y. Liu, *et al.* Angew. Chem. Int. Ed. **51**, 5629 (2012)
5. C. Chen, Y. Kang, Z. Huo, *et al.* Science **343**, 1339 (2014)
6. M. P. Andrews, S. C. O'Brien, J. Phys. Chem. **96**, 8233 (1992)
7. F. Wang, L. Sun, W. Feng, *et al.* Small **6**, 2566 (2010)
8. E. Bauer, J. H. van der Merwe, Phys. Rev. B.: Condens. Matter. **33**, 3657 (1986)
9. F. R. Fang, D. Y. Liu, Y. F. Wu, *et al.* J. Am. Chem. Soc. **130**, 6949 (2008)

10. S. A. Chambers, Surf. Sci. Rep. **39**, 105 (2000)
11. A. W. K. Liu, M. B. Santos, Eds. Thin films: heteroepitaxial system, Series on Directions in Condensed Matter Physics, World Scientific, **15**, p27 (1999)
12. H. Gerisher, C. W. Tobias, Eds. Advances in electrochemistry and electrochemical engineering. Wiley-Interscience, New York, **11**, 125 (1978)
13. R. Johnston, J. Wilcoxon, Eds. Metal nanoparticles and nanoalloys. Elsevier, Oxford, p8 (2012)
14. H. Lipson, Ed. The study of metals and alloys by X-ray powder diffraction methods. University College Cardiff Press, Cardiff, Wales, p7 (2001)
15. Z. Jiang, Q. Zhang, C. Zong, *et al*. J. Mater. Chem. **22**, 18192 (2012)
16. Z. Xu, E. Lai, S. Shao-Horn, *et al*. Chem. Commun. **48**, 5626 (2012)
17. Y. W. Lee, M. Kim, S. Kang, *et al*. Angew. Chem. Int. Ed. **50**, 3466 (2011)
18. M. K. Carpenter, T. E. Moylan, R. S. Kukreja, *et al*. J. Am. Chem. Soc. **134**, 8535 (2012)
19. S. I. Choi, S. U. Lee, W. Y. Kim, *et al*. Appl. Mater.Interfaces **4**, 6228 (2012)
20. A. R. Denton, N. W. Ashcroft, Phys. Rev. A **43**, 3161 (1991)
21. L. Zhang, J. Zhang, Q. Kuang, *et al*. J. Am. Chem. Soc. **133**, 17114 (2011)
22. Y. Ma, W. Li, E. C. Cho, *et al*. ACS Nano **4**, 6725 (2010)
23. M. Jin, H. Zhang, J. Wang, *et al*. ACS Nano **6**, 2566 (2012)
24. S. Xie, S. Choi, N. Lu, *et al*. Nano Lett. **14**, 3570 (2014)
25. A. Q. Zhang, D. J. Qian, M. Chen, Eur. Phys. J. D **67**, 231 (2013)
26. C. Zhu, J. Zeng, J. Tao, *et al*. J. Am. Chem. Soc. **134**, 15822 (2012)
27. J. Chen, B. Wiley, Z. Li, *et al*. Adv. Mater. **17**, 2255 (2005)
28. L. Durán Pachón, M. Thathagar, F. Hartl, *et al*. Phys. Chem. Chem. Phys. **8**, 151 (2006)
29. B. Lim, M. Jiang, P. H. C. Camargo, *et al*. Science **324**, 1302 (2009)
30. V. R. Stamenkovic, B. Fowler, B. S. Mun, *et al*. Science **315**, 493 (2007)
31. S. U. Son, Y. Jiang, J. Park, *et al*. J. Am. Chem. Soc. **126**, 5026 (2004)
32. H. Bonnemann, R. A. Brand, W. Brijoux, *et al*. Appl. Organometal. Chem. **19**, 790 (2005)
33. R. D. Rutledge, W. H. Morris, M. S. Wellons, *et al*. J. Am. Chem. Soc. **128**, 14210 (2006)
34. I. Robinson, S. Zacchini, L. D. Tung, *et al*. Chem. Mater. **21**, 3021 (2007)
35. S. H. Sun, C. B. Murray, D. Weller, *et al*. Science **287**, 1989 (2000)
36. S. H. Sun, Adv. Mater. **18**, 393 (2006)
37. M. Chen, J. P. Liu, S. Sun, J. Am. Chem. Soc. **126**, 8394 (2004)
38. M. Chen, J. Kim, J. P. Liu *et al*. J. Am. Chem. Soc. **128**, 7132 (2006)
39. C. Wang, Y. Hou, J. Kim, S. Sun, Angew. Chem. Int. Ed. **46**, 6333 (2007)
40. Y. D. Li, L. D. Li, H. W. Liao, *et al*. J. Mater. Chem. **9**, 2675 (1999)
41. Q. Liu, Z. Yan, N. L. Henderson, *et al*. J. Am. Chem. Soc. **131**, 5720 (2009)
42. D. Xu, Z. Liu, H. Yang, *et al*. Angew. Chem. Int. Ed. **48**, 4217 (2009)
43. J. W. Hong, Y. W. Lee, J. Kim, *et al*. Chem. Commun. **47**, 2553 (2011)
44. A. X. Yin, X. Q. Min, Y. W. Zhang, *et al*. J. Am. Chem. Soc. **133**, 3816 (2011)
45. L. Kuai, X. Yu, S. Wang, *et al*. Langmuir **28**, 7168 (2012)
46. Y. W. Lee, J. Kim, Z. H. Kim, *et al*. J. Am. Chem. Soc. **131**, 17036 (2009)
47. B. Lim, J. Wang, P. H. C. Camargo, *et al*. Angew. Chem. Int. Ed. **48**, 6304 (2009)
48. Y. Liu, M. Chi, V. Mazumder, *et al*. Chem. Mater. **23**, 4199 (2011)
49. C. Wang, H. Yin, R. Chan, *et al*. Chem. Mater. **21**, 433 (2009)
50. S. E. Skrabalak, J. Chen, Y. Sun, *et al*. Acc. Chem. Res. **41**, 1587 (2008)
51. J. Chen, J. M. McLellan, A. Siekkinen, *et al*. J. Am. Chem. Soc. **128**, 14776 (2006)
52. S. E. Skrabalak, J. Chen, L. Au, *et al*. Adv. Mater. **19**, 3177 (2007)
53. Y. Sun, Y. Xia, Nano Lett. **3**, 1569 (2003)
54. J. Chen, B. Wiley, J. M. McLellan, *et al*. Nano Lett. **5**, 2058 (2005)
55. Y. Sun, Y. Xia, J. Am. Chem. Soc. **126**, 3892 (2004)
56. R. Newman, K. Sieradzki, Science **263**, 1708 (1994)
57. L. Martinez, M. Segarra, M. Fernandez, *et al*. Metall. Trans. B **24B**, 827 (1993)
58. H. Zhang, M. Jin, J. Wang, *et al*. J. Am. Chem. Soc. **133**, 6078 (2011)

59. N. M. Shahjamali, M. Bosman, S. Cao, *et al.* Adv. Funct. Mater. **22**, 849 (2012)
60. X. Peng, Adv. Mater **15**, 459 (2003)
61. Y. Xiong, Y. Xia, Adv. Mater. **19**, 3385 (2007)
62. Y. Xia, Y. Xiong, B. Lim, *et al.* Angew. Chem. Int. Ed. **48**, 60 (2009)
63. S. E. Habas, H. Lee, V. Radmilovic, *et al.* Nat. Mater. **6**, 692 (2007)
64. J. Li, Y. Zheng, J. Zeng, *et al.* Chem. Eur. J. **18**, 8150 (2012)
65. B. Lim, H. Kobayashi, T. Yu, *et al.* J. Am. Chem. Soc. **132**, 2506 (2010)
66. L. Wang, Y. Nemoto, Y. Yamauchi, J. Am. Chem. Soc. **133**, 9674 (2011)
67. W. Chen, R. Yu, L. Li, *et al.* Angew. Chem. Int. Ed. **49**, 2917 (2010)

Chapter 4
Interactions of Metallic Nanocrystals with Small Molecules

Ran Long and Yujie Xiong

Abstract Metallic nanocrystals have been widely used in heterogeneous catalysis and biomedical applications. The essence of these applications is more or less related to interactions of metallic nanocrystals with molecules such as O_2, H_2, and CO. This chapter summarizes the progress on the related research with a focus on fundamentals. Two typical small molecules, O_2 and H_2 are highlighted to demonstrate the mechanisms for metal–molecule interactions, followed by brief introduction to their applications. Notably, charge transfer process plays a central role in the interactions. Acquiring this information, one can rationally tune the performance of metal nanocrystals in catalysis and biomedicine by tailoring their parameters.

4.1 Introduction

Catalytic reactions such as CO oxidation and hydrogenation have received wide interest due to increasing energy concerns. Metallic nanomaterials have been commonly used as catalysts in these reactions. Heterogeneous catalysis essentially occurs through a process of interfacial reactions. Thus the surface states of catalysts play an important role in determining the species adsorption and molecular activation, and in turn, hold promise to tailoring reaction activity and selectivity in catalysis. In this regard, we can learn a lot from surface science, which community has extensive experience in surface–molecule interactions on flat surface. However, it is well known that as the size of particles shrinks, their boundary effects will become more dominant. In particular, metallic nanomaterials have demonstrated a range of fascinating properties that are distinct from those appearing in bulk materials, basically due to its unique surface/interface states. Such difference of heterogeneous catalysts from single crystals has been commonly recognized in surface science as "material gap". To better understand the nature of metallic nanocatalysts, it is imperative to investigate the interactions of these nanocrystals with small molecules

Y. Xiong (✉) · R. Long
School of Chemistry and Materials Science, University of Science
and Technology of China, Hefei 230026, Anhui, P. R. China
e-mail: yjxiong@ustc.edu.cn

© Springer International Publishing Switzerland 2015
Y. Xiong, X. Lu (eds.), *Metallic Nanostructures,* DOI 10.1007/978-3-319-11304-3_4

from the viewpoint of fundamentals. In parallel, molecular activation serves as a key process to biomedicine such as cancer therapy and nanoparticles toxicity. For instance, reactive oxygen species (ROS), whose generation relies on O_2 activation, are known to be capable of reducing cell viability. Thus the surface–molecule interactions also hold the key to designing agents for cancer treatment and safe nanomaterials for commercialization.

Given the extensive research in related fields, it seems impractical to include all the activities in this chapter. In order to highlight the key facts, here we mainly outline two small molecules' interactions (O_2 and H_2) with metallic nanocrystals. From the perspective of fundamentals, the mechanisms involved are not limited to surface science, quantum confinement effect, and molecular dynamics. It is anticipated that the discussions will in turn facilitate further applications other than catalysis and biomedicine.

4.2 Molecule of O_2

Oxidation reactions are an important category of chemical transformation in organic chemistry. The oxidative species includes a variety of molecules and compounds, among which molecular oxygen (O_2) is the one that can readily meet the criteria for green chemistry and low costs. In principle, molecular O_2 can be activated to form various ROS species including superoxide (O_2^-), hydrogen peroxide (H_2O_2), hydroxyl radicals ($OH\bullet$), and singlet oxygen ($^1O_2$). Each species may play an important role in chemical reactions and biological systems. The corresponding electron spinning configurations for species, along with their specific scavengers that can effectively inhibit the production of species, are summarized in Fig. 4.1.

It is well known that the ground state of O_2 is in a triplet form, while most organic compounds are singlet. In the organic systems, chemical reactions between singlet

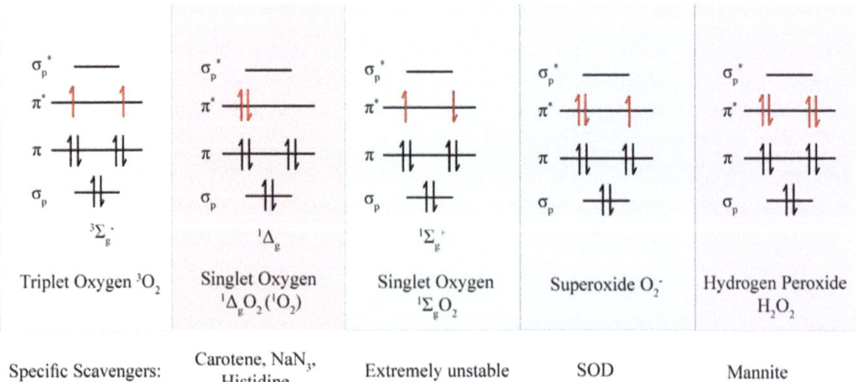

Fig. 4.1 Molecular orbital diagrams of O_2 molecule and ROS species, and several commonly used specific scavengers

organic molecules and triplet O_2 ($^3\Sigma$) that produce new singlet compounds are generally forbidden by the Wigner's spin selection rule, making the organic oxidations kinetically slow [1, 2]. For this reason, the excitation of inert ground triplet O_2 to highly reactive singlet O_2 is a key step in many oxidation reactions. In recent years, there has been a significant progress to discover that metallic nanoparticles exhibit excellent catalytic activities in various oxidation reactions such as epoxidation [3] and CO [4], hydrocarbon [5], alcohol [6], and glucose oxidations[7] in the presence of molecular oxygen [6, 8, 9]. The interactions between O_2 and metal surface play a central role in a variety of chemical and biological systems, but are not yet fully understood despite several decades of research in surface science.

Among various model systems, chemical/physical behavior of O_2 molecule over metal surface is the particular one that has been widely investigated. Numerous different theories have been proposed to address the related issues such as the diffusion of "hot atoms" [10], atom-repelling via adsorption [11], and spin selection rule [12]. Meanwhile, it has been recognized that the states of adsorbed O_2 are varied because factors such as the types, sizes, surface facets, and charge states of metals can affect the adsorption process. In this section, we will use three types of metallic nanocrystals as conceptual models to discuss recent progress on nanocrystal-O_2 interactions.

4.2.1 Au–O_2 Interactions

It has been identified for long time that bulk gold is very inactive towards oxygen activation unless sufficient energy can be exotically provided to break the O=O (double bond) [13]. However, when the gold materials downsize to the nanoscale or sub-nanoscale, they become an effective catalyst for oxidation reactions at low temperature using molecular O_2. This finding was first achieved on low-temperature CO oxidation by Haruta in 1987 [14]. Following this important discovery, a large number of experimental and theoretical studies have been undertaken, attempting to elucidate the nature of this catalytic activity. With an insightful saturation chemisorption investigation, Salisbury and Whetten found that the adsorbed O_2 can be activated by the electrons that are donated by Au clusters, which plays a key role in oxidation reactions [15]. This conclusion is now generally accepted by researchers. In this section, we will highlight some detailed mechanisms for the interactions of molecular O_2 with Au clusters or nanocrystals, hopefully in turn facilitating related research in the future.

4.2.1.1 Au Clusters

Based on the charge states of Au clusters, it can be divided into three types: cationic clusters (positively charged), neutral clusters, and anionic ones (negatively charged). Experimental and theoretical studies have revealed that the activation of adsorbed O_2 can be affected by the charge states of Au clusters. Specifically, the activation mainly occurs on neutral and anionic clusters, whereas no interaction

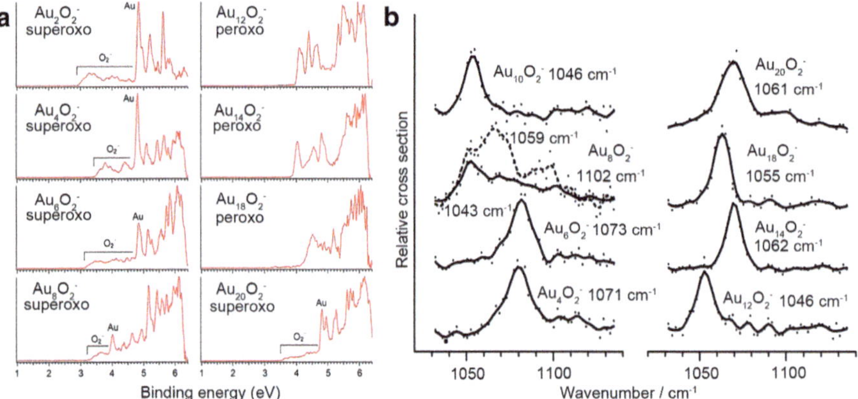

Fig. 4.2 **a** PES of $Au_nO_2^-$ ($n=2, 4, 6, 8, 12, 14, 18, 20$) [17, 18]. **b** IR-MPD spectra of $Au_nO_2^-$ ($n=4, 6, 8, 10, 12, 14, 18, 20$) [19]

Fig. 4.3 Structures of anionic $Au_nO_2^-$ with even atom numbers acquired by theoretical simulations for the lowest energy [17]

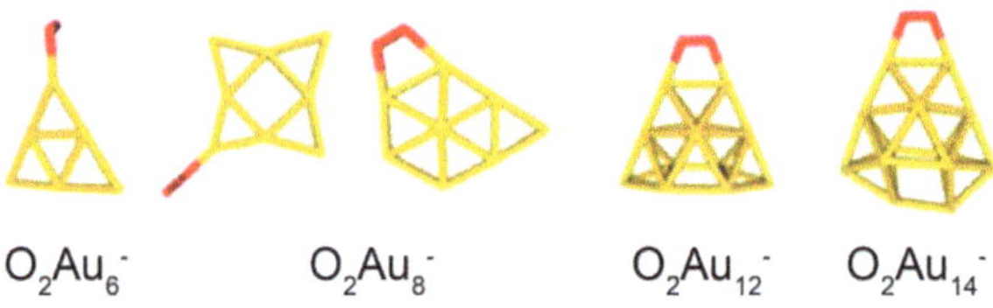

between Au_n^+ and O_2 has been experimentally observed [16]. Besides the charge states, the even–odd number (i.e., magic number) of clusters has significant impact on their reactivity with O_2. As a result, the interactions between Au clusters and O_2 become quite complicated.

Anionic Clusters Huang and Wang presented a systematic photoelectron spectroscopy (PES) study for the interactions between molecular O_2 and Au_n^- clusters (Fig. 4.2, $n=2, 4, 6$). Their observations show that both the electron-bonding energies (E_b) and vibrational frequencies ($v(O–O)$) of adsorbed O_2 are distinguished from free molecular O_2 when $n=2, 4, 6$. In that work, the inertness of odd-sized Au clusters ($n=1, 3, 5, 7$) towards O_2 was also observed for the first time. Based on the findings, Pal and Zeng further elucidated that there exist two modes of chemisorbed O_2 on even-sized anionic Au_n^- clusters: the superoxo (O_2^-) and peroxo (O_2^{2-}) states (see Fig. 4.2 and 4.3) [17]. For the clusters of $n=2, 4, 6, 20$, O_2 is adsorbed in the superoxo form via one-electron charge transfer from anionic clusters to O_2. However, $n=8$ is the onset of the even-sized anionic clusters from the superoxo to the peroxo forms via chemisorption. When $n=10, 12, 14, 18$, the O–O bond is more elongated than that in $n=2, 4, 6, 20$, indicating that the peroxo form of adsorbed O_2 is more activated than the superoxo form.

Besides the PES, another direct evidence of O_2 activation on Au clusters is infrared multiple-photon dissociation (IR-MPD) spectroscopy provided by Woodham and Fielicke (see Fig. 4.2b) [19]. Based on the IR-MPD data, they have been able

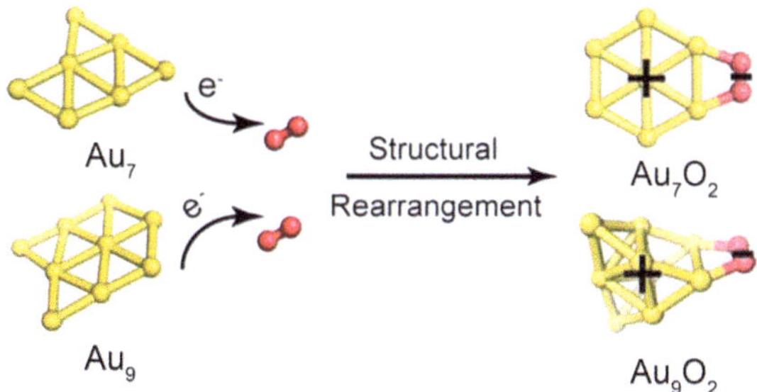

Fig. 4.4 Structures of anionic Au_nO_2 with odd atom numbers acquired by theoretical simulations for the lowest energy [20]

to establish an approximate anticorrelation between electron affinity (EA) and vibrational frequencies (v(O–O)). In a word, the clusters with low EA are more likely to donate electron into the π^* orbital of O_2, resulting in a weaker O–O bond and a lower v(O–O). It is worth mentioning that an anionic Au cluster only can accommodate one O_2 molecule, no matter whether it is of even or odd atom numbers [15].

Neutral Clusters The presence of ROS on neutral Au clusters arises from an electron transfer from the Au_n to the oxygen, similarly to the findings observed for anionic clusters. Woodham and Fielicke have presented spectroscopic evidence for the formation of odd-sized Au_nO_2 ($n=7, 9, 11, 21$) [20]. The electron transfer from Au_n leads to the formation of a superoxo (O_2^-) species. In stark contrast, the even-sized Au_n clusters have been found largely unreactive with molecular O_2 when $n=4, 10, 12$. Only slightly activated O_2 was captured upon even-sized Au_nO_2 complexation.

The charge transfer from Au_n to O_2 not only impacts on the states of O_2, but also induces significant rearrangement on the Au clusters (see Fig. 4.4). Density functional theory (DFT) calculations suggest that after the rearrangement process, the neutral Au cluster more resembles a cationic structure. This rearrangement mechanism (along with binding and electron transfer) highlights the importance of structural flexibility to O_2 activation by Au clusters.

Taken together, both the charge state and even-odd effect are major parameters in the O_2 activation by Au clusters. Table 4.1 outlines some related results as a reference for further investigations. Due to the complexity of this $Au_n–O_2$ system, early-stage theoretical calculations failed to explain such results with sharp turnovers caused by subtle structure/charge difference of Au clusters. However, after

Table 4.1 Different forms of adsorbed oxygen on Au_n clusters (Data collected from [17, 20])

Even-sized clusters				Odd-sized clusters
Neutral clusters	Unreactive			Superoxo ($n=7, 9, 11, 21$)
Anionic clusters	Superoxo ($n=2–6, 20$)	Superoxo to peroxo ($n=8$)	Peroxo ($n=10–18$)	Unreactive

Fig. 4.5 TOF (s^{-1}) *versus* particle size (Data collected from [24–30])

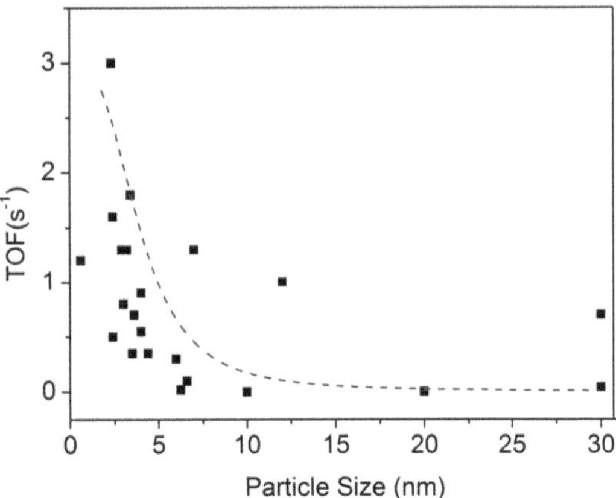

the tremendous efforts on theoretical calculations to benchmark work functions, the research community can now explain and predict the adsorption behavior of molecular O_2 on Au clusters [18, 21–23].

4.2.1.2 Au Nanocrystals

As the model system switches to nanoscale crystals, the mechanisms of molecule–metal interactions become quite different from those for clusters. In principle, the difference may originate from the facts: (1) a number of nanocrystal sizes are above the regime for quantum confinement effect; (2) the nanocrystals are enclosed by certain surface facets that affect the adsorption modes of molecules. In general, the parameters listed below play important roles in determining the activation of O_2 molecules on Au nanocrystals.

Size At large sizes (approximately > 10 nm), Au nanocrystals behaves like bulk Au (i.e., unreactive, or extra energy required) in the activation of oxygen. The catalytic activity of Au nanocrystals in oxidation reactions is promoted rapidly with its size reduction (see Fig. 4.5). As the nanocrystals have a fixed and long-range ordered structure, they cannot undergo a thorough restructuring during Au–O_2 interaction like the Au clusters. As a result, the O_2 activation process on the nanocrystals would be very different, making the large-size Au relatively inactive. For this reason, tremendous efforts have been devoted to tune the sizes of Au nanocrystals for catalytic oxidation reactions.

Light Raviraj and Hwang reported that molecular O_2 could be activated to $^1O_2$ on the surface of Au and Ag nanocrystals by illumination of light [31, 32]. In comparison, Au and Ag nanocrystals are unable to promote the formation of $^1O_2$ in the absence of light (see Fig. 4.6). Given the importance of photons to O_2 activation, a "localized surface plasmon resonance (LSPR) sensitization" mechanism has been proposed as shown in Fig. 4.7 [33]. Similar findings were also brought about by

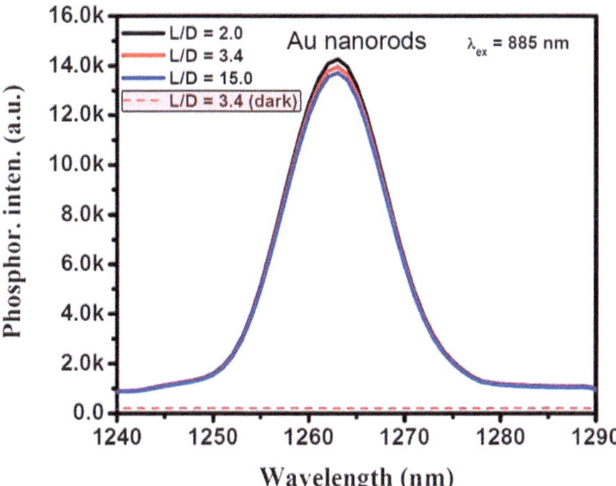

Fig. 4.6 Phosphorescence emission of $^1O_2$ by sensitization of Au nanorods at 885 nm (*solid lines*) and in dark (*dash line*) [32]

Huang and Tian [34]. Nevertheless, there have been some other reports to exclude the necessity of light in the Au-mediated O_2 activation. For instance, they performed oxidation reactions in the presence of molecular O_2 and Au nanocrystals, and the participation of light was not emphasized as a necessary element [24–30]. In addition to the energy transfer through LSPR coupling depicted in Fig. 4.7, it is quite possible that the LSPR photothermal effect can convert the light into heat promoting the O_2 activation. It seems that this light effect is still an unfinished story, and more benchwork research is needed to deepen understanding on the underlying mechanisms.

Support Au nanocrystals supported on various materials, often present unexpected catalytic properties. The remarkable catalytic behavior may arise from strong electronic interactions between Au and supports. Turner and Lambert synthesized a series of Au nanoparticls from 1.4 to 30 nm that were supported on boron nitride

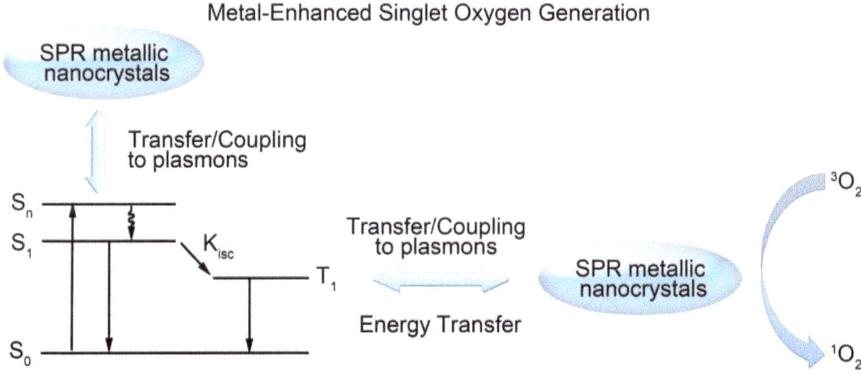

Fig. 4.7 Schematic for surface-plasmon-enhanced $^1O_2$ generation [33]

Table 4.2 Catalytic data for styrene oxidation using O_2 and supported Au. Mean particle size of Au was calculated by counting particles in HRTEM images. All reactions were carried out at 100 °C in toluene [35]

Catalyst	Mean Au size (nm)	Conversion yield (%)	Selectivity (%)		
			Benzaldehyde	Styrene epoxide	Acetophenone
Au/BN	1.6	19.2	82.3	14.0	3.9
Au/SiO$_2$	1.5	25.8	82.1	12.0	5.7
Au/C	17	No reaction	–	–	–

(BN), silicon dioxide (SiO_2) or carbon (C). The catalytic behavior by these three supports is briefly summarized in Table 4.2, showing selective oxidation of styrene by molecular O_2. It turns out that the Au nanoparticles with extremely small sizes that have been supported on chemically and electronically inert materials possess high capabilities of adsorbing and activating O_2 for selective oxidation. It is believed that this oxidation process is initiated through dissociation of O_2 to yield O adatoms. Interestingly, a sharp size threshold in catalytic activity has been identified: the nanoparticles with diameters above 2 nm are completely inactive [35]. In fact, the integration of Au nanoparticles with supports has been widely used in various oxidation systems. Figure 4.8 illustrates support-dependent activities of Au nanocrystals in CO oxidation. The semiconductor supports have shown superior enhancement to the dielectric materials [36]. In principle, charge transfer may occur between semiconductor and metal so that the charge density of Au nanocrystals becomes tunable to improve O_2 activation.

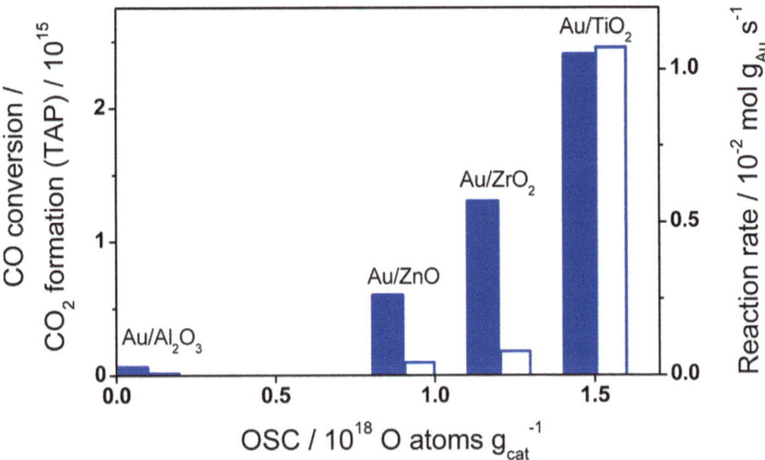

Fig. 4.8 CO conversion (*filled bars*) and reaction rates (*empty bars*) of different supported Au nanocrystals [36, 37]

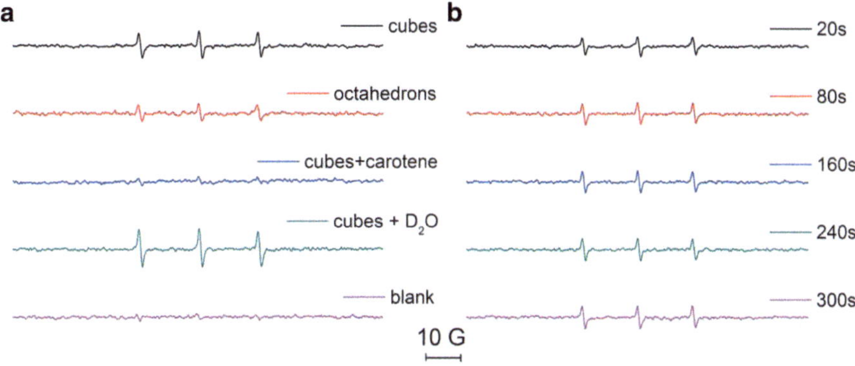

Fig. 4.9 ESR spectra of 4-oxo-TMP solution and Pd nanocrystals mixture in different conditions: **a** nanocubes in H$_2$O, octahedrons in H$_2$O, nanocubes in the presence of carotene (^{1}O$_2$ scavenger), and nanocubes in H$_2$O with 50-μL D$_2$O (which can prolong ^{1}O$_2$ lifetime), respectively; **b** nanocubes under irradiation of UV light source (ν < 400 nm) for different time [38]

4.2.2   Pd–O$_2$ Interactions

4.2.2.1   Surface Facet

It is known that Pd nanoparticles exhibit excellent catalytic performance in various oxidation reactions. Like the Au system, the activation of O$_2$ on the surface of Pd nanocrystals should have a strong correlation with their surface states. A recent work by Xiong et al. indicates that the surface facets have huge impact on their efficiency in O$_2$ activation, using single-faceted Pd nanocrystals as a model system.

In that work, the O$_2$ activation was probed by electron spin resonance (ESR) spectroscopy, a common way to determine the type of ROS in the presence of specific radical trapping agents. Specifically, 2,2,6,6-tetramethyl-4-piperidone hydrochloride (4-oxo-TMP) was used as an ^{1}O$_2$-sensitive trapping agent, which can produce a stable nitroxide radical 4-oxo-TEMPO, to identify the production of singlet oxygen. The 4-oxo-TEMPO has characteristic signals with a 1:1:1 triplet of g = 2.0055 in ESR spectroscopy. Figure 4.9a clearly shows that ^{1}O$_2$-analogous species can be formed on the surface of Pd nanocrystals. Such formation of ^{1}O$_2$-analogous species has also been proven by the control experiments in the presence of D$_2$O (enhancer) or carotene (quencher). Interestingly, the experiment results show that the ^{1}O$_2$-analogous species is preferentially formed on {100} facets of Pd nanocrystals rather than {111} facets (see Fig. 4.9a).

This finding highlights the opportunity for tuning O$_2$ activation via facet engineering, followed by further investigations to elucidate the mechanism behind. As

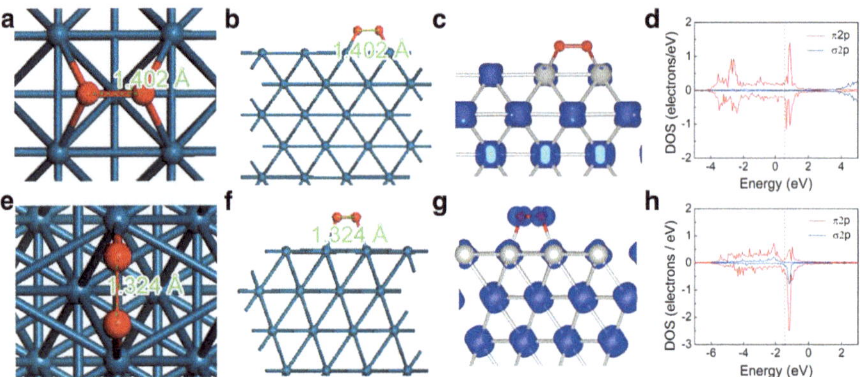

Fig. 4.10 **a** Top view, **b** side view, **c** spin charge density, and **d** projected density of states (PDOS) diagram of O_2 on Pd{100} facet. **e–h** The case of O_2 adsorbed on Pd{111} [38]

mentioned in Sect. 4.2.1.2, the "LSPR sensitization" mechanism has been tentatively employed as an explanation for $^1O_2$ generation in the Au and Ag systems [31, 32]. In order to assess whether the LSPR plays a role in the case of Pd, Xiong et al. employed the 4-oxo-TMP ESR probing technique to characterize the samples under irradiation of different light sources. As shown in Fig. 4.9b, illumination of light does not promote the signals for $^1O_2$, suggesting that the LSPR sensitization should not be responsible for the finding in the Pd system.

Instead, their theoretical simulations reveal that this O_2 activation is actually caused by electron transfer from Pd nanocrystal surface to molecular O_2 during chemisorption. As illustrated by Fig. 4.10, surface facets may alter the chemisorption state of molecular O_2 on Pd surface and, in turn, cause the O_2 activated to different levels. The surface facets are thus a key parameter for tuning efficiency in generating $^1O_2$-analogous species. This species behaves like singlet oxygen physically and chemically but has higher charge density due to the acceptance of electrons. Enabled by this tunable O_2 activation, the {100}-enclosed Pd nanocubes have been identified as more active catalysts for oxidation reactions as compared with the {111}-enclosed Pd octahedrons [38]. Given the key role of ROS in reducing viability, this facet control also provides a knob to optimize the performance of metallic nanocrystals in cancer treatment.

4.2.2.2 Surface Charge State

Since Pd surface donates electrons for the production of $^1O_2$-analogous species, the charge state of Pd surface should be a critical parameter to the O_2 activation efficiency. Xiong et al. proposed a strategy for tailoring the Pd charge state through metal-semiconductor hybrid configuration. In the configuration, Pd nanocrystals enclosed by {100} facets are deposited on TiO_2 supports. Driven by the Schottky junction, the photoexcited semiconductor supports can supply electrons for Pd

Table 4.3 Simulation results for 25%-coverage oxygen on Pd {100} (parallel configuration) in the presence of different additional electrons or holes [39]

25% Coverage, parallel	Charge -2	Charge 0	Charge $+1$	Charge $+2$
Total energy/eV	-883.357	-883.482	-882.833	-882.029
Local magnetic moment $_{average}/\mu_B$	0.008	0.004	0.002	0.001
O–O distance $_{average}/\text{Å}$	1.396	1.406	1.411	1.415
Obtained electrons per O_2	0.64	0.69	0.72	0.74

{100} surface under illumination, increasing the Pd electron density. As indicated by theoretical simulations, appropriate quantities of additional electrons brought to the Pd {100} surface promote its activity in O_2 activation (see Table 4.3). Based on this mechanism, an enhancement of O_2 activation has been observed on TiO_2-supported Pd nanocubes [39].

Interestingly, the light intensity turns out to affect the amount of electrons supplied to the Pd {100} surface, with a turnover point at certain intensity [39]. Accordingly, the efficiency of O_2 activation shows the same trend. This feature suggests that too strong UV illumination reduces the number of transferred electrons from TiO_2 to Pd. Ultrafast spectroscopy reveals that plasmonic effect of Pd can generate hot electrons on metal nanocrystals, which are in turn injected into the conduction band of the semiconductor at the metal-semiconductor interface. This injection of hot electrons constitutes a reverse flow of electrons from Pd to TiO_2, thus lowering the electron density of the Pd surface as a side effect. As Pd is a metal with relatively weak plasmonic effect, the migration of hot electrons becomes more dominant under strong illumination. This case informs us that use of semiconductor supports should be an effective approach to tuning O_2 activation but light intensity should be precisely modulated for optimal efficiency. As a result, the optimized O_2 activation under appropriate illumination has enabled significantly improved catalytic efficiency in glucose oxidation [39].

4.2.3 Ag–O_2 Interactions

Epoxidation of ethylene with molecular O_2 catalyzed by Ag is one of the most important and successful industrial reactions. The importance of this reaction to chemical engineering has attracted tremendous attention to study

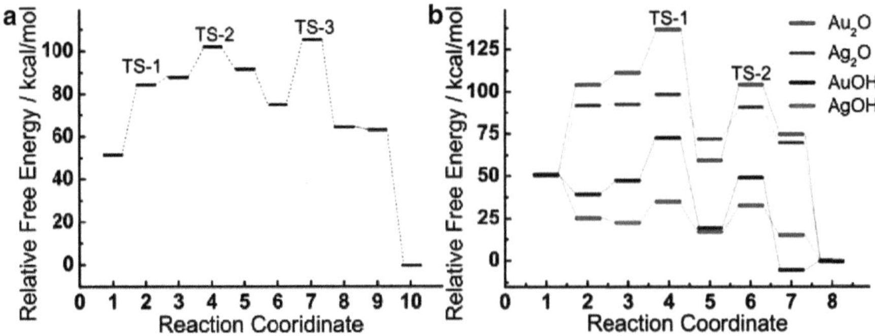

Fig. 4.11 Activation energy pathway for PATP oxidation process by **a** $^3O_2$ and **b** AgOH/Ag$_2$O [34]

the nature of this reaction. Santen et al. has figured out that the efficiency of Ag-catalyzed ethylene epoxidation with molecular O_2 relies on the state of adsorbed O_2 [40]. Raman spectroscopy provides a versatile tool to examine the behavior of adsorbed O_2 on the surface of Ag films. Pettenkofer et al. found that there are at least three different types of oxygen species on the Ag surface under illumination of laser light, with O–O vibrational bands around 1050, 700, and 630 cm^{-1} [41].

Huang and Tian further proposed that Ag oxide/hydroxide is the key element in $^3O_2$ activation during oxidation reactions [34]. Raman spectra monitoring oxidation of *p*-aminothiophenol (PATP) in different chemical environment have confirmed such hypothesis. Theoretically, DFT calculation results show that the thermodynamic energy for the PATP oxidation process by $^3O_2$ is over -60 kcal mol^{-1}. In comparison, the activation barriers for the PATP oxidation process by AgOH and Ag$_2$O are 7.3 and 9.5 kcal mol^{-1}, respectively (see Fig. 4.11). Given that Ag is a metal that can be easily oxidized, such reaction pathway hypothesis should be quite reasonable. Moreover, they pointed out that, both energy transfer and local heating effects raised by the LSPR of metal nanocrystals should have played a critical role.

4.2.4 Summary for Metal–Oxygen Interactions

As outlined above, molecular oxygen can be activated to ROS on metal clusters or nanocrystals during a spontaneous process via adsorption. For this reason, some metal clusters or nanocrystals have been used as highly efficient and selective oxidation catalysts at low temperatures. In the cases of Au cluster and Pd nanocrystal, charge transfer occurs between metal and molecular O_2, responsible for the O_2 activation. This feature provides more versatility to tune the activation efficiency via materials design. Here, this section will be ended with a brief summary of adsorbed O_2 on the common metals (see Table 4.4).

Table 4.4 Different situations for adsorbed oxygen on metallic nanocrystals

		Adsorbed oxygen	Light	Ref.
Pd	Nanocrystals	$^1O_2$-analogous	Unnecessary	[38, 39]
Pt	Nanocrystals	$^1O_2$	Necessary	[42]
Ag		peroxo O_2^{2-}		[43]
	Nanocrystals	$^1O_2$	Necessary	[42]
	Nanocrystals	$^2O_2^-$	Necessary	[34]
Au	Neutral clusters (odd-sized)	Superoxo O_2^-		[20]
	Anionic clusters (even-sized except Au_8^-)	Superoxo O_2^-		[17]
	Anionic clusters (Au_8^-)	Peroxo O_2^{2-}		[17]
	Nanocrystals	$^1O_2$	Necessary	[42]
	Nanocrystals	$^2O_2^-$	Necessary	[34]
Zn		Peroxo O_2^{2-}		[43]
Cd		Peroxo O_2^{2-}		[43]
Ti		$-OOH$		[43]
Zr		$-OOH$		[43]
Co	Nanocrystals	O_2^-	Unnecessary	[44]
Cr		$-OOH$		[43]
Mo		$-OOH$		[43]
W		$-OOH$		[43]

4.3   Molecule of H_2

Catalytic hydrogenation with molecular H_2 includes a large group of addition reactions of hydrogen at unsaturated bonds, which are among the simplest transformations in organic chemistry and extensively involved in various processes. Palladium is the only element in group-10 transition metals that can absorb molecular H_2 exothermically [45]. Hydrogen dissociative adsorption on the Pd can easily occur at room temperature and under ambient pressure. For this reason, this section is mainly focused on Pd nanostructures and their alloys from the viewpoint of H_2-metal interactions.

4.3.1   Pd–H_2 Interactions

In this section, we start with a brief description for the interactions between molecular H_2 and single-facet Pd surface. With the concepts from traditional surface

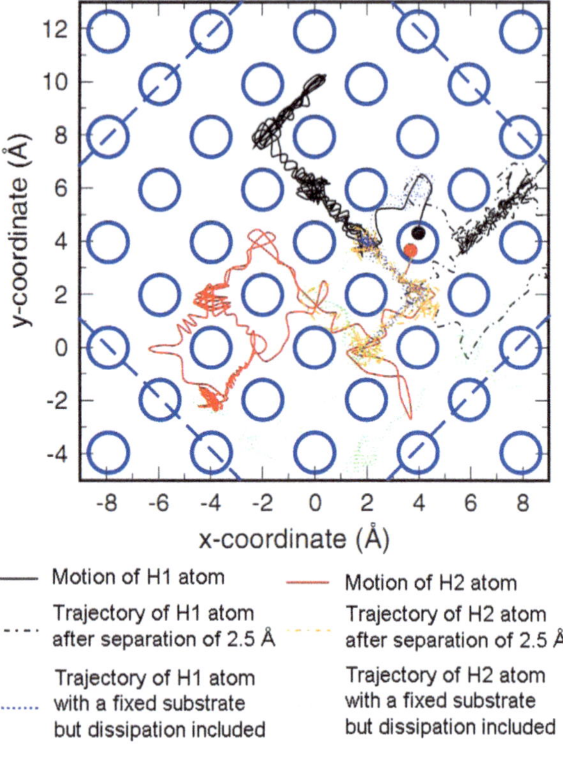

Fig. 4.12 Calculated motion and trajectory of H atoms after dissociative adsorption on clean Pd{100} surface [50]

Table 4.5 Binding energies (E_b) and adsorption sites of H on Pd single-facet surface [48]

	E_b/eV	Site
Pd{111}	−2.86	fcc
Pd{100}	−2.81	Hollow
Pd{110}	−2.75	Short bridge

science in mind, it would be more straightforward to understand the case of Pd nanocrystals.

Dissociative Adsorption The dissociative adsorption of molecular H_2 on Pd surface is the first step (and most of the time, the crucial step) in heterogeneous catalysis. In practice, there is almost no barrier for the dissociative adsorption of molecular H_2 on Pd surface, which makes this step easily occur at room temperature [46, 47]. The calculated binding energies and site preferences of H on Pd surface are listed in Table 4.5. Among the three basic crystal facets, Pd {111} shows the highest surface atom density and the lowest surface energy, as well as exhibits the highest selectivity for synthesis of H_2O_2 from H_2 and O_2 [48]. Like the O_2 activation, this feature once again demonstrates that the performance of a catalyst highly depends on its surface structure.

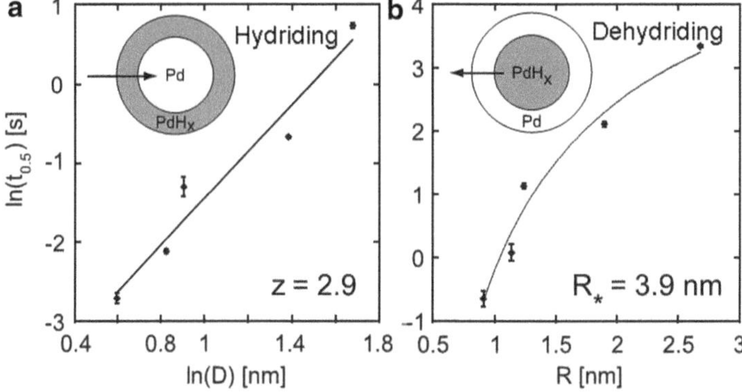

Fig. 4.13 Power-law for $t_{0.5}$ (half of the maximum) and D (mean diameter of Pd nanocrystals): **a** hydriding process, and **b** dehydriding process [51]

Trajectory of Dissociated H Atoms After the first step of adsorption, the extra energy gained from the dissociative adsorption process leads to the formation of "hot" atoms (i.e., atoms with energy much higher than external thermal energy). The diffusive motion of "hot" atoms can significantly affect catalytic reaction pathway, since it represents the way of reactants to reach the surface of catalysts [49, 50]. Groß et al. has performed ab initio molecular dynamics (AIMD) simulations to reveal the mean free path of the dissociated H atoms on clean Pd{100} surface [50]. The calculation results show that the mean displacement of a single H atom on Pd{100} is about 7 Å. In addition to acquiring information for H diffusion, this picture of atomic motion also illustrates the importance of single-atom alloy surface to separation of H atoms (see Sect. 4.2.2).

Formation of PdH It is well known that Pd not only interacts with H_2 through surface adsorption, but also incorporates dissociated H atoms in lattice and form palladium hydride phase (PdH). The hydride formation is a result from attractive H-H interactions [51]. As compared with Pd, PdH possesses quite different physical and chemical properties, enabling wider applications. The phases of hybrids depend on H_2 pressure: α-phase PdH can be produced under low H_2 pressure, while PdH prefers to form a β-phase as the pressure becomes higher [52]. At moderate temperature and ambient pressure, the transition from α to β-phase starts at 1–2 % H_2, leading to an increase in strain energy due to the mismatch of lattice constants. As demonstrated in Sect. 4.2.3, this lattice mismatch also facilitates the application of Pd nanocrystals in H_2 sensors.

As the model system switches to nanocrystals, a kinetics study has revealed that the hydriding and dehydriding processes occurring on Pd nanocrystals are strongly dependent on their particle sizes (see Fig. 4.13). The hydriding process follows the function of $\ln(t_{0.5}) = z \ln(D)$ where $t_{0.5}$ is half of maximal hydriding degree and D represents mean diameter of Pd nanocrystals. The dehydriding process is even more complicated, which involves growth of newly formed metallic shells and shrinkage of hydrides. During dehydriding, H atoms diffuse from the hydride cores to the outside, which are still attracted by the metallic shells. As this research proceeded, Kasemo et al. have modified the function of dehydriding to be $\ln(t_{0.5}) = \ln(CR) - R_*/R$ where C is a constant and R stands for nanocrystal radius [51]. From both equations for hydriding and dehydriding, one can see that these two reversal processes exhibit size-dependent behavior. For this reason, the size control provides a knob for tuning the performance of Pd nanocrystals in H_2 activation.

4.3.2 Alloys with Isolated Pd Atoms

Using single-facet metal as a model system, Sykes et al. identified interesting behavior of molecular H_2 on Pd single-atom alloy (SAA) surface. It is well known that there is no notable barrier for the H_2 dissociative adsorption step on Pd surface; however, the adsorbed H atoms have very strong binding to the surface. In sharp contrast, the dissociative adsorption of H_2 on Cu surface is relatively inert. Taking the relative merits of Pd and Cu surface, the isolated Pd SAA surface offers low dissociation barrier due to the presence of Pd atom and at the same time, has hydrogen atoms weakly bounded to the Cu surface. As a result, the H atoms activated by Pd can spill over onto the Cu surface [53, 54]. Based on this feature, Pd SAA can be utilized a bifunctional surface, which not only boosts catalytic efficiency in hydrogenation but cost-effectively reduces usage of noble metals.

4.3.3 Applications

The fundamental research for interactions between H_2 and metal surfaces in surface science sheds more light on underlying mechanisms for metallic nanocatalysts, which in turn can help form designing rules for high-performance heterogeneous catalysts or facilitate other applications.

4.3.3.1 Catalytic Reactions

Due to its active interaction with molecular H_2, Pd has been a common catalytic material in hydrogenation reactions. Since the reaction occurs at the surface of catalysts, the activity and selectivity of a catalytic reaction is strongly correlated with the shape and size of the nanocrystals involved in catalysts, which is more or less relevant to their surface states.

Kiwi-Minsker et al. investigated three types of Pd nanocrystals with uniform shapes in the hydrogenation of 2-methyl-3-butyn-2-ol (MBY): nanocube with {100}, octahedron with {111}, and cuboctahedron with both {100} and {111} facets [55]. Catalytic data suggested that two types of active sites were involved in the catalysis: the atoms on the planes (σ_1 sites) and the ones at edges (σ_2 sites; see Fig. 4.14 a). As these two types of atoms have different coordination numbers, the

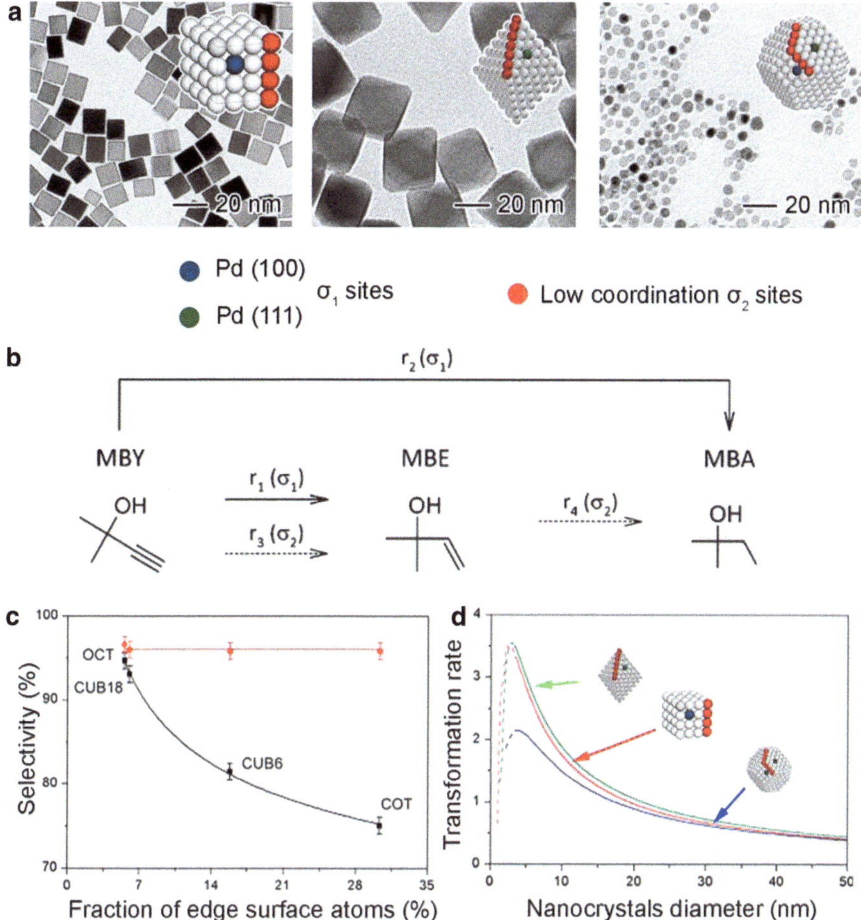

Fig. 4.14 **a** TEM images of 18-nm Pd nanocubes, 31-nm Pd octahedrons and 5.5-nm Pd cuboctahedrons. **b** Schematic for hydrogenation of MBY. **c** Selectivity of MBY as a function of edge atoms. **d** Dependence of MBY conversion rate on size and shape of Pd nanocrystals [55]

reaction activity and selectivity are varied. It turns out that semihydrogenation preferentially occurs at the σ_1 sites to produce 2-methyl-3-butyn-2-ol MBY, while the σ_2 sites promote over hydrogenation for formation of 2-methyl-3-buten-2-ol MBE (see Fig. 4.14b and 4.14c). Following a two-site Langmuir–Hinshelwood mechanism, the data fitted with kinetic modeling can well depict the activity and selectivity of reactions. Overall, shape and size control of nanocrystals provides a brand new model for catalysis research, beyond the traditional concepts in surface science given the unique roles of corner and edge atoms in activating molecules.

When the reactions move up to the industrial scale, the catalysts are mostly required to be mounted on supports. For this reason, supported catalysts have been

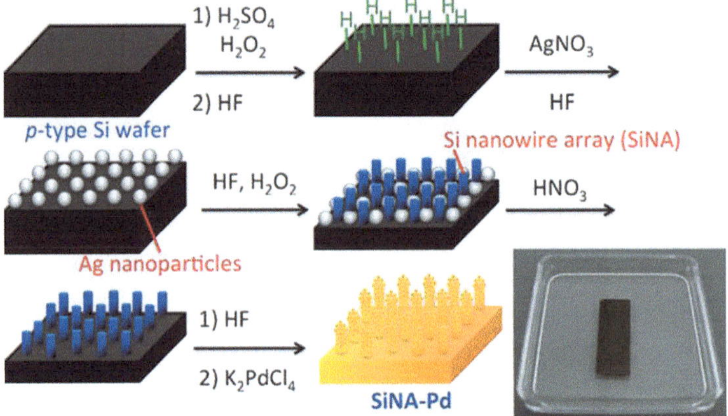

Fig. 4.15 Schematics for fabricating Si nanowire array-supporting Pd catalyst [56]

widely investigated by research community. Yamada et al. developed a catalyst made of silicon nanowire array-stabilized Pd nanoparticles (SiNA-Pd; see Fig. 4.15). Such a SiNA-Pd structure showed high catalytic activity and reusability in hydrogenation of alkene and nitrobenzene. Strikingly, the activity of this supported catalyst reached a turnover number (TON) as high as 2,000,000 [56]. Kobayashi et al. has also developed an efficient system for hydrogenation reaction using a microchannel reactor based on the Pd catalysts. Such a microreactor substantially reduces the time and space needed for organic synthesis, boosting the atomic economy. The hydrogenation reactions in this system turn out to proceed to the desired products within 2 min [57]. There is no doubt that this approach should be extendable to other multiphase reactions.

4.3.3.2 H$_2$ Sensing

Development of H$_2$ sensor with low detection limit is crucial for safe hydrogen-based applications. The spontaneous dissociation of H$_2$ on Pd surface, together with the huge difference between Pd and PdH, makes Pd an excellent material for H$_2$ sensing. Specifically, the sensor may work electrically or optically.

Electrical Sensors As described above, PdH can be form upon the dissociative adsorption of H$_2$ on Pd surface. Such hybrids have very different conductivity from Pd, so the conductance of materials gradually changes when exposed to H$_2$. Based on this feature, the conductance can be probed to monitor the presence of H$_2$, making a hydrogen sensor. In a typical case, Crego-Calama et al. used conventional microfabrication techniques (deposition and etching under angles (DEA)) to fabricate Pd nanowires on Si substrate. The nanowires show a reversible response to H$_2$ at concentrations as low as 27 ppm (see Fig. 4.16) [58]. When H$_2$ is turned on, the conductance of Pd nanowire starts to decrease and finally reaches a steady value for a given H$_2$ concentration. In comparison, no signal has been observed for

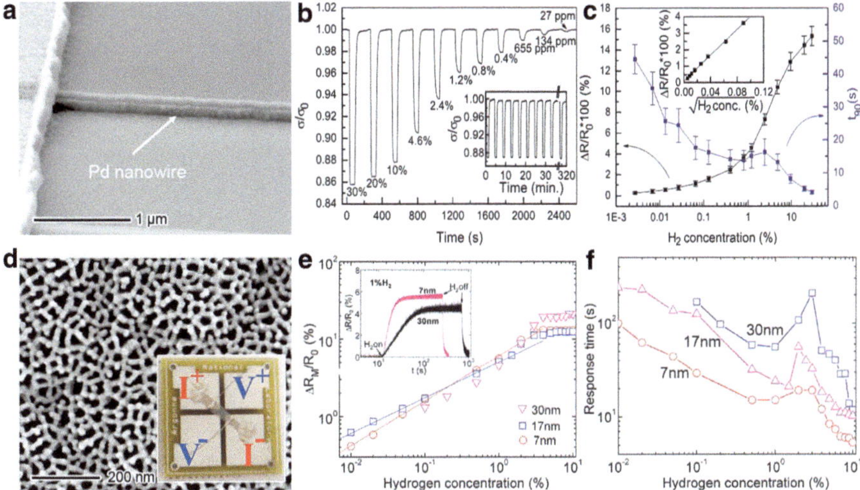

Fig. 4.16 **a** SEM image of a Pd nanowire with 50–80 nm diameter in sensor device. **b** Conductance change of a single Pd nanowire when exposured to H_2 at concentrations ranging from 30 % to 27 ppm. The *inset* shows the reversibility of a single Pd nanowire for 80 cycles. **c** Averaged response time from six nanowires. The *inset* shows a linear fit to the steady-state response for square root of H_2 concentration [58]. **d** SEM image of Pd nanowire network. The inset is the photograph of a sensor sample. H_2 concentration-dependent **e** resistance change and **f** response time of of Pd network with different thickness [59]

O_2, validating the application of this sensor in real environment. The Pd nanowire-based sensor possesses high response speed and sensitivity, and as such, the networks of ultrathin (< 10 nm) Pd nanowires can be utilized as a high-sensitivity and fast-response sensor as demonstrated by Zeng and Xu (see Fig. 4.16d–4.16f) [59].

During the formation of PdH, the Pd materials also expand in volume. This feature offers an opportunity of designing a "on-off switch" H_2 sensor based on the Pd nanoparticle assembly. Favier and Penner demonstrated a Pd nanowire-based sensor for the H_2 detection [60]. This nanowire is composed of densely packed Pd nanoparticles, and as such, the conductive nanowire becomes to have high resistance after exposed to H_2, as the contact between the nanoparticles gets loose after volume expansion (i.e., mode I). After the first cycle, the resistance of nanowire is reduced once exposed to H_2, simply because the volume expansion can shrink the gap between the nanoparticles. The detection limit for "mode I" and "mode II" sensors is 0.5 % and 2.25 % H_2, respectively. Despite the lower sensitivity, the "mode II" sensor can produce a "wait state" where no power or noise is generated in the absence of H_2. This is a huge advantage to further industrial applications. By designing different structures, this working mechanism has been extended to other H_2 sensors by Villanueva and Favier [61].

Optical Sensors The structural changes during Pd–H_2 interactions not only change the conductivity of Pd, but also tune its LSPR optical properties. However, spectral signals arising from Pd-H interactions are too weak to be detected given Pd is not an ideal metal for LSPR (see Fig. 4.17a-up). Liu and Alivisatos demonstrated a

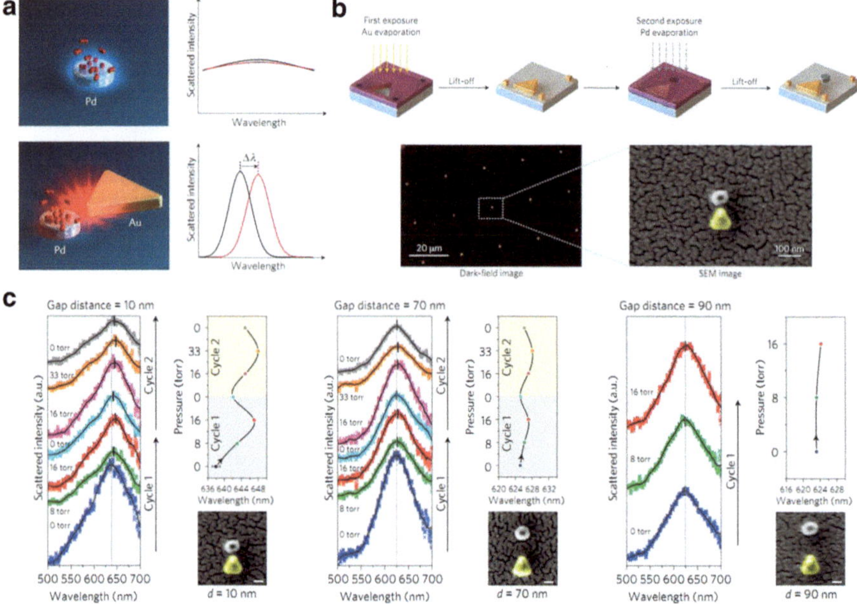

Fig. 4.17 **a** Schematic for antenna-enhanced single-particle H_2 sensing. **b** Fabrication of antenna-enhanced nanostructures using double electron-beam lithography with double lift-off. **c** Optical-scattering measurements of a single Pd–Au triangle antenna nanostructure on H_2 exposure. The results show that adsorption peak shift is dependent on separation **d** between the Au antenna and the Pd particle [62].

resonant antenna-enhanced single-particle H_2 sensor, working along a mechanism similar to tip-enhanced phenomenon [62]. The Au antenna and Pd particle can be facilely manufactured through a lift-off process involved in two separate nanofabrication cycles (see Fig. 4.17b). The distance between Au antenna and Pd particle can be controlled by tuning fabrication parameters. It turns out that the distance between Au antenna and Pd particle is the key factor to optimizing sensitivity. The smaller the distance, the greater the single shift (see Fig. 4.17c). Certainly, the shape of Au is also influential to the signal information due to shape-dependent LSPR property. This design of antenna-enhanced single-particle sensor has been able to in-situ monitor catalytic reactions.

Nevertheless, collecting antenna-enhanced plasmonic signals usually requires complicated devices. To simplify the process, Chiu and Huang designed and synthesized Au–Pd core-shell nanocrystals with various shapes to yield visible optical evolution during Pd–H interactions (see Fig. 4.18) [63]. In the designed nanostructures, the Au cores act as a nanoantenna to enhance the plasmonic signals (i.e., LSPR peak shift with H_2 adsorption). The Au–Pd core-shell nanocrystals show nearly reversible spectral shifts during hydriding and dehydriding cycles. As a consequence, it appears to be a promising and reusable H_2 sensor for various applications. Notably, this optical sensor is H_2-specific, as no spectral change can be

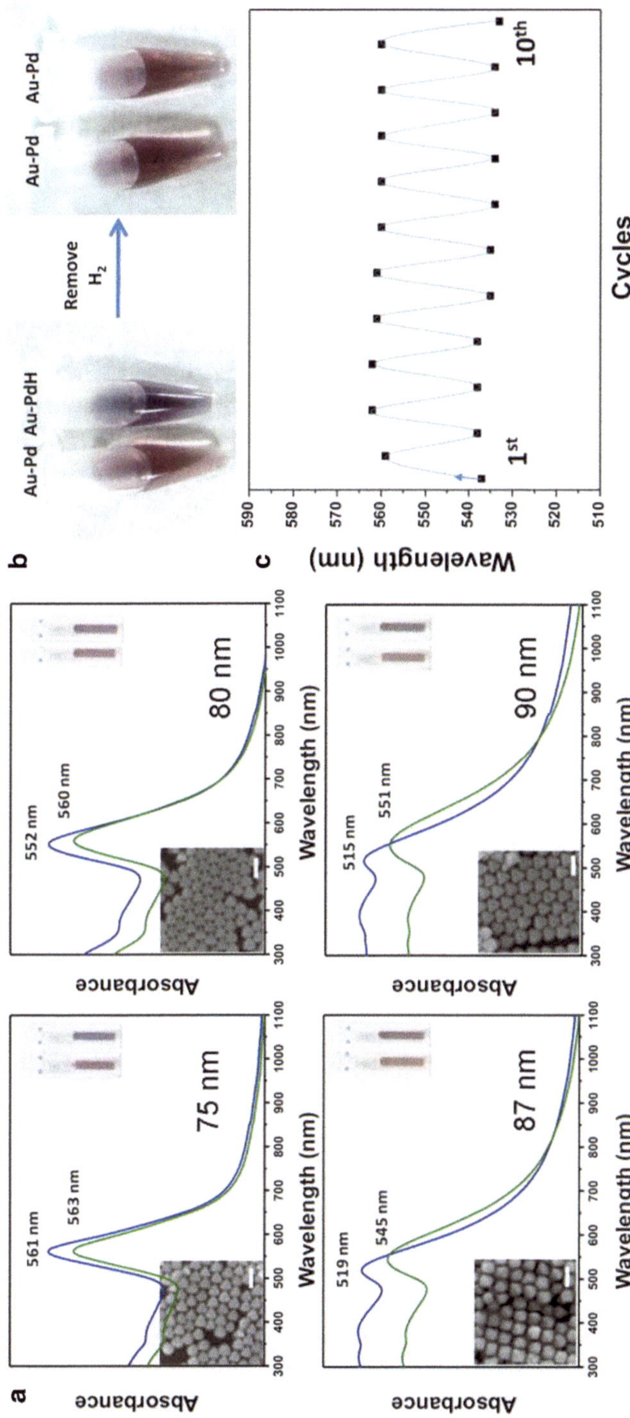

Fig. 4.18 **a** UV-vis absorption spectra of Au–Pd core-shell nanocrystals before (*blue*) and after (*green*) the H₂ exposure. **b** Photographs of suspension of Au–Pd core-shell nanocrystals before and after H₂ exposure. **c** Spectral reversibility of Au–Pd core-shell nanocrystals during H₂ adsorption and desorption cycles [63]

observed when exposed to O_2 or CO. As compared with the "top-down" approach, this design coming up with "bottom-up" synthesis represents a simpler technique to address the demand for H_2 sensing.

4.4   Conclusion and Outlook

To summarize, the interactions between small molecules and metallic nanocrystals have been discussed in this chapter. Understanding on these interactions at molecular scale would open new avenues to the design of materials with specific functions. Taking Pd–H system for example, researchers have developed a variety of applications based on the hydriding and dehydriding processes. In the applications, lattice changes and optical properties along with the processes are fully utilized. The insights into the underlying mechanisms also highlight the importance of controlled synthesis of metallic nanocrystals/clusters, which should hold the promise for tuning their behavior in molecular activation to boost various applications.

4.5   Appendices

4.5.1   A. Synthetic Methods

Pd Nanocubes Covered by {100} Facets   About 0.105 g of poly(vinyl pyrrolidone) (PVP, M.W. = 55,000), 0.060 g of L-ascorbic acid and 0.100–0.600 g of KBr were dissolved in 8 mL of deionized water at room temperature. The solution was placed in a 3-neck flask (equipped with a reflux condenser and a magnetic Teflon-coated stirring bar) and heated in air at 80 °C for 5 min. Meanwhile, 0.065 g of potassium palladium(II) chloride (K_2PdCl_4) was dissolved in 3 mL of deionized water at room temperature. The Pd stock solution was then injected into the flask through a syringe pump. Heating of the reaction at 80 °C was continued in air for 3 h.

Pd Octahedrons by {111} and Pd Cuboctahedrons by {111}/{100}   About 0.105 g of poly(vinyl pyrrolidone) (PVP, M.W. = 55,000), 0.060 g of citric acid and 0.060 g of L-ascorbic acid were dissolved in 8 mL of deionized water at room temperature. The solution was placed in a 3-neck flask (equipped with a reflux condenser and a magnetic Teflon-coated stirring bar) and heated in air at 120 °C for 5 min. Meanwhile, 0.065 g of potassium palladium(II) chloride (K_2PdCl_4) was dissolved in 3 mL of deionized water at room temperature. The Pd stock solution was then injected into the flask through a syringe pump at 5 mL/h-octahedrons, 360 mL/h-cuboctahedrons. Heating of the reaction at 120 °C was continued in air for 3 h. The samples were washed with acetone and then with ethanol and water three times to remove most of the PVP and other molecules by centrifugation.

Fig. 4.19 ESR spectra of
a DMPO/OH, **b** DMPO/
O_2^- [66], **c** 4-oxo-TEMPO
(4-oxo-TMP/$^1O_2$) [38]

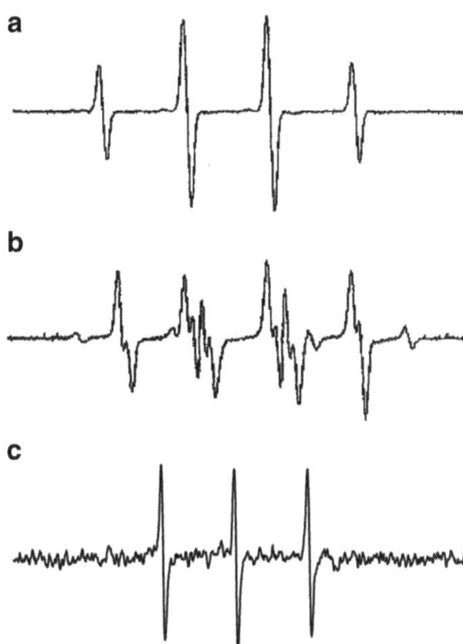

4.5.2   B. ESR Measurement

Reactive oxygen species (ROS) are transient species that can be identified by using radical trapping agents (spin traps). Spin traps react with transient ROS in solution to yield stable products—spin adducts, which can be readily observed by ESR spectroscopy. 5,5-dimethyl-1-pyrroline N-oxide (DMPO) is particularly useful for identifying oxygen-centered radicals (superoxide radical anion and hydroxyl radicals), as the formed spin adducts have characteristic ESR parameters and can be readily distinguished from other radicals (see Fig. 4.19a and 4.19b) [64]. 4-oxo-TMP has been commonly used as a sensitive probe for $^1O_2$ that produces stable nitroxide radical 4-oxo-TEMPO (see Fig. 4.19c) [38, 65].

Measurement Protocol as an Example A 50 μL of aqueous suspension of nanocrystals was mixed with 500 μL of spin-trap solution (4-oxo-TMP, DMPO). The solution was then characterized with ESR spectroscopy at 20 °C. The measurements can be performed in various chemical environment and incident light depending on experimental needs.

References

1. Wahlen, J., De Vos, D. E., Jacobs, P. A., et al., Adv Synth Catal. **346**, 152 (2004)
2. Kovalev, D., Fujii, M., Adv Mater. **17**, 2531 (2005)
3. Hikazudani, S., Mochida, T., Yano, K., et al., Catal Commun. **12**, 1396 (2011)

4. Jin, M., Liu, H., Zhang, H., et al., Nano Res. **4**, 83 (2010)
5. Reetz, M. T., Westermann, E., Angew Chem Int Ed Engl. **39**, 165 (2000)
6. Enache, D. I., Edwards, J. K., Landon, P., et al., Science. **311**, 362 (2006)
7. Besson, M., Lahmer, F., Gallezot, P., et al., J Catal. **152**, 116 (1995)
8. Kesavan, L., Tiruvalam, R., Ab Rahim, M. H., et al., Science. **331**, 195 (2011)
9. Wittstock, A., Zielasek, V., Biener, J., et al., Science. **327**, 319 (2010)
10. Brune, H., Wintterlin, J., Behm, R. J., et al., Phys Rev Lett. **68**, 624 (1992)
11. Komrowski, A., Sexton, J., Kummel, A., et al., Phys Rev Lett. **87**, 246103 (2001)
12. Behler, J., Delley, B., Lorenz, S., et al., Phys Rev Lett. **94**, 036104 (2005)
13. Gottfried, J. M., Schmidt, K. J., Schroeder, S. L. M., et al., Sur Sci. **511**, 65 (2002)
14. Haruta, M., Kobayashi, T., Sano, H., et al., Chem Lett. **16**, 405 (1987)
15. Salisbury, B. E., Wallace, W. T., Whetten, R. L., Chem Phys. **262**, 131 (2000)
16. Cox, D. M., Brickman, R., Creegan, K., et al., Z Phys D Atom Mol Cl. **19**, 353 (1991)
17. Pal, R., Wang, L. M., Pei, Y., et al., J Am Chem Soc. **134**, 9438 (2012)
18. Huang, W., Zhai, H. J., Wang, L. S., J Am Chem Soc. **132**, 4344 (2010)
19. Woodham, A. P., Meijer, G., Fielicke, A., Angew Chem Int Ed Engl. **51**, 4444 (2012)
20. Woodham, A. P., Meijer, G., Fielicke, A., J Am Chem Soc. **135**, 1727 (2013)
21. Ding, X., Li, Z., Yang, J., et al., J Chem Phys. **120**, 9594 (2004)
22. Roldan, A., Ricart, J. M., Illas, F., et al., Phys Chem Chem Phys. **12**, 10723 (2010)
23. Zhao, Y., Khetrapal, N. S., Li, H., et al., Chem Phys Lett. **592**, 127 (2014)
24. Comotti, M., Della Pina, C., Matarrese, R., et al., Angew Chem Int Ed Engl. **43**, 5812 (2004)
25. Schubert, M., Hackenberg, S., van Veen, A. C., et al., J Catal. **197**, 113 (2001)
26. Okumura, M., Nakamura, S., Tsubota, S., et al., Catal Lett. **51**, 53 (1998)
27. Haruta, M., Tsubota, S., Kobayashi, T., et al., J Catal. **144**, 175 (1993)
28. Kang, Y. M., Wan, B. Z., Catal Today. **26**, 59 (1995)
29. Park, E. D., Lee, J. S., J Catal. **186**, 1 (1999)
30. Kozlov, A. I., Kozlova, A. P., Liu, H. C., et al., Appl Catal A-Gen. **182**, 9 (1999)
31. Vankayala, R., Sagadevan, A., Vijayaraghavan, P., et al., Angew Chem Int Ed Engl. **50**, 10640 (2011)
32. Vankayala, R., Kuo, C.-L., Sagadevan, A., et al., J Mater Chem B. **1**, 4379 (2013)
33. Zhang, Y., Aslan, K., Previte, M. J., et al., J Fluoresc. **17**, 345 (2007)
34. Huang, Y. F., Zhang, M., Zhao, L. B., et al., Angew Chem Int Ed Engl. **53**, 2353 (2014)
35. Turner, M., Golovko, V. B., Vaughan, O. P. H., et al., Nature. **454**, 981 (2008)
36. Widmann, D., Behm, R. J., Accounts Chem Res. **47**, 740 (2014)
37. Widmann, D., Liu, Y., Schuth, F., et al., J Catal. **276**, 292 (2010)
38. Long, R., Mao, K., Ye, X., et al., J Am Chem Soc. **135**, 3200 (2013)
39. Long, R., Mao, K., Gong, M., et al., Angew Chem Int Ed Engl. **53**, 3205 (2014)
40. Vansanten, R. A., Kuipers, H. P. C. E., Adv Catal. **35**, 265 (1987)
41. Pettenkofer, C., Pockrand, I., Otto, A., Sur Sci. **135**, 52 (1983)
42. Vankayala, R., Sagadevan, A., Vijayaraghavan, P., et al., Angew Chem Int Ed Engl. **50**, 10640 (2011)
43. Trimm, D. L. *Design of Industrial Catalysts*; Elsevier Scientific Publishing Company: Amsterdam-Oxford-New York, 1980.
44. Leonard, S., Gannett, P. M., Rojanasakul, Y., et al., J Inorg Biochem. **70**, 239 (1998)
45. Okuyama, H., Siga, W., Takagi, N., et al., Sur Sci. **401**, 344 (1998)
46. Maiti, A., Gee, R., Maxwell, R., et al., J Phys Chem B. **110**, 3499 (2006)
47. Staykov, A., Kamachi, T., Ishihara, T., et al., J Phys Chem C. **112**, 19501 (2008)
48. Tian, P., Ouyang, L., Xu, X., et al., Chinese J Catal. **34**, 1002 (2013)
49. Blanco-Rey, M., Juaristi, J. I., Diez Muino, R., et al., Phys Rev Lett. **112**, 103203 (2014)
50. Groß, A., Phys Rev Lett. **103**, 246101 (2009)
51. Langhammer, C., Zhdanov, V. P., Zoric, I., et al., Phys Rev Lett. **104**, 135502 (2010)
52. Tripodi, P., McKubre, M. C. H., Tanzella, F. L., et al., Phys Lett A. **276**, 122 (2000)
53. Kyriakou, G., Boucher, M. B., Jewell, A. D., et al., Science. **335**, 1209 (2012)
54. Tierney, H. L., Baber, A. E., Kitchin, J. R., et al., Phys Rev Lett. **103**, 246102 (2009)

55. Crespo-Quesada, M., Yarulin, A., Jin, M., et al., J Am Chem Soc. **133,** 12787 (2011)
56. Yamada, Y. M., Yuyama, Y., Sato, T., et al., Angew Chem Int Ed Engl. **53,** 127 (2014)
57. Kobayashi, J., Mori, Y., Okamoto, K., et al., Science. **304,** 1305 (2004)
58. Offermans, P., Tong, H. D., van Rijn, C. J. M., et al., Appl Phys Lett. **94,** 223110 (2009)
59. Zeng, X. Q., Latimer, M. L., Xiao, Z. L., et al., Nano Lett. **11,** 262 (2011)
60. Favier, F., Walter, E. C., Zach, M. P., et al., Science. **293,** 2227 (2001)
61. Villanueva, L. G., Fargier, F., Kiefer, T., et al., Nanoscale. **4,** 1964 (2012)
62. Liu, N., Tang, M. L., Hentschel, M., et al., Nat Mater. **10,** 631 (2011)
63. Chiu, C. Y., Huang, M. H., Angew Chem Int Ed Engl. **52,** 12709 (2013)
64. Bilski, P., Reszka, K., Bilska, M., et al., J Am Chem Soc. **118,** 1330 (1996)
65. Konaka, R., Kasahara, E., Dunlap, W. C., et al., Free Radical Bio Med. **27,** 294 (1999)
66. Noda, Y., Anzai, K., Mori, A., et al., Biochem Mol Biol Int. **42,** 35 (1997)

Chapter 5
Plasmonic Nanostructures for Biomedical and Sensing Applications

Samir V. Jenkins, Timothy J. Muldoon and Jingyi Chen

Abstract Noble metal nanostructures are appealing for therapeutic, diagnostics, and sensing applications because of their unique optical properties and biocompatibility. The collective oscillation of electrons in a metal nanostructure resonates with particular wavelengths of light, generating the localized surface plasmon resonance (LSPR). The LSPR results in absorption and scattering of incoming photons. Absorption leads to photothermal generation of heat, photoluminescence, and quenching of fluorophores in close proximity. Scattering results in reflected photons and its amplification of the local electromagnetic field can enhance fluorescence, phosphorescence, and Raman scattering. These optical properties make plasmonic nanostructures ideal candidates for theranostic applications including light-induced thermal therapy, drug/gene delivery, and biomedical imaging. The use of plasmonic nanostructures for ex vivo detection of chemicals and biomolecules are also discussed in this chapter.

5.1 Introduction

Nanomaterials are broadly defined as materials that have at least one dimension in the range of 1–100 nm and possess unique properties distinct from those of their bulk counterparts [1]. Noble metal nanomaterials (particularly Au and Ag) have long been known to exhibit brilliant colors as a result of their interaction with light, known as localized surface plasmon resonance (LSPR). This phenomenon arises from the resonant oscillation of conduction electrons confined in a metal particle with an incident electromagnetic wave (photons), resulting in strong scattering and/ or absorption at the resonant wavelength. Historically, Au and Ag nanoparticles (AuNPs and AgNPs) have been (unknowingly) used to stain glass. The Cup of

J. Chen (✉) · S. V. Jenkins
Department of Chemistry and Biochemistry,
University of Arkansas, Fayetteville, AR 72701, USA
e-mail: chenj@uark.edu

T. J. Muldoon
Department of Biomedical Engineering,
University of Arkansas, Fayetteville, AR 72701, USA

© Springer International Publishing Switzerland 2015
Y. Xiong, X. Lu (eds.), *Metallic Nanostructures,* DOI 10.1007/978-3-319-11304-3_5

Lycurgus, a famous fourth century Roman relic currently housed in the British Museum, is made of dichroic glass containing Au and Ag colloids. This cup changes color depending on its orientation, appearing red by transmitted light and green by reflected light, an effect now known to arise from the optical properties of the Au and AgNP contained in the glass.

In addition to their plasmonic properties, Au and Ag colloids have a long history of biomedical use. Ancient Indian, Chinese, and Egyptian medicine utilized Au-based ingestible substances as therapeutics for arthritis [2, 3]. Such treatment relied on the oral administration of Au salts, which ultimately formed colloids in situ in the synovial fluid [4]. Gold-based therapeutics were also unsuccessfully applied to treat tuberculosis, lupus, epilepsy, and migraines [3]. Many ancient civilizations transported water in Ag containers, unknowingly tapping into the antimicrobial properties of the metal. While Ag is toxic in large doses, these applications typically resulted in low enough doses to prevent severe adverse effects. A condition known as argyria did develop in some people who consumed excessive amounts of Ag. The Ag had deposited in the skin, generating a colloid that ultimately caused the patients to turn blue [5].

Modern study of AuNPs began with Michael Faraday in 1857. Faraday synthesized ruby-red Au colloids [6] that are still on display in the Faraday Museum. Since Faraday's synthesis of stable Au colloids, the understanding of the plasmonic materials has advanced significantly, and they have found diverse applications ranging from catalysis to chemical sensing and biomedicine. The plasmonic properties are strongly correlated to the size and shape of the particle. This chapter focus on the plasmonic properties of noble metal nanostructures especially Au nanostructures and their biomedical and sensing applications. Particular attention is paid to the strong absorption, scattering, and photoluminescence phenomena that Au nanostructures exhibit, followed by discussion of their applications in therapeutics, diagnostics, and sensing.

5.2 Optical Properties

5.2.1 Localized Surface Plasmon Resonance (LSPR)

When metal NP are smaller than the wavelength of incident light, the oscillation of the conduction electrons can resonate with incident electromagnetic wave and give rise to the LSPR (Fig. 5.1) [7]. The effect of the LSPR is often referred to as extinction and is the sum of two components: absorption and scattering [8]. In 1908, Gustav Mie developed a theory to describe the interaction between spherical particles and light [9]. The Mie theory represents the exact solution to Maxwell's equations and describes a plane wave interacting with a metal sphere with radius R having the same dielectric functions as its bulk counterpart. If the boundary conditions are specified, the extinction cross-section (σ_{ext}) of a sphere can be obtained. Once can derive the extinction efficiency factor (Q_{ext}) which is defined as ratio of

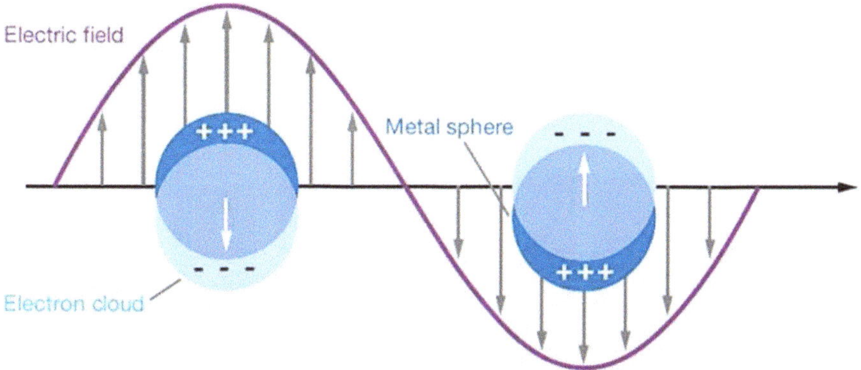

Fig. 5.1 Schematic illustration of LSPR phenomenon. (Reproduced from reference [7])

σ_{ext} to the physical cross-sectional area (πR^2). Under the condition that $R<<\lambda$, σ_{ext} can be determined from Eqn. 1.

$$\sigma_{ext}(\lambda) = \frac{24\pi^2 R^3 \varepsilon_m^{3/2}}{\lambda} \frac{\varepsilon_i(\lambda)}{\left[\varepsilon_r(\lambda)+2\varepsilon_m\right]^2 + \varepsilon_i(\lambda)^2} \tag{1}$$

where ε_m is the dielectric constant of the embedding medium and ε_r and ε_i are the real and imaginary components of the metal dielectric function, respectively. Resonance occurs when the condition $\varepsilon_r(\lambda) = -2\varepsilon_m$ is satisfied. All metal nanoparticles exhibit LSPR. In the case of coinage metals (Cu, Ag, and Au) the resonant condition is met in the visible region of the spectrum, accounting for their brilliant color. Typically, Ag nanospheres have an LSPR at ~400 nm; while the LSPR of Au and Cu nanospheres is located at ~520 and ~600 nm, respectively [10, 11]. The position of the LSPR peak is sensitive to the size of nanospheres. For example, increasing the diameter of Au nanospheres from 15–110 nm resulted in an LSPR shift from ~520 to ~585 nm (Fig. 5.2) [12, 13].

In addition to size, the position of the LSPR is dependent on the shape of the nanoparticles. For biomedical applications, the tissue transparent windows are the spectral regions where light penetration through tissue is the greatest, located in the near infrared (NIR) region (Fig. 5.13). The first NIR window ranges between 650 and 900 nm where the penetration depth is 1–2 cm with some tissue autofluorescence that may interfere with the optical signal [14]. The second NIR window exists between 1,000 and 1,350 nm at which autofluorescence is significantly reduced, although the availability of detectors is rather limited [15]. By altering the shape, the optical properties of the Au nanostructures can be tuned to the NIR window, while maintaining their small size.

Nanoshells, nanorods, and nanocages are three representative Au nanostructures with tunable LSPR in the NIR peaks that have been extensively for biological applications. Gold nanoshells, developed by Halas et al. [16], are made by depositing a thin layer of Au on a silica core with small AuNPs adsorbed. The LSPR is less

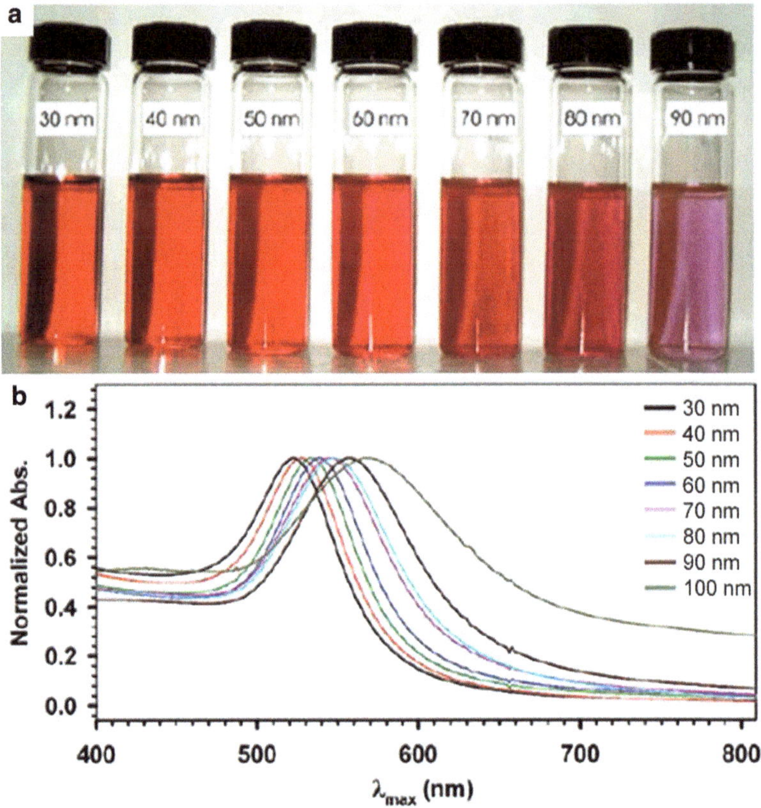

Fig. 5.2 Dependence of LSPR on diameter of spherical AuNPs: **a** photograph of AuNPs of different diameters and **b** corresponding extinction spectra. (Modified from reference [13])

red-shifted as the shell becomes thicker. Gold nanorods, developed by Murphy [17] and El-Sayed [18], have a longitudinal mode of the LSPR that shifts to longer wavelengths as their aspect ratio increases (Fig. 5.4a) [19]. Xia et al. [20] developed the synthesis of Au nanocages using galvanic replacement between Ag nanocubes and chloroauric acid. As the nanocubes are hollowed out, the optical spectra gradually shift to the red and reach the NIR region (Fig. 5.4b) [21, 22]. The optical spectra of these more complex geometries are typically calculated using the discrete dipole approximation. Readers may refer to an excellent review by Schatz et al. [23] on the calculation of optical spectra of plasmonic nanostructures of with arbitrary geometries.

5.2.2 *Absorption*

Absorption by metal nanoparticles at their LSPR gives rise to three interesting phenomena: nonradiative relaxation of the plasmon, photoluminescent emission

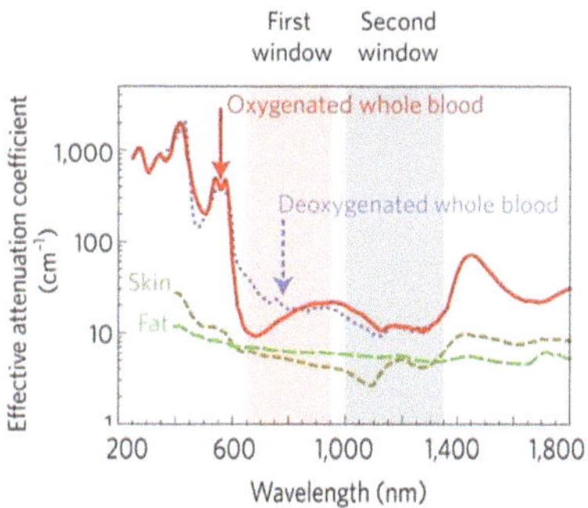

Fig. 5.3 Spectral locations of the first and second tissue transparent windows as well as the attenuation coefficients of skin, fat, and oxygenated or deoxygenated whole blood. (Reproduced from reference [15])

(5.2.3), and quenching of adjacent fluorophores (5.3.3). Nonradiative relaxation results in the generation of heat, a process which has been reviewed by Govorov et al. [24]. Basically, the excited electronic oscillation relaxes by releasing energy in the form of heat dissipated to the surroundings. In fact, most light absorbers generate heat via nonradiative relaxation. Because of their extremely large absorption coefficients, plasmonic nanomaterials can convert the photon energy into heat energy at very low concentration, several orders of magnitude lower than that of other materials [25]. The temperature profile over time can be described by Eqn. 2 [26].

$$T(t) = T_0 + A/B(1 - e^{-Bt}) \qquad (2)$$

where T_0 is the initial temperature, A is the rate of energy absorption (°C/s) and B is the first order rate constant for heat dissipation (s⁻¹).

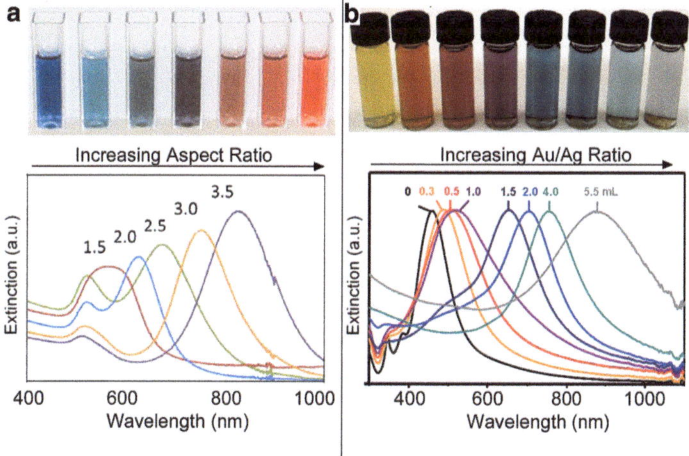

Fig. 5.4 Tunable LSPR maxima: **a** color photograph and extinction spectra of Au nanorods with increasing aspect ratio; and **b** color photograph and extinction spectra of Au/Ag nanocages with increasing ratio of Au/Ag. (Modified from references (**a**) [19] and (**b**) [21, 22])

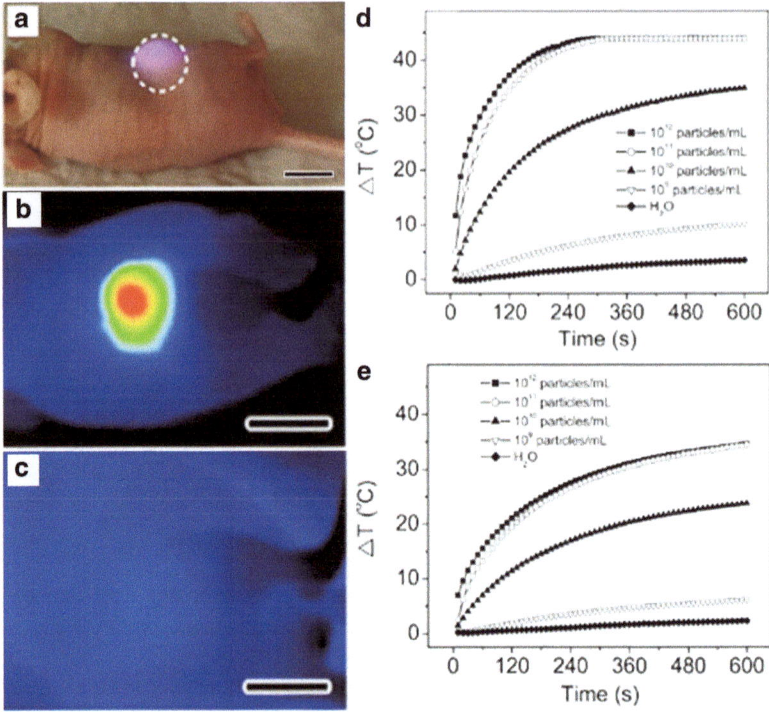

Fig. 5.5 Photothermal effect of Au nanocages: **a** photograph of a mouse undergoing laser irradiation of tumor; thermal image during irradiation; **b** with and **c** without Au nanocages; temperature response profile as a result of varying concentration of Au nanocages at laser flux of **d** 1.0 and **e** 0.5 W/cm². (Modified from reference [28])

The relaxation time of the LSPR is size dependent. For example, the relaxation following absorption occurs on the order of 10 and 380 ps for 4 and 50 nm AuNP, respectively [27], at which time point another photon can be absorbed. The photothermal effect can be varied by the nanoparticle concentration (more precisely, the optical density at the excitation wavelength), the optical flux, and the irradiation time. For instance, a power density of 1 W/cm² is capable of increasing the temperature of a 2 nM suspension of Au nanocages by more than 40 °C in less than 5 min, while a 0.2 nM suspension irradiated at 0.5 W/cm² has not reached a 30 °C temperature increase after 10 min irradiation (Fig. 5.5) [28]. The photothermal effect will be further discussed in the context of acoustic (5.4.2.1) and thermal imaging (5.4.2.2), hyperthermal therapy (5.4.1.1), and on-demand drug release (5.4.1.2).

5.2.3 Photoluminescence

An interesting, but not yet well understood plasmonic phenomenon is photoluminescence. Distinct from scattering, the photon is absorbed by the LSPR, and subsequently emitted in a short period. The wavelength of the emitted photon is slightly red-shifted relative to the excitation wavelength. The first reports of photoluminescence from bulk Cu and Au were published in 1969 [29], Shortly thereafter it was found that roughened metal surfaces emitted more frequently than smooth surfaces such that two-photon luminescence could be achieved [30]. Plasmonic photoluminescence is particularly appealing for bioimaging because it bypasses the photobleaching problems of organic fluorophores and blinking issues of quantum dots [31]. The luminescence efficiency of this process is quite low, with a milestone efficiency of 10^{-4} (i.e. 1 photon emitted for every 10^4 absorbed) for some nanoscale particles, while emission from bulk metals is practically nondetectable ($\sim 10^{-10}$) [32].

Two plausible mechanisms of photoluminescence have been proposed: retardation of nonradiative decay or enhancement of radiative decay [33]. For metallic structures, d band holes are generated by the excitation, then recombine with sp electrons in the metal [29]. This recombination releases a particle plasmon that then decays by emitting a photon. This luminescence intensity, however, has been shown to be independent of the size of particles and in some situations can be more easily detected than the scattered signal. It is worth noting that *nonplasmonic* metal clusters have a high photoluminescence yield, comparable to quantum dots. Readers may refer to a thorough review on photoluminescence of metal clusters by Dickson et al. [34].

5.2.4 Scattering

All objects scatter light to some extent. Lord Rayleigh first described the elastic scattering light by small molecules [35]. No change in wavelength is observed, and the scattering intensity is proportional to the diameter of the object to the sixth power and inversely proportional to the wavelength of light to the fourth power (i.e. $I \propto d^6, \lambda^{-4}$). Rayleigh scattering applies to objects that are significantly smaller than the wavelength of the incident light (e.g. oxygen and water molecules in the sky). Tyndall scattering is similar to Rayleigh scattering, but applies to larger objects with a diameter approaching the wavelength of light [36]. Both Rayleigh and Tyndall scattering are more efficient with shorter wavelengths of light.

Plasmonic scattering is also wavelength dependent; and the LSPR maximum determines the region of maximum scattering [37]. Scattering by the LSPR can be visualized using darkfield microscopy with the particles typically appearing as the color of the scattered light, that is, AgNP are cyan [38], AuNP are green [39], Au nanorods are yellow or red [40]. Hyperspectral imaging records the scattered spectrum of light at a particular point on a microscope image, and can be used to generate a map that records both the position and LSPR profile of nanoparticles. These maps have been used to distinguish various geometries of particles and to monitor particle agglomeration [39].

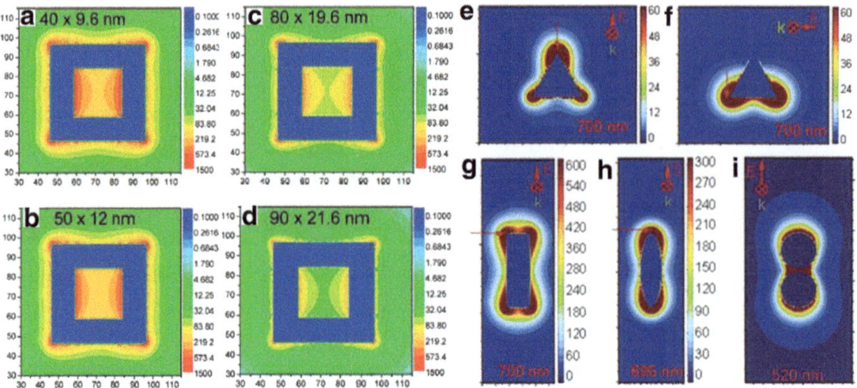

Fig. 5.6 Calculated electromagnetic field enhancement of various geometries: **a–d** hollow Au nanoframes of varying edge length and wall thickness; **e, f** Ag nanoprisms under different polarizations of light; **g** cylindrical and **h** tapered Ag nanorod; and **i** Ag dimer separated by 2 nm. (Modified from references (**a–d**) [41] and (**e–i**) [42])

5.3 Surface Effects

The oscillations of the LSPR allow the elastic absorption and reemission of the photon (i.e. scattering). The motion of electrons changes the charge density around the nanostructure causing enhancement of the electromagnetic field. Simulations of this "near-field" have shown significantly greater field enhancement in areas of high curvature (Fig. 5.6) [41, 42]. The corners of nanocubes [43], the tips of nanorods [44], and the void between two metal surfaces [45] have been indicated to be the sites of maximum enhancement on their respective structures. Agglomeration of AuNP results in significant field enhancement at the points where particles abut, which is largely the result of plasmon coupling [46]. Similarly, the interior and the pores of hollow Au nanocages and nanoshells exhibit orders of magnitude enhancement of the electromagnetic field. The interior of concentric structures (e.g. nanorattles or nanomatrioshka) displays extraordinary local field enhancement, particularly when the core is adjacent the side wall [47]. The enhanced electromagnetic fields can interact with near-field molecules giving rise to a variety of optical phenomena including surface-enhanced Raman scattering (SERS), surface-enhanced fluorescence (SEF), and nanometal surface energy transfer (NSET).

5.3.1 Surface-Enhanced Raman Scattering (SERS)

The LSPR can enhance Raman signal of molecules on the surface by several orders of magnitude, a phenomenon known as SERS. Raman scattering by a molecule, first observed in 1928 [48, 49], is an inelastic process wherein the energy of the

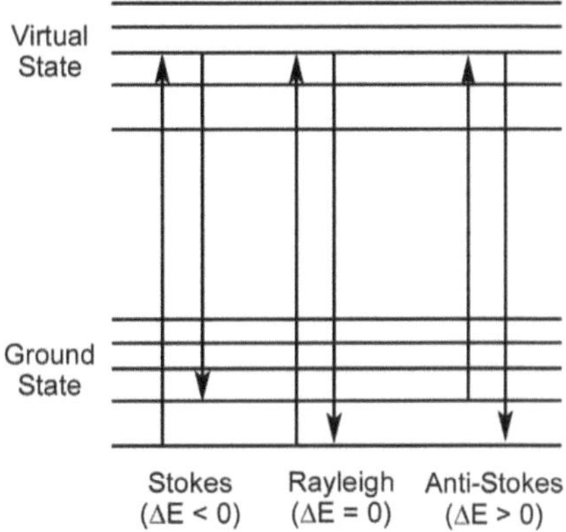

Fig. 5.7 Schematic illustration of scattering processes that utilize the transient virtual excited state of the molecules. The energy change relative to the excitation wavelength is included in parenthesis

scattered photon (E_{sca}) is different from that of the absorbed photon (E_{abs}). This energy difference, known as Raman shift, corresponds to changes in the vibrational state of the molecule, and can be classified as Stokes shift ($E_{sca} < E_{abs}$) or anti-Stokes shift ($E_{sca} > E_{abs}$) (Fig. 5.7) [50]. The Stokes and anti-Stokes spectra are mirror images, with the Stokes spectrum having higher intensity because of the larger population of the ground vibrational state than an excited vibrational state in a sample [50, 51]. Raman shifts, both Stokes and anti-Stokes shifts, are typically reported as wavenumber (cm⁻¹) because this unit is linearly proportional to energy in a photon. The Raman shift can be defined by Eqn. 3.

$$\Delta v = (1 / \lambda_0 - 1 / \lambda) \cdot (10^7 \ \text{nm/cm}) \tag{3}$$

where Δv is the Raman shift (in cm⁻¹), λ_0 is the excitation wavelength, and λ is the wavelength of the scattered photon (both in nm).

Generally, the Raman cross section of a molecule is extremely small, and thus the signal is extremely weak relative to Rayleigh scattering. In 1974, Fleischmann et al. [52] observed that molecules on roughened Ag surfaces exhibited significantly enhanced Raman scattering. Since then, SERS has become a powerful analytical tool that has been extensively investigated for many applications, particularly chemical and biological detection [53]. There are two primary mechanisms responsible for SERS: chemical and electromagnetic enhancement. Chemical enhancement refers to the process of charge transfer from the plasmonic metal to the molecule, which better enables the vibrational transitions necessary for Raman shifts. Such enhancement has been shown to enhance Raman signal by up to a factor of 10^2 [54]. The electromagnetic component contributes the majority of Raman enhancement, typically enhancing the electric field as a total of E^4 [55]. In this scenario, the plasmonic metal acts as a two-way antenna: increasing incident light flux on the molecule thereby increasing the number of transient absorptions and amplifying the scattered Raman signal [56]. SERS enhancement is inversely correlated with distance to the 10th power ($1/d^{10}$) [57], and thus the enhancement decreases rapidly away from the

Fig. 5.8 Surface-enhanced
Raman spectroscopy:
a schematic illustration
of agglomeration-based
activation of SERS signal
with **b** typical, reversible pH
responsive SERS spectra;
c distance dependence of
SERS enhancement factor of
Ag island films. (Modified
from references (**a**–**b**) [58]
and (**c**) [59])

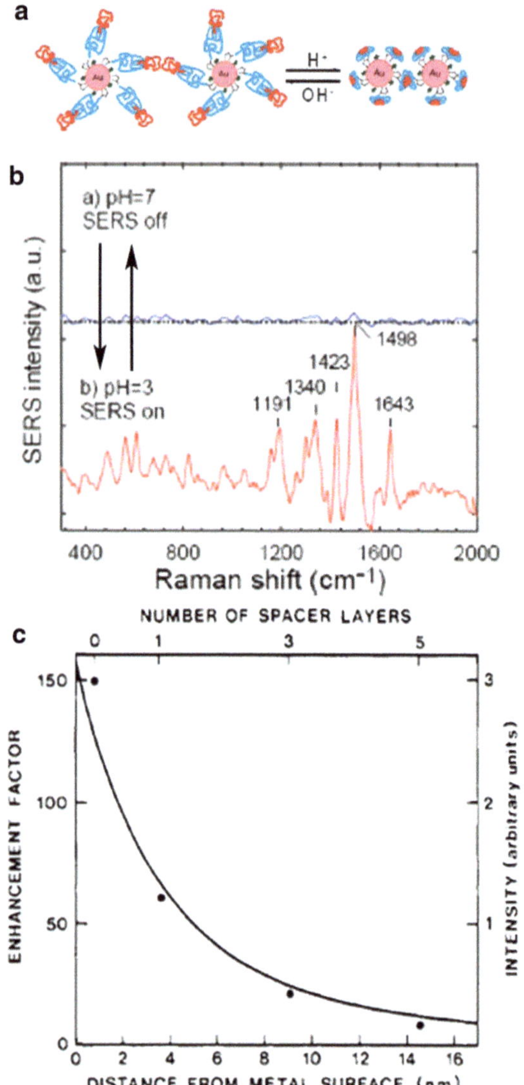

metal surface (Fig. 5.8) [58, 59]. Therefore, maximum enhancement is observed for
surface adsorbates. Because Raman scattering requires a change in the polarizabil-
ity of a molecular vibration to be observed and the interaction of scattered photons
with the LSPR is orientation dependent, SERS can be used to probe the orientation
of Adsorbed molecules on the surface [60].

Raman enhancement occurs best when a molecule resides in a "hot spot," an area
of high local enhancement of the electromagnetic field [61]. The location of these
hot spots corresponds to the regions of calculated maximum field enhancement in

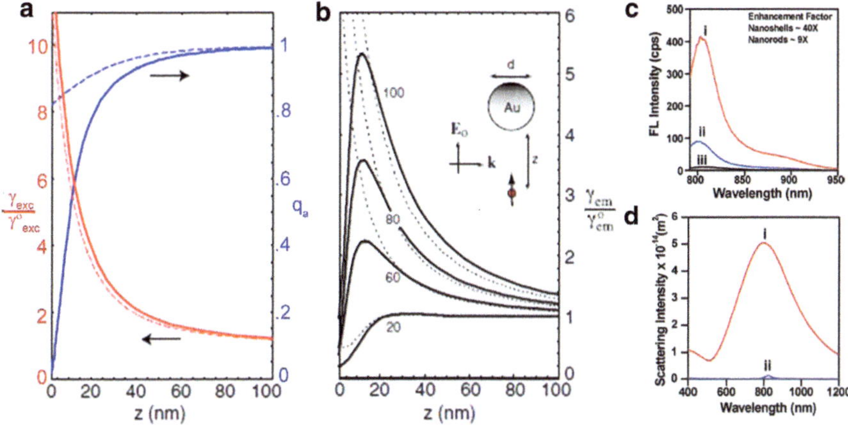

Fig. 5.9 Fluorescence enhancement: **a** sample calculated distance dependence of fluorescence (*left axis*) and quenching (*right axis*); **b** additive fluorescence quantum yield based on combination of enhancement and quenching; **c** fluorescence signal from fluorophore/albumin conjugated to Au nanoshell *(i)* Au nanorod *(ii)* and without Au *(iii)*; **d** calculated scattering intensity of Au nanoshell *(i)* and Au nanorod *(ii)*. (Modified from References (**a**, **b**) [67] and (**c**, **d**) [71])

a nanostructure or an array of nanostructures [62, 63]. For example, agglomeration points are areas of both high curvature and nearby metal surfaces, and under appropriate conditions can generate hot spots with enhancement factors on the order of 10^{15} [64]. Such hot spots have enabled the detection of single molecules residing within them [65].

5.3.2 *Surface-Enhanced Fluorescence (SEF)*

Similar to SERS, localized enhancement of electromagnetic field on the metal surface results in the enhancement of fluorescence from fluorophores in close proximity, known as SEF. This phenomenon was first observed with organic dyes on roughened metal films [66]. This process competes with quenching processes (5.3.3) at short distances but is dominant at a longer distance range. Consequently, enhancement is observed over a distance range of up to $d = 100$ nm from the metal surface, while the fluorescence is primarily quenched at distance close to the surface, d less than the critical distance ~ 20 nm (Fig. 5.9a and b) [67].

The plasmon can interact with a fluorophore in several ways to enhance emission. First, the LSPR can increase the excitation incidence, by increasing the optical flux at the fluorophore thereby increasing the frequency of excitation events. Second, the LSPR can increase the rate of radiative decay, causing emission to occur more readily from the excited state of the fluorophore. Third, the LSPR can engage in far-field coupling with the emitted photon, increasing the likelihood of emission [68]. Functionally, the plasmonic structure serves as a bidirectional antenna that increases the number of photons absorbed and the number of photons emitted [69]. The spectral position of the LSPR plays an important role in fluorescence enhancement. Generally, overlap between the *emission* wavelength and the LSPR provides the maximum enhancement. The LSPR overlapping with the excitation wavelength results in slight enhancement while its red-shifting from the emission wavelength results in reduction of enhancement [70]. Other studies have shown that the enhancement factor increases with the size (more precisely, the scattering cross-section) of the nanostructure (Fig. 5.9c and d) [71]. Theoretical calculations have shown that the longitudinal orientation of the fluorophore results in the maximum vector sum of the metal dipole and fluorophore transition dipole, which leads to an increase of the emitting dipole [72]. In contrast, transversal and perpendicular orientations of the fluorophore cause antiparallel orientation with the metal dipole, resulting in the reduction of emission.

In addition to fluorescence enhancement, the LSPR can also enhance the quantum efficiency of triplet state phenomena. Plasmonic metal enhancement of phosphorescence was demonstrated in 2004 using AgNPs deposited on a layer of phosphors on glass substrate [73]. This enhanced phosphorescence is the result of the LSPR stabilizing the triplet state of the phosphor and enabling radiative decay, similar to fluorescence enhancement. This phenomenon has also been shown to enhance the phosphorescent signal from the decay of singlet oxygen. Relaxation of singlet oxygen to its ground triplet state leads to phosphorescence at ~ 1270 nm, and utilization of plasmonic structures can enhance the signal from singlet oxygen, leading to lower detection limits [74]. Singlet oxygen itself is typically generated through triplet-triplet annihilation of an excited photosensitizer molecule [75]. As a result, the LSPR can effectively enhance the generation of singlet oxygen by the adjacent photosensitizers, improving efficacy of photodynamic therapy [76, 77]. Similar to fluorescence quenching, close proximity between the particle and the phosphor results in quenching by the LSPR, and subsequent loss of emission [78]. A distance dependence similar to that of fluorescence enhancement has also been observed [79]. Various studies have further demonstrated that the maximum stabilization of the triplet state can be achieved by matching the LSPR with the wavelength of phosphorescent emission [80, 81].

5.3.3   *Nanometal Surface Energy Transfer (NSET)*

Analogous to the distance-dependence quenching of fluorophores observed in Förster resonance energy transfer (FRET) [82], NSET was first observed in 1969 [83] and then applied to metal nanoparticles by Strouse et al. in 2005 [84]. For NSET to occur, the metal surface is treated as a two-dimensional array of dipoles, while the fluorophore is treated as a point dipole. The quenching efficiency of a metal surface is distance dependent with a $1/d^4$ loss. The value d_0 for NSET is defined as the distance at which quenching efficiency is 50 % and is described by Eqn. 4.

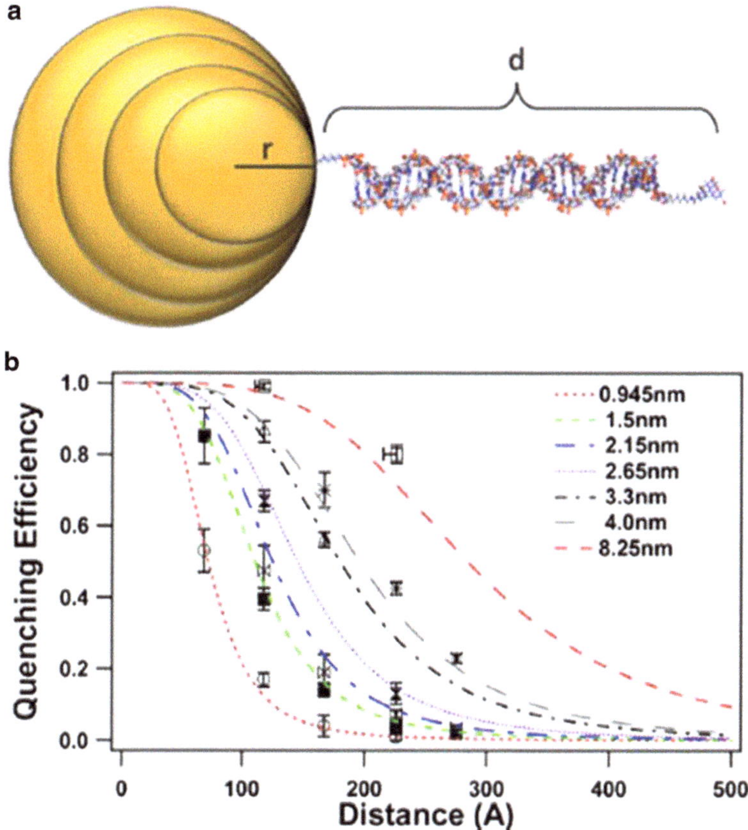

Fig. 5.10 Fluorescence quenching efficiency is both distance and size dependent: **a** schematic for distance and size change; **b** calculated *(lines)* and experimental *(points)* quenching efficiency for varying AuNP radius, "r", and separation from fluorophore, "d." (Reproduced from reference [86])

$$d_0 = \left(\frac{0.525 \cdot c^3 \cdot \Phi_D}{\omega^2 \cdot \omega_f \cdot k_f} \right)^{\!1/4} \tag{4}$$

where c is the speed of light, Φ_D is the donor quantum efficiency, ω is the donor electronic transition, ω_f is the Fermi frequency for the (bulk) metal, and k_f is the Fermi wave vector of the (bulk) metal. Further experiments revealed that the mechanism for NSET does not involve an appreciable suppression of the radiative rate of the fluorophore, but rather an increase in the rate of energy transfer [85]. The fluorescence quenching efficiency is both distance and size dependent (Fig. 5.10) [86]. Additionally, the early studies of NSET used 1.5 nm AuNPs, which are too small to exhibit LSPR. Larger particles have a plasmon, and their quenching efficiency has been shown to have a $1/d^6$ distance dependence, the same as that of FRET [87]. This distance dependence is due to the LSPR and the fluorophore both behaving as point dipoles and the d_0 can be described by Eqn. 5

$$d_0 = 0.211 \cdot \left[\kappa^2 \cdot \Phi_D \cdot n^{-4} \cdot J(\lambda) \right]^{\frac{1}{6}} \tag{5}$$

where κ^2 is the relative orientation of the dipoles and is taken to be 2/3 due to rapid averaging, Φ_D is the quantum yield of the free donor, n is the refractive index of the medium and $J(\lambda)$ is the overlap integral between the normalized donor emission and acceptor extinction coefficient. Quenching by NSET yields a larger d_0 value than that for traditional FRET systems. This increased d_0 value allows the two objects to function as optical rulers on a longer scale than those based on FRET. There have been numerous demonstrations of the use of nanoparticle-fluorophore systems to measure distances at the nanoscale [88].

5.3.4 Surface Chemistry

Functionalization of the surface of noble metal NP has become a very important topic for their use in biomedical and sensing applications. An excellent review by El-Sayed and Murphy has summarized the recent advancement of surface chemistry on Au nanostructures [21]. In this subsection, a brief discussion will focus on Au-thiolate chemistry and its application for in vivo targeting.

The Au-S bond is among the strongest bonds to the Au surface with roughly the same strength as that of the strongest hydrogen bond (~40 kcal/mol) [89]. Whitesides et al. pioneered the work in self-assembled monolayers on Au films using alkanethiols [90, 91]. Other sulfur-containing molecules, such as disulfides and thioethers, can also be used to functionalize the Au surface. Using heterobifunctional linker molecules with one thiol terminus (e.g. HS-linker-X), a variety of functional groups can be introduced to the Au surface via the other terminus. The thiol serves as an anchor to the nanoparticle platform while the variable terminus acts as a functional handle for further chemical modification. Additionally, mixed monolayers can be attached to the Au surface for inclusion of multiple components in controllable compositions [92, 93]. Poly(ethylene glycol) (PEG) is one of the most common surface ligands because of its ability to repel proteins and increase particle circulation lifetimes [94]. Heterobifunctional PEG allows for a thiol group to serve as an anchor to an Au nanoparticle and the other exposed terminus to function as a handle for further conjugation with a fluorophore, therapeutics, marker, or targeting moiety.

For in vivo applications, one of the grand challenges of nanomedicine is successful delivery of the nanomaterial to the desired site. Two primary delivery mechanisms can be applied: passive targeting and active targeting. The primary distinction between the two is the use of targeting moieties. For instance, in passively targeted systems, no ligand is utilized to direct materials to a particular site [95]. For tumors,

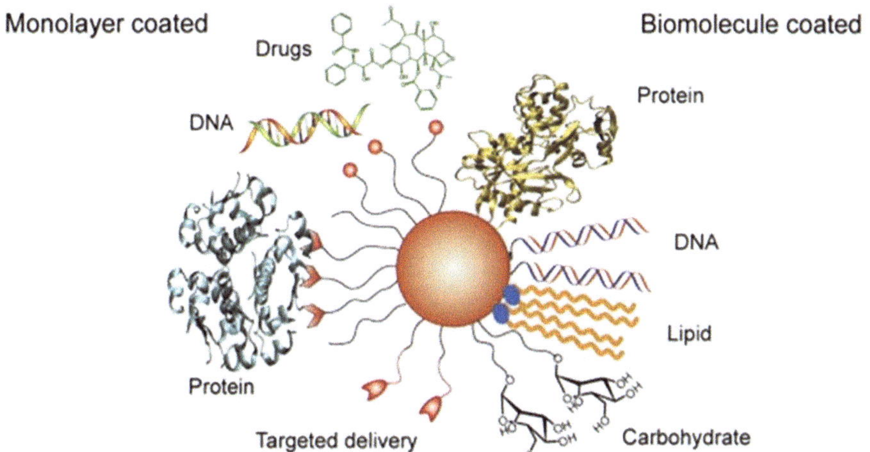

Fig. 5.11 Schematic of various targeting moieties. (Reproduced from reference [98])

the leaky vasculature allows for increased uptake of nanoparticles with diameters typically below 200 nm [96]. The inflammation usually associated with tumors leads to reduced clearance of the materials from the tumor, and thus increases retention of the materials [97]. Together these phenomena are known as the enhanced permeability and retention (EPR) effect. Particles clearance from the bloodstream typically occurs through the spleen and the liver; therefore these organs can be passively targeted as well.

In contrast to passive targeting, active targeting uses specific recognition moieties to promote the association with particular cells (Fig. 5.11) [98]. Biomolecules, such as peptides, antigens, aptamers, and antibodies, have been utilized as targeting ligands [99]. Targeting may lead to a significant increase in tumor uptake thereby decreasing particle concentration in other organs [100]. Targeting typically involves interaction of the ligands with specific receptors expressed at high concentration on the target cell surface. The high local density of the ligand on a particle's surface enables strong multidentate binding to several receptors to facilitate receptor-mediated endocytosis of nanoparticles [101, 102]. In addition to endocytosis, cell penetrating peptides can shuttle the conjugated nanoparticles through the membrane to the cytosol without being encapsulated in a vesicle [103]. It is worth noting that the formation of a protein corona in biological systems could mask the targeting moieties to an extent, though they are often not fully compromised [104].

5.4 Biomedical and Sensing Applications

Plasmonic nanostructures, in particular Au nanostructures, are potential candidates as theranostic agents due to unique their optical properties, facile surface modification and bioinertness. "Theranostics" is a portmanteau of therapeutics and diagnostics; theranostic materials possess both of these capabilities [105]. Most studies have concentrated on oncological applications which has been covered in the book chapter by Chen [106]. In this context, nononcological applications of plasmonic nanostructures are the primary focus. This section is divided into three subsections: therapeutics, biomedical imaging and sensing. Each subsection will start with the basis of each application, followed by examples.

5.4.1 Therapeutics

Plasmonic nanostructures are particularly appealing for therapeutic applications due to their capability of light-to-heat conversion. Plasmonic nanostructure mediated photothermal therapy, chemotherapy, and gene therapy will be discussed in this subsection.

5.4.1.1 Photothermal Therapy

Plasmonic nanomaterials can efficiently convert the absorbed photons to phonons through lattice vibrations, generating heat that can be dissipated to their surroundings. The large absorption cross-section at their resonance frequency makes Au nanostructures ideal photothermal agents [107]. Targeting these nanostructures to cells or tissue and then irradiating the region allows controllable localized heating [108]. Increasing temperature around mammalian cells can induce cell death *via* two different mechanisms: necrosis and apoptosis. Necrosis is typically induced at a temperature greater than 50 °C, and this process is known as thermal ablation [109]. The extremely high heat causes the direct cell killing and tissue coagulation [110]. When the temperature mildly increases to 42–47 °C (body temperature being 37 °C), many tumor cells have been shown to begin undergoing apoptosis, programmed cell death [109]. Hyperthermia elevates the cell temperature only high enough to induce apoptosis, whereas ablation causes significant protein denaturation that in turn causes the cells to die. Additionally, photothermal therapy combining with other therapy such as chemotherapy could further improve the treatment efficacy through the synergistic effect (Fig. 5.12a and b) [111].

In photothermal therapy, the thermal effect is generated by the plasmonic particles upon light irradiation at their resonance wavelength. By controlling the nanoparticle concentration and the irradiation dose, the temperature can be finely tuned to maximize therapeutic effect and minimize damage to healthy tissue. The photothermal effect was first demonstrated on tumors by Halas et al. using Au nanoshells [112, 113]. Other Au nanostructures such as nanorods [114, 115], nanocages, [116, 117] nanostars [118], have been applied to different cell lines and *in vivo* models for photothermal treatment of cancer. The photothermal effect is typically generated by a NIR continuous wave (c.w.) laser while an ultrafast pulse laser introduces alternative complications that may cause necrotic cell death such as

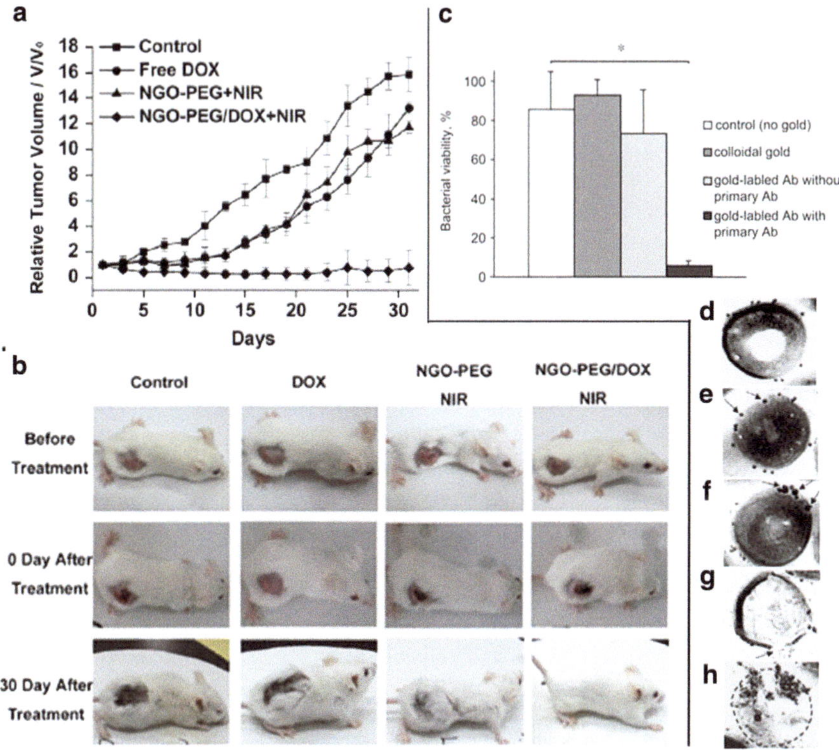

Fig. 5.12 Chemotherapy delivery: **a** synergistic effect of photothermal and chemo therapies based on tumor size; **b** photographs of mice under various treatment conditions; **c** chart of bacterial viability after photothermal treatment under various AuNP formulations; TEM images of *S. aureus* targeted with various AuNPs **d** before treatment; after 0.5 J/cm² irradiation using **e** nonclustered and **f** clustered AuNPs; and after 3.0 J/cm² irradiation using **g** nonclustered and **h** clustered AuNPs. (Modified from references (**a, b**) [111] and (**c–h**) [120])

cavitation or nanobubbles around the plasmonic nanoparticles [119]. Photothermal destruction of bacterial cells, pioneered by Zharov and Smeltzer, has been demonstrated against both gram negative and gram positive bacterial strains using AuNPs and Au nanorods (Fig. 5.12c–h) [120, 121].

5.4.1.2   Chemotherapy

The facile surface chemistry of plasmonic nanostructures makes them ideal candidates for drug delivery. Drug molecules can be incorporated on the nanoparticle surface covalently *via* chemical bonds or noncovalently through intermolecular interactions. A vast library of covalent conjugation strategies developed for biochemical labeling is readily applied to these materials [122]. Amide coupling is among the most commonly-used method to covalently attach the drug molecule to the handle of the bifunctional linker on the particle surface. "Click chemistry" is another particularly appealing conjugation route, due to the orthogonality of the functional groups involved [123]. Additionally, these covalent linkages can be designed to respond to a variety of external or internal stimuli such as irradiation [124], pH change [125, 126], or exposure to biomolecules [117], enabling controlled release [127]. Covalent bonds offer relatively stable conjugates with low leakage during transport in the circulatory system. However, the pharmacokinetics and efficacy of drug molecules could be compromised because their structures may have been permanently modified by the conjugation. Noncovalent strategies offer the advantage of no chemical modification of the drug, however, they are more prone to leakage due to the equilibrium-based nature of the binding [128]. The drug molecules are incorporated to the surface coating on the NP through van der Waals forces including electrostatic forces, hydrogen bonding, hydrophobic interactions, and π-π interactions. Readers may refer to an excellent review by Rotello et al. [129] on the topics of drug delivery using AuNPs.

Assisted by the photothermal effect, the drug payload can be released controllably by NIR light. Halas et al. [130] demonstrated the Au nanoshell embedded into heat-responsive, poly(N-isopropylacrylamide) (pNiAAm) hydrogel could control the release of drugs by NIR irradiation. Instead of a gel, a monolayer coating of thiol-anchored pNiAAm serves as a gate to keep drugs inside the hollow Au nanocage (Fig. 5.13a) [131]. The release can then be regulated by the NIR light through the photothermally induced phase change of the polymer from hydrophilic, brush-like conformation to hydrophobic, collapsed state. Collapse of the polymer exposes the pores allowing drug molecules to diffuse outward. Similarly, drugs can be embedded in a phase-change material within the nanocage void, which transitions from solid to liquid at a specific temperature (Fig. 5.13b) [132, 133]. Photo-induced heating mediated by the nanocages causes the solid interior to melt and the drugs to diffuse outward. Alternatively, drugs can be loaded into the interior of vesicles composed of Au nanostructure tethered polyelectrolyte capsules [134, 135], or liposomes [136, 137]. Irradiation of the Au nanostructures results in photothermal rupture of the vesicles to release the encapsulated drug (Fig. 5.13c) [138], or photothermal activation of chemical reaction to cleave the drug molecules (Fig. 5.13d) [139]. Other studies have demonstrated that Au structures can enhance photodynamic therapeutic efficacy of conjugated photosensitizers by increasing delivery of the drug to the tumor site and/or enhancing generation of therapeutic singlet oxygen [77, 140].

In addition to anticancer drugs, AuNPs have been utilized for the delivery of antibiotics. Typically, AuNPs are conjugated with an antibiotic through an available amine group which provides sufficiently strong interactions to densely pack on the AuNP surface, although the active portion of the antibiotic must be exposed to the environment. It has been suggested that therapeutic efficacy could be reduced as a consequence of agglomeration due to the presence of multiple amino groups in the antibiotic [141]. The close-packing of the antibiotic vancomycin on the particle surface mimics the mechanism of action of the drug by removing the entropic

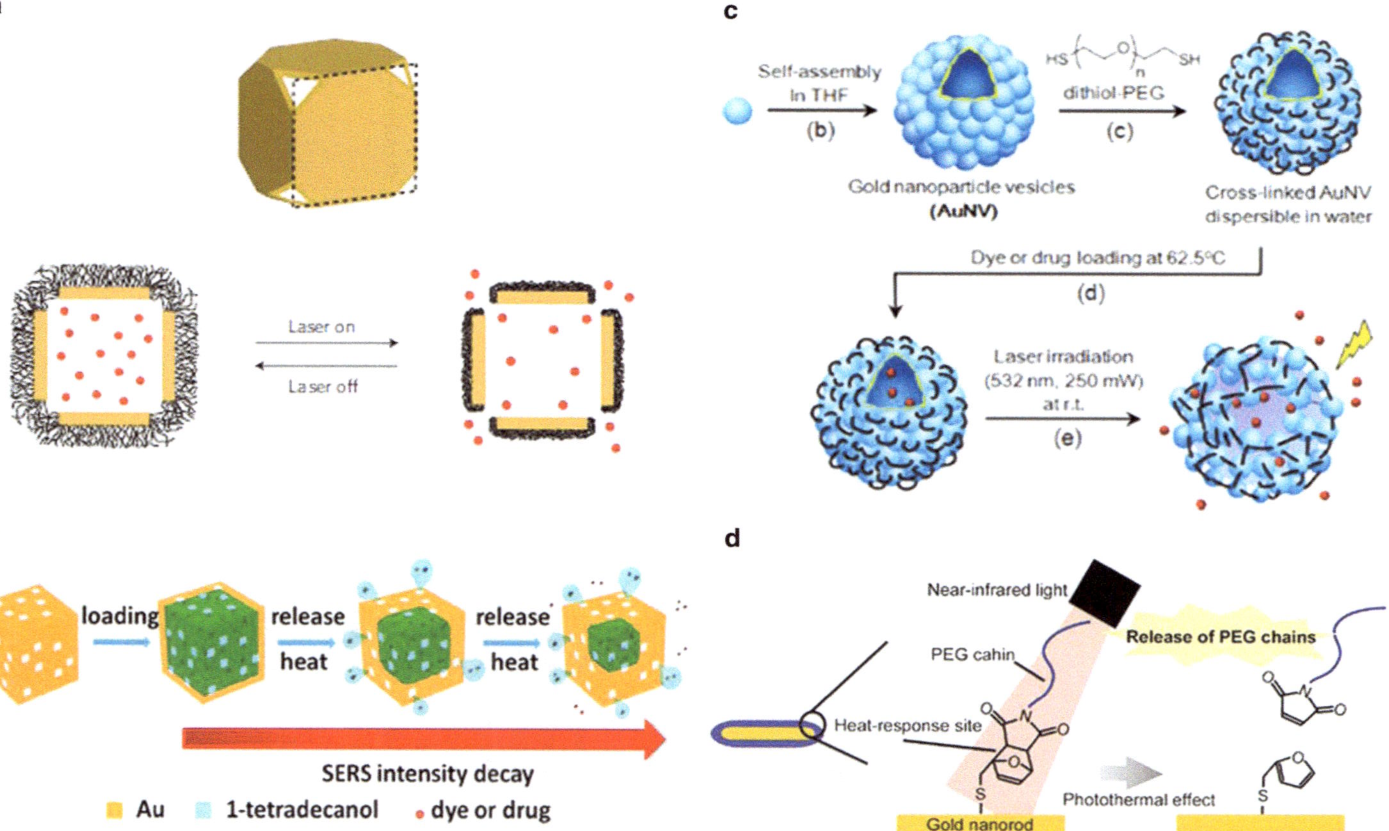

Fig. 5.13 Photothermal drug release: **a** from Au nanocages conjugated with heat responsive smart polymer; **b** from Au nanocages loaded with phase change material and a drug; **c** from rupture of AuNP vesicle loaded with a drug payload; **d** following retro Diels-Alder reaction. (Modified from references (**a**) [131], (**b**), [133], (**c**) [138], and (**d**) [139])

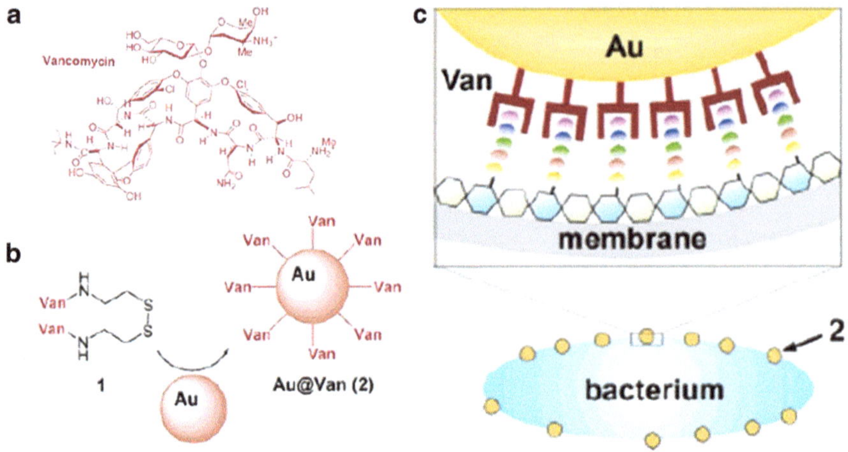

Fig. 5.14 Schematic illustration of the conjugation of vancomycin to Au nanoparticles and its bactricidal mechanism of action: **a** structure of vancomycin; **b** illustration of conjugation via cysteamine; and **c** illustration of binding mechanism of the conjugated particle with bacterial membrane. (Modified from reference [142])

contribution to the activation (Fig. 5.14) [142]. Multiple vancomycins attached to the particle surface facilitate oligomerization of the antibiotic at the cell membrane, thereby lowering its minimum inhibitory concentration. Other reports have shown improved efficacy of quinolone [143], aminoglycosides ("-mycins") [144], and ampicillin [145] against gram negative and gram positive strains when they were conjugated onto the AuNP surface. Even in the case of small molecules (e.g. pyrimidine and derivatives) which lack intrinsic antibacterial activity, conjugation to AuNPs has been shown to turn them into potent bactericides [146]. The antibiotic cefaclor has also been utilized as reducing and capping agent for synthesizing AuNPs while its β-lactam ring remains intact and thus antimicrobially active [147]. Replacing the amine with a thiol for anchoring to the AuNP surface can be advantageous because the orientation of the attached antibiotics could be more tightly controlled, which is paramount for its efficacy [141].

5.4.1.3 Gene Therapy

Gold nanostructures have been widely-used to deliver genetic material, commonly DNA (deoxyribonucleic acid) and siRNA (small interfering ribonucleic acid), for gene therapy. The two primary conjugation strategies for genetic molecules are to anchor oligonucleotides to the nanoparticle surface through a terminal thiol or to tether anionic DNA/RNA strands to the cationic particles by the electrostatic interactions (Fig. 5.15) [148, 149]. The former yields a perpendicular orientation of the genetic molecules to the nanoparticle surface while a parallel orientation is found on the NP surface in the latter. Gold nanoparticles provide a significant increase in

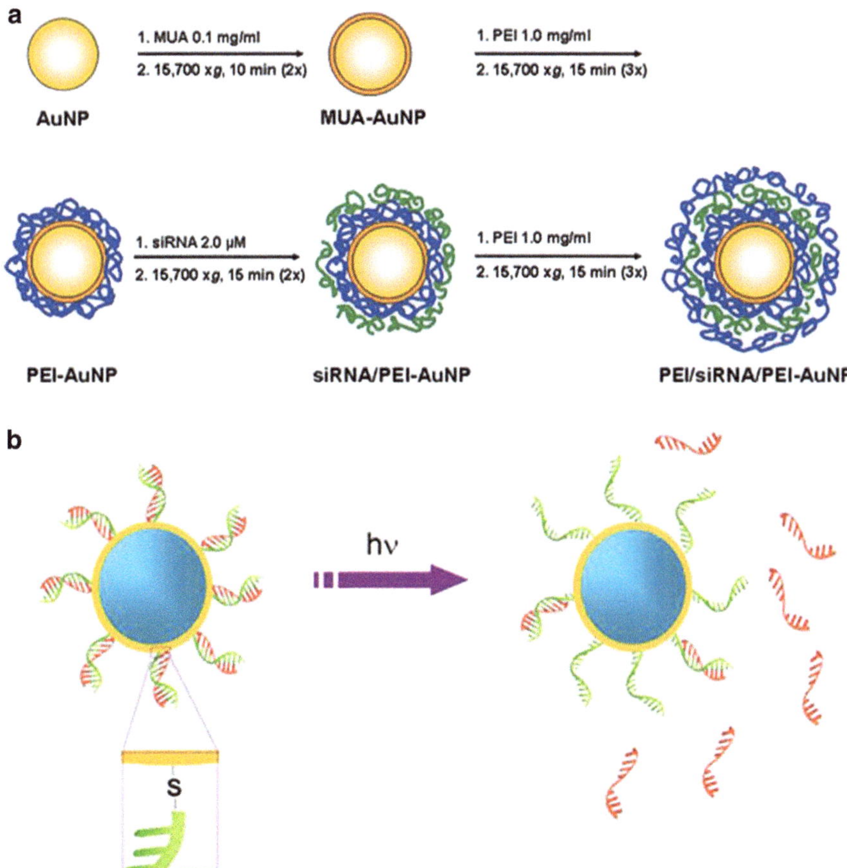

Fig. 5.15 Gene conjugation strategies: **a** oligonucleotides are layered on the Au surface using electrostatic interactions; **b** oligonucleotides are anchored via a thiol chain, PT heating causes the release of single-stranded oligonucleotides. (Modified from references (**a**) [148] and (**b**) [149])

transfection efficiency relative to free DNA possibly due to increased resistance to degradation by DNAse [150] and being targetable to the nucleus [151]. Readers can refer to a comprehensive review by Rotello et al. [152] on AuNPs for nucleic acid delivery.

In addition to its function as a platform, AuNPs are more advantageous over other carriers because their plasmonic properties enable controlled release of nucleic acid by NIR light. The ubiquity of nucleases in biological samples has long prevented the delivery of "naked" genetic material. The light-to-heat conversion by AuNPs increases the conjugate temperature above the melting point of the attached double-stranded DNA, leading to the release of a single strand for transfection [149]. Further, multiple DNA oligonucleotides could be selectively released from surface of Au nanorods with different LSPR wavelengths by simply varying the wavelength of incident light [153]. Aside from DNA delivery, AuNPs have shown successful delivery of siRNAs, which are a class of double-stranded RNA molecules (20–25 base pairs) that interfere with protein expression and subsequently turn various functions off and on. In an addiction model, an Au nanorod/siRNA conjugate was able to effectively shut down the positive response to stimulus normally present in dopaminergic cells [154]. Similarly, Au nanoshells conjugated with siRNA were able to arrest the production of green fluorescent protein in cells following photothermal release [155].

5.4.2 Biomedical Imaging

Plasmonic nanostructures are particularly attrative as constrast agents for biomedical imaging because of their nonblinking and nonphotobleaching optical properties. In this subsection, we discuss the use of plasmonic nanostructures for molecular imaging by six optical imaging modalities: photoacoustic tomography, photothermal imaging, darkfield microscopy, optical coherence tomography, Raman mapping, and multiphoton luminescence.

5.4.2.1 Photoacoustic Tomography (PAT)

The capability of strong light absorption makes plasmonic nanostructures particularly compelling as contrast agents for photoacoustic (PA) imaging. In PA imaging, the irradiation light is absorbed and converted to thermal energy, resulting in thermal expansion of the objects and generation of mechanical (acoustic) waves that can be detected using an ultrasonic probe (Fig. 5.16a) [156]. PAT combines the high lateral resolution of optical imaging and the deep information resultant from ultrasonic imaging (Fig. 5.16b) [156]. The PA signal is proportional to the light absorption ability of the objects, therefore, tissue with different absorption capability show inherent PAT contrast. To enhance contrast in the region of interest both endogenous and exogenous contrast agents are considered. Although endogenous contrast agents (e.g. blood) can reveal some biological aspects associated with a disease such as O_2 diffusion rate, exogenous contrast agents provide additional information that is essential for disease diagnosis particularly in early stages. Gold nanostructures possess tunable, large absorption cross-sections in the tissue transparent window (i.e. NIR region), allowing significant enhancement of PA contrast compared with the tissue alone.

Early demonstrations showed that Au nanoshells [157] and nanocages [158] could be used as PA contrast agents to image rat brain (Fig. 5.16c–e). Nanocages and nanorods have the advantage over the nanoshells for contrast enhancement due to their relatively large absorption cross-section [159]. Nanocages have also been demonstrated as optical tracers for PA sentinel lymph node mapping that could reach a depth of 33 mm below the skin surface [153, 160]. Such techniques could potentially be translated to the clinical setting for staging breast cancer. Nanocages conjugated

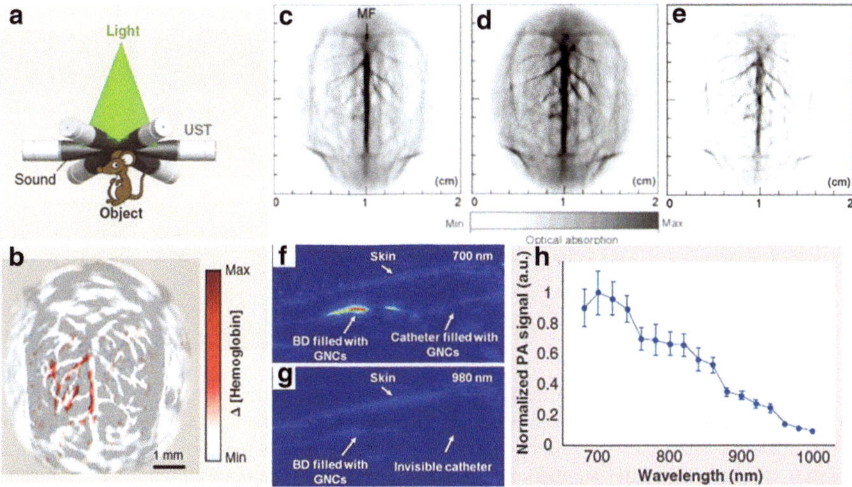

Fig. 5.16 Photoacoustic tomography: **a** schematic illustration of photoacoustic tomography setup with **b** sample tracking of blood flow in a brain. Photoacoustic images of brain **c** before and **d** after I.V. AuNP injection and **e** subtracted images. Photoacoustic images of mouse bladder following intraurethral injection of Au nanocages with excitation at **f** 700 nm and **g** 980 nm **h** as well as the wavelength dependence of PA signal. (Modified from references (**a**, **b**) [156], (**c**–**e**) [157], and (**f**–**h**) [164])

with tumor targeting peptides could enable molecular PAT of tumors such as melanomas for early cancer diagnosis [161]. Alternatively intravascular PAT showed the successful imaging of atherosclerotic plaque in rabbit arteries using AuNPs [162, 163]. PAT has also been applied for cystography (bladder imaging) following intraurethral administration of the nanocages as contrast agents (Fig. 5.16f–h) [164]. This administration route could lead to nanoparticle elimination during the first urination, effectively nullifying any toxicity risk. In-depth mechanisms and recent advancements of PAT can be found in the excellent review by Wang et al. [165].

5.4.2.2   Photothermal Imaging

The capability of light-to-heat conversion makes plasmonic nanostructures as appealing contrast agents for photothermal (PT) imaging, which relies on the photothermal effect of the analytes upon irradiation. Photothermal detection was first introduced by Kitamori et al. [166] using the thermal lens effect to detect trace amount of small molecule light absorbers, possibly to the level of single molecule. The technique was then further advanced by coupling the thermal lens with polarization interference to image AuNPs with diameters down to 2.5 nm [167]. PT imaging overcomes the limitation of Rayleigh scattering-based imaging (e.g. darkfield microscopy, 5.4.2.3), which typically requires particle size large than 40 nm.

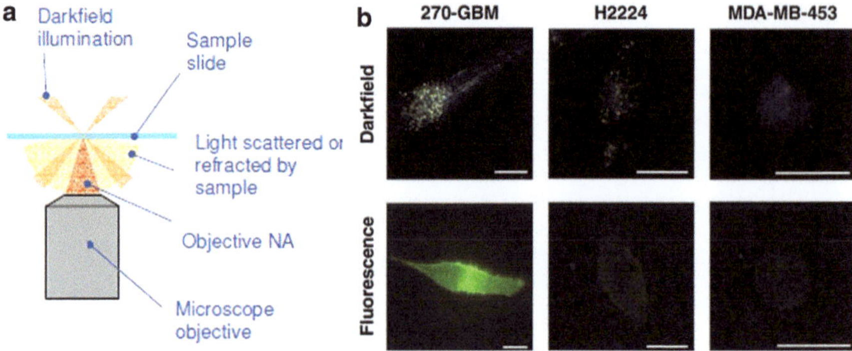

Fig. 5.17 Darkfield microscopy: **a** schematic illustration of darkfield microscope set up; and **b** images of three different cell lines with high, low, or no expression of EGFR: darkfield images were acquired after incubation with anti-EGFR labeled Au nanoparticles, and fluorescence images were acquired after incubation with labeled anti-EGFR antibody. (Modified from references (**a**) [172] and (**b**) [175])

Because of the large absorption cross-section of a AuNP, at least two orders of magnitude higher than that of a small molecule absorber, photon energy is efficiently converted to heat at extremely low particle concentration. Thus, irradiation doses well below the safe standard for medical lasers can be used for PT imaging [168].

PT imaging, consisting of a PT probe and sensitizer, is ideal to trace metal nanoparticles within the cells. This technique was pioneered by Zharov et al. [107]. Typically, two laser pulses are used with an initial pulse followed by a probe ~ 50 ns later. Subtraction of these two signatures has been used to image AuNPs in biomimetic media. Photothermal imaging can be coupled with flow cytometry, and this technique has been used to visualize 20-nm AuNPs in the lymphatic system [169]. This imaging modality could be adapted to monitor circulating cells targeted with AuNP for diagnostic or therapeutic purposes.

5.4.2.3 Darkfield Microscopy

The tunable optical properties and favorable surface chemistry of Au nanostructures make them attractive agents for immunotargeted labeling of cells and tissues [170, 171]. An additional imaging modality, darkfield microscopy, is particularly sensitive to the strong optical scattering of plasmonic nanostructures. Darkfield microscopy is a technique which employs the use of a specially designed condenser annulusto block the central portion of the illumination beam normally used in the conventional bright field configuration (Fig. 5.17a) [172]. The resulting ring of illumination will, if configured properly, illuminate the sample such that the cone of light falls outside the cone of acceptance of the objective lens; therefore, this method of imaging is useful only for relatively low numerical aperture objectives. A nonscattering sample appears black like the background; however, should the incoming photons undergo a scattering event the angle of the scattered photon may fall within the acceptance cone of the objective lens. Scattering materials appear as discrete, bright objects against a black background; this high signal-background ratio is ideal for detection of relatively low numbers of scattering particles [172].

Darkfield microscopy of targeted gold nanostructures has been successfully demonstrated using antibody-directed binding to selected cellular targets. Extracellular receptors, such as epidermal growth factor receptor (EGFR), are highly overexpressed in certain epithelial cancers, such as squamous carcinoma of the oral cavity and breast adenocarcinoma [173]. Several groups have successfully demonstrated darkfield imaging of immortalized human cancer cell lines using targeted Au nanostructures as scattering optical contrast agents (Fig. 5.17b) [174, 175]. This method is highly promising as a molecular imaging technique to specifically detect abnormally overexpressed cellular receptors in dysplastic cells, a potential method for early detection of cancers in epithelial tissues.

5.4.2.4   Optical Coherence Tomography (OCT)

Optical coherence tomography is a scattering-based optical imaging modality that is also referred to as "optical ultrasound." OCT works by reading the optical signal backscattered by opaque samples using an interferometer. This technique utilizes the wavelength of the light to remove background scattering noise and only read photons that are scattered 180° toward the source (Fig. 5.18a and b) [176]. Compared with ultrasound, OCT can provide high resolution images of tissue (axial image pixel resolution as high as 1.6 μm) using nonionizing radiation without invasive procedures. However, the technique is limited by the penetration depth of light through tissue. The theory underlying OCT and its clinical applications have been discussed in a thorough review by Boppart et al. [177].

Plasmonic nanostructures are appealing as contrast agents for OCT because their strong light scattering capability in the NIR region. The scattering cross-section of Au nanostructures (e.g. nanocages) is five orders of magnitude larger than that of dye molecules. Many Au nanostructures with immune-labeling have been demonstrated to enhance OCT contrast at the cellular level for cancer cell differentiation including Au nanoshells, nanocages, and nanorods [178–181]. In addition to cancer diagnosis, OCT has been widely-used in ophthalmology, because it enables high spatial resolution to visualize the back of the eye [182]. Using contrast agents such as Au nanorods, the OCT image contrast of the eyes in living mice could be improved nearly 50-fold higher than the control mice, providing more information of the ocular structures (Fig. 5.18c and d) [183]. Therefore, the use of plasmonic nanostructures in OCT can potentially overcome the limitation of intrinsic tissue scattering, providing structural information at the region of interest.

Fig. 5.18 Optical coherence tomography: **a** Schematic of setup with **b** sample depth-based output; **c** AuNP concentration dependence of contrast; and **d** optical coherence tomography images of mouse cornea in vivo with varying concentration of Au nanorods. (Modified from references (**a**, **b**) [176] and (**c**, **d**) [183])

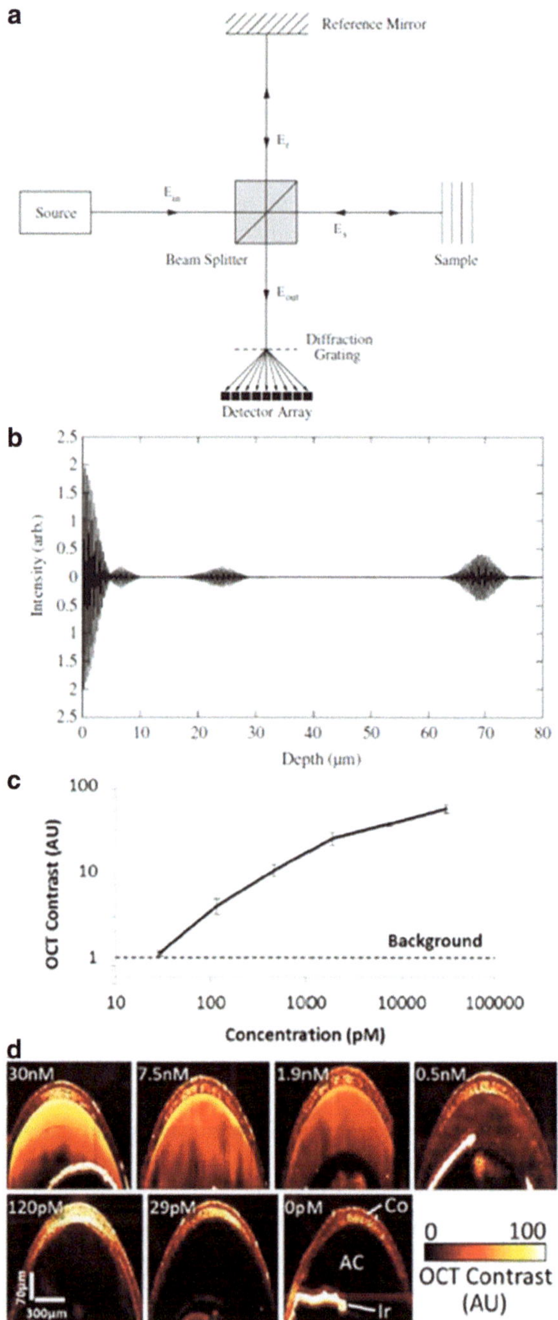

5.4.2.5 Raman Imaging and Mapping

Enhancement of the Raman scattering signal by plasmonic nanostructures means that Raman mapping can be utilized with greater sensitivity than fluorescence microscopy [184]. The two primary imaging methodologies are "label-free" and "labeled," which refer a structure without or with a Raman reporter conjugated to the particle, respectively (Fig 5.19a–e) [185]. Using a label-free system allows the detection of various endogenous components by their unique Raman signals. However, it may be compromised by overlapping Raman signatures from different components in a living system. Therefore, the labeled system is widely-used with Raman reporters having a known spectral signature or "fingerprint."

A SERS label for Raman imaging and mapping contains a metal core, adsorbed reporter, and targeting agent. This basic strategy has been used to identify and image cancer cells in vitro and in vivo [186]. SERS is suitable for multiplexed imaging because unique reporters and affinity agents can be coupled to the metal core for identification of different biomarkers in an entity [187]. On the other hand, SERS has also been applied to track the vesicles containing plasmonic nanostructures as they are taken up and trafficked within living cells (Fig. 5.19f–k) [188].

5.4.2.6 Multiphoton Luminescence

Multiphoton irradiation is advantageous over traditional single photon illumination for biological imaging because of the deeper tissue penetration at longer wavelengths, low autofluorescence from the biological sample and reduced photobleaching if a dye molecule is used [189]. Compared with higher order events, two-photon luminescence is well established and understood. For small molecule fluorophores, two-photon absorption is effectively a simultaneous occurrence at which two photons of the appropriate energy must be absorbed within the same 10^{-18} s to reach the excited state and emit fluorescence [189]. In the case of plasmonic nanostructures, two-photon luminescence occurs by emitting photons after a sequential, two-photon absorption process with a longer-lived intermediary state, on the order of 1 ns [190]. Two-photo luminescence has been used to track Au nanostructures in biological system [191–193]. Two-photon excitation at ~ 800 nm may overlap with the LSPR of the Au nanostructures, causing photothermal damage. The use of three-photon [194] and four-photon [195] schemes have been demonstrated to generate photoluminescence while overcome the photothermal effect.

5.4.3 Sensing Applications

Plasmonic nanostructures are also widely used for *ex vivo* chemical and biological sensing applications because of their unique optical properties. In this subsection, we discuss four plasmonically mediated, optical-based sensing techniques: agglomeration-, LSPR-, Raman-, and fluorescence-based detection.

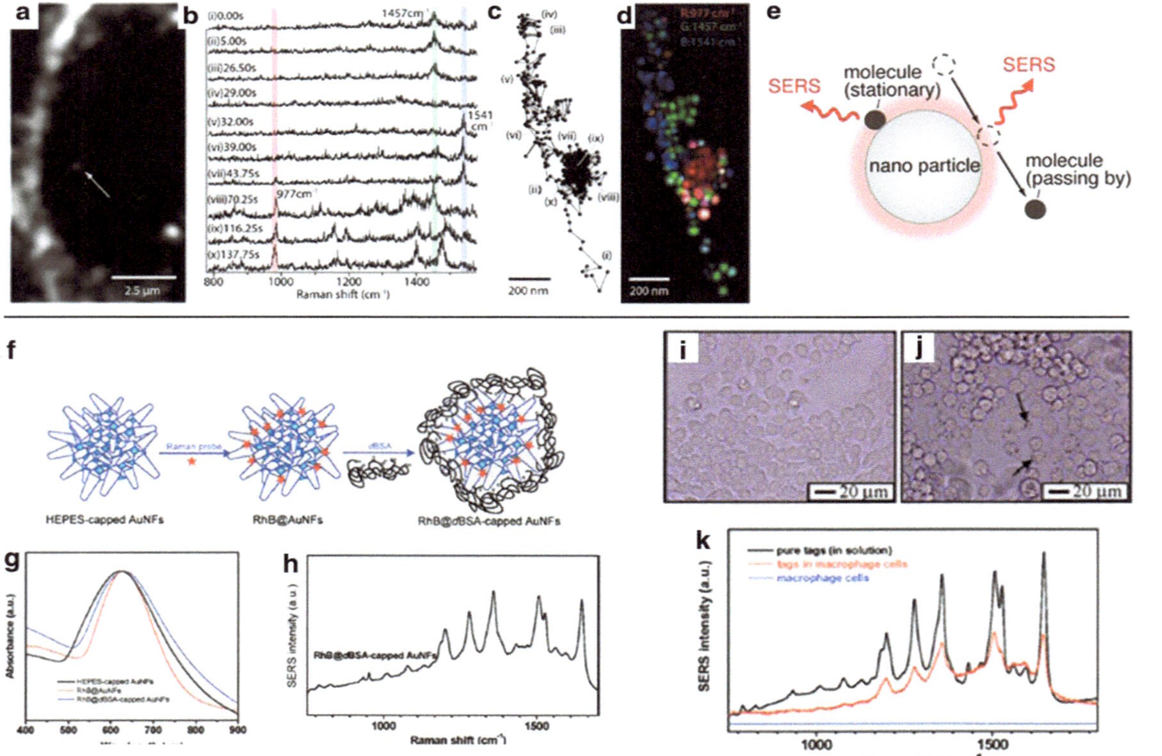

Fig. 5.19 Raman imaging: pH Indicative label free Raman detection of Au nanoparticles in endosomes shown on **a** darkfield **b** with spectral mapping **c** tracking the particle, and **d** mapping the pH; **e** schematic of the origin of label-free SERS signal; **f** schematic of fabrication of SERS label using Au nanostars, fluorophores, and protein with **g** UV-Vis and **h** SERS spectra as well as brightfield images of macrophages incubated **i** without and **j** with Raman active Au nanostars and **k** corresponding spectra. (Modified from references (**a**–**e**) [185] and (**f**–**k**) [188])

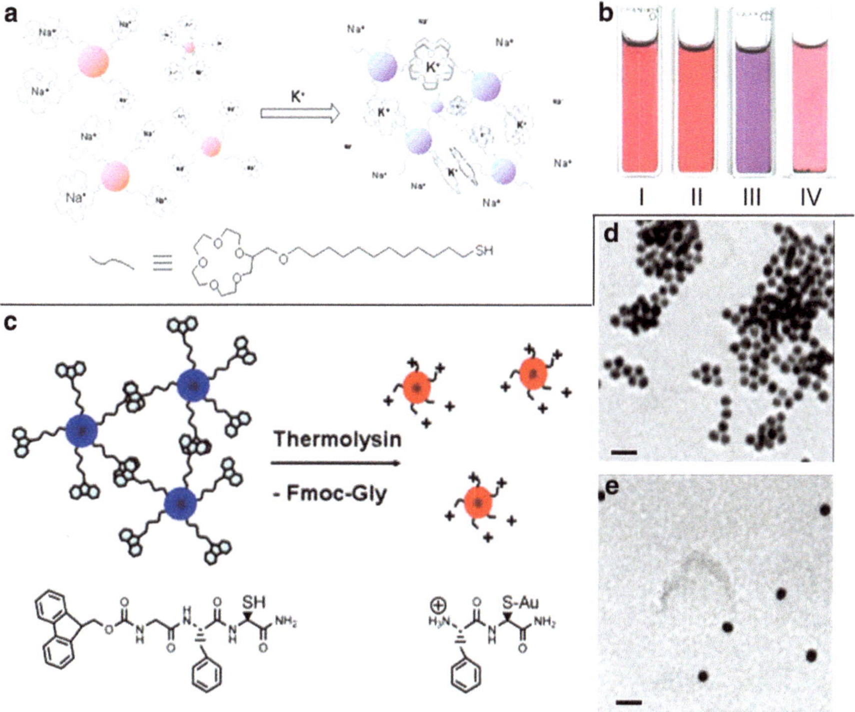

Fig. 5.20 Agglomeration-based sensing: **a** schematic of AuNPs conjugated with 15-crown-5 ether as a selective K$^+$ sensor; **b** color photos of the sensor incubated (I) without ions, (II) with Na$^+$, and with Na$^+$ and K$^+$ (III) immediately after and (IV) 1 day after addition of K$^+$; **c** schematic of thermolysin responsive AuNP assembly where the presence of thermolysin causes release of AuNPs from the assembly; TEM images of the assembly **d** without and **e** with thermolysin. (Modified from references (**a, b**) [197] and (**c–e**) [200])

5.4.3.1 Agglomeration-Based Detection

The LSPR of AuNP shifts dramatically as particles agglomerate which can be detected visually or by a simple colorimetric technique. This method was first introduced by Mirkin et al. [196] for the screening of particular oligonucleotides using AuNP which were conjugated with the complementary DNA strands of the analytes. Following this conceptual demonstration, various types of chelating ligands were conjugated to the surface of AuNPs for detection of corresponding analytes. Basically, the presence of a particular analyte causes the ligands on different AuNPs to bind the same analyte, resulting in the formation of a nanoparticle network and a subsequent shift of the plasmonic peak. In addition to multidentate binding, careful selection of chelating ligands is required to increase the sensitivity and selectivity.

The agglomeration-based sensing technique has then been widely used to detect different types of physiological important analytes in aqueous solution. For example, AuNPs functionalized with 15-crown-5 were used to detect K$^+$ and a number of other cations in the concentration range of µM to mM (Fig. 5.20a and b) [197].

AuNPs terminated with carboxylic acid group were employed to detect a number of toxic heavy metal ions in aqueous solution [198]. More complex ligands were considered for the detection of anions and small molecules [199]. In addition to small molecule ligands, biomolecules can be used as chelating compounds (Fig. 5.20c–e) [200]. Thymine bases in DNA can base pair with each other in the presence of Hg^{2+} ions, enabling a selective sensor for Hg^{2+} [201]. The reverse action, AuNP deagglomeration, was also demonstrated for colorimetric detection. A system of networked hydrogen bonds on AuNPs was disrupted in the presence of Pb^{2+}, leading to a quantifiable deagglomeration of the nanoparticles [202]. Aptamers have demonstrated particular applicability for detection of small molecules with high selectivity [203], even in complex environment as a "dipstick" to test blood for the presence of particular small molecules [204]. Apart from detection of ions and small molecules, proteins can be identified using this colorimetric technique by functionalizing AuNPs with carbohydrates [205] or specific, enzyme-cleavable sequences [206].

5.4.3.2   LSPR-Based Detection

Rather than monitor the significant shift between primary and agglomerated AuNPs, small shifts in the LSPR can be used to detect binding of small molecules, based on the change of the local dielectric constant surrounding the nanoparticles (Fig. 5.21a–c) [207]. Typically these sensors exist in the form of nanoparticles immobilized on a substrate and conjugated with binding moieties. In solution, AuNPs are typically conjugated with antibodies and as ligands bind, a shift in the plasmon is observed [208]. The signal can be amplified by interactions between the LSPR and the metal film used as a substrate for surface plasmon resonance measurements [209]. Utilization of two unique antibodies can significantly enhance signal as a result of significant increase of the plasmonic shift. In this case, the primary antibody is bound to the surface of a thin gold film which is exposed to ligands and then to a secondary antibody conjugated to AuNPs [210]. A similar strategy was employed using complementary nucleotides, with half the complement conjugated to the film and the other half attached to the nanoparticles [211].

Instead of exploiting the LSPR shift that results from binding, the LSPR shift as a result of morphological changes can also be utilized for chemical sensing of peroxides and other small molecules. The peroxides are scavenged, acting as a reducing agent for metal ions which can be deposited on preexisting metal sites on a substrate [212]. The growth patterns vary with peroxide concentrations, resulting in significant shifts in the LSPR. The plasmonic shift is inversely proportional to the concentration of H_2O_2 concentration, thereby providing extremely high sensitivity and low detection limits (Fig. 5.21d) [213]. A similar system was exploited that utilized the generation and growth of AuNPs as a response to different neurotransmitters [214].

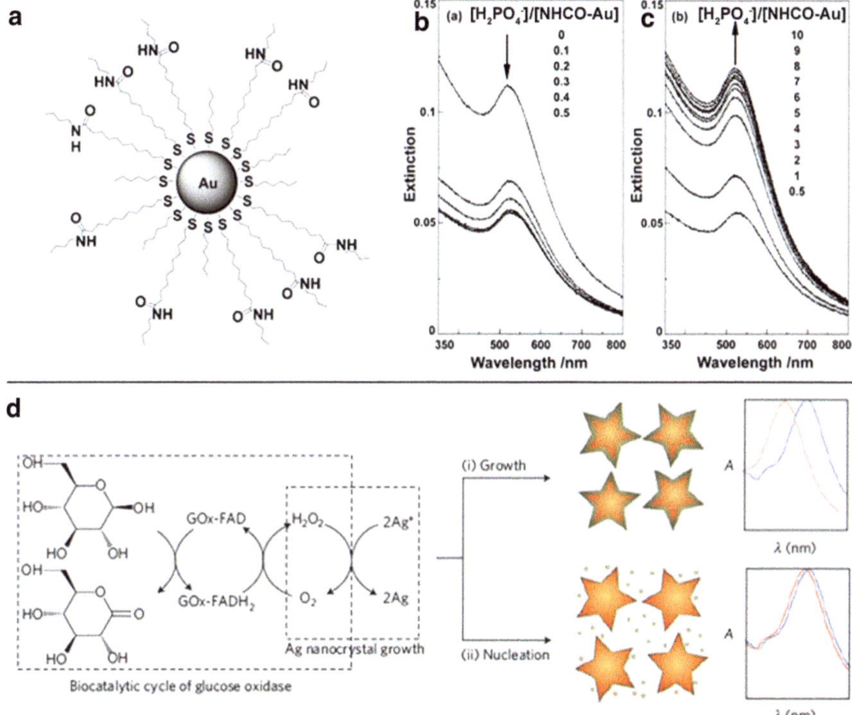

Fig. 5.21 Localized surface plasmon resonance-based sensing: **a** schematic illustration of AuNP surface coating for use as an anion sensor; **b, c** LSPR shift of the sensor in the presence of increasing concentrations of $H_2PO_4^-$; **d** catalytic cycle resulting in hydrogen peroxide generation and reduction of Ag^+ to Ag metal; at *(i)* low concentration, Ag plates the Au nanostars, resulting in a clear shift in LSPR, while at (ii) high concentration, Ag self-nucleates, forming independent AgNPs that do not appreciably affect the LSPR of Au nanostars. (Modified from references (**a–c**) [207] and (**d**) [213])

5.4.3.3 Raman-Based Sensing

Plasmonic nanostructures are particularly apt for Raman-based sensing because of the massive enhancement factor that can be achieved with precisely shaped particles. When a Raman-active dye is adsorbed to the surface of the plasmonic nanoparticles, fluorescence is quenched, minimizing its competing signal. Raman-based detection can be performed with label-free assays or Raman reporters. Label-free detection is often utilized for the detection of small molecules. The analytes induce agglomeration of nanoparticles, which in turn results in the generation of hot spots and a detectable Raman signal from the analytes. This methodology has been used with AuNPs to monitor the pH of intracellular compartments [215]. Similar

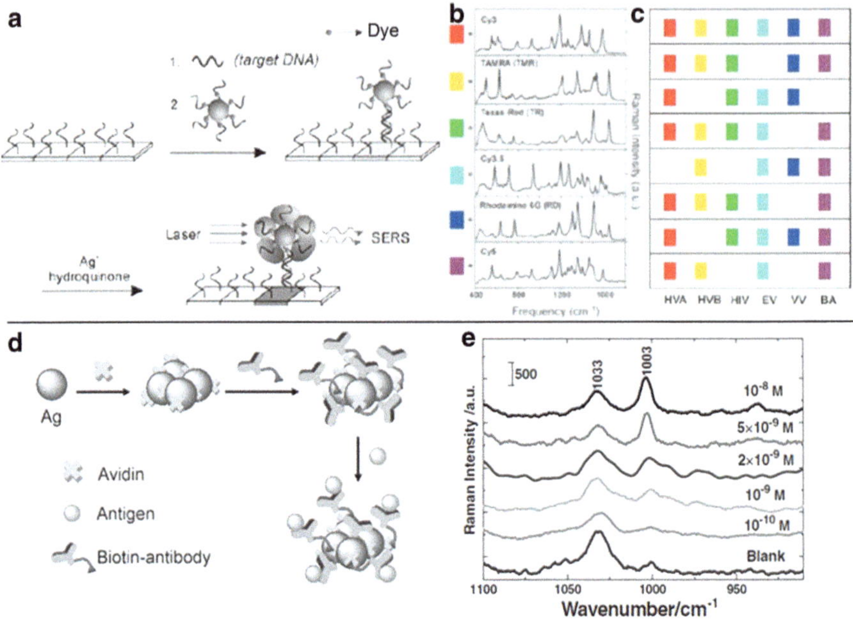

Fig. 5.22 Raman-based detection: **a** schematic of Raman-based chemical nose system using **b** various dyes with unique SERS spectra and **c** analysis of various pathogens to determine the presence of these genes simultaneously from a complex sample; **d** schematic of label-free system using antibody coated Ag agglomerates where **e** ligand binding results in new, concentration-dependent Raman signal. (Modified from references (**a–c**) [219] and (**d**) [221])

applications have been demonstrated with AgNPs which provide greater Raman enhancement due to the strong scattering properties of Ag, however, the instability of Ag nanostructures remains a concern [216].

Raman reporters can be conjugated to plasmonic nanoparticle surface for multiplexed detection. Characteristic peaks of Raman reporters are well-known and their vibrational fingerprints have been catalogued [217]. Several strategies have been demonstrated for detection of biomolecules. For example, Raman-labeled oligonucleotides adsorbed to the AuNP surface yield strong SERS signal of the reporter for oligonucleotide detection [218]. Gold nanoparticles conjugated with single-stranded oligonucleotides through a Raman reporter as a linker were used as SERS probes to bind substrates through the complementary strand, followed by detecting the Raman signals after enhancement (Fig. 5.22a–c) [219]. By utilizing distinct Raman reporters, this methodology has been demonstrated to quantify up to eight oligonucleotide simultaneously [220]. Proteins are often detected using a sandwich approach wherein a Raman reporter is conjugated to the surface of a particle (or film), and a primary antibody conjugated above the reporter (Fig. 5.22d and e) [221]. A secondary antibody is conjugated to free nanoparticles, protein immobilization by the primary antibody is followed by binding of the nanoparticle-bound secondary antibody, which generates a hot spot that enhances the Raman signal [222].

5.4.3.4 Fluorescence-Based Sensing

Gold nanoparticle-fluorophore-based sensing can be categorized two ways: "conformation-based" and "release-based" sensing. Conformation-based sensing results from the quenching of fluorophores near the particle surface through a FRET-like process [84]. Because the quenching efficiency is distance dependent, designing the system so the fluorophore-to-nanoparticle spacing changes in the presence of a particular molecule is paramount for effective sensing. Typically, a self-complementary oligonucleotide conjugated to a fluorophore is utilized to provide a linker of known length. Binding with DNA causes rigidity of the oligonucleotide and separation from the Au surface, resulting in the recovery of fluorescence (Fig. 5.23a) [223]. Most fluorescence-based detection, however, relies on the deconjugation of the fluorophore from the nanoparticle surface.

Release from the nanoparticle surface can be achieved in many ways, depending on the type of the conjugation. For example, conjugating quantum dots to biotinylated AuNPs through streptavidin results in separation of the conjugate in the presence of avidin, and recovery of fluorescence [224]. Alternatively, a fluorophore labeled oligonucleotide could be conjugated to a mostly complementary strand on the AuNPs surface, but then displaced by a strand with higher binding affinity [225]. Similarly, green-fluorescence protein (GFP) can be noncovalently conjugated to AuNP; and competitive binding between analyte proteins and the GFP results in GFP displacement and the recovery of fluorescence (Fig. 5.23b) [226]. These techniques were multiplexed toward the development of a "chemical nose." The nose consisted of AuNPs that had been monolayer coated with six different cationic termini followed by introduction of an anionic, fluorescent polymer that bound electrostatically. Introduction of proteins led to competitive binding with the fluorescent polymer, but different proteins showed different affinities for different proteins, resulting in a chemical fingerprint that can be used to identify proteins [227].

5.5 Conclusion and Outlook

Noble metal nanostructures interact with incident light and give rise to LSPR at particular wavelengths. Their strong plasmonic properties, together with their biocompatibility, make noble metal nanostructures, in particular Au nanostructures, appealing for biomedical and sensing applications. The LSPR is the sum of absorption and scattering of the incoming photons. Absorption leads to the photothermal effect, photoluminescence, and quenching of fluorophore in close proximity, while scattering results in emitted photons and amplifies the local electromagnetic field, enhancing fluorescence, phosphorescence, and Raman scattering.

The optical properties make plasmonic nanostructures ideal candidates for theranostic applications. Early studies have demonstrated Au nanostructures with LSPR in the NIR region, especially nanoshells, nanocages, and nanorods, for molecular imaging, photothermal therapy, and drug delivery. Recent advancements emphasize

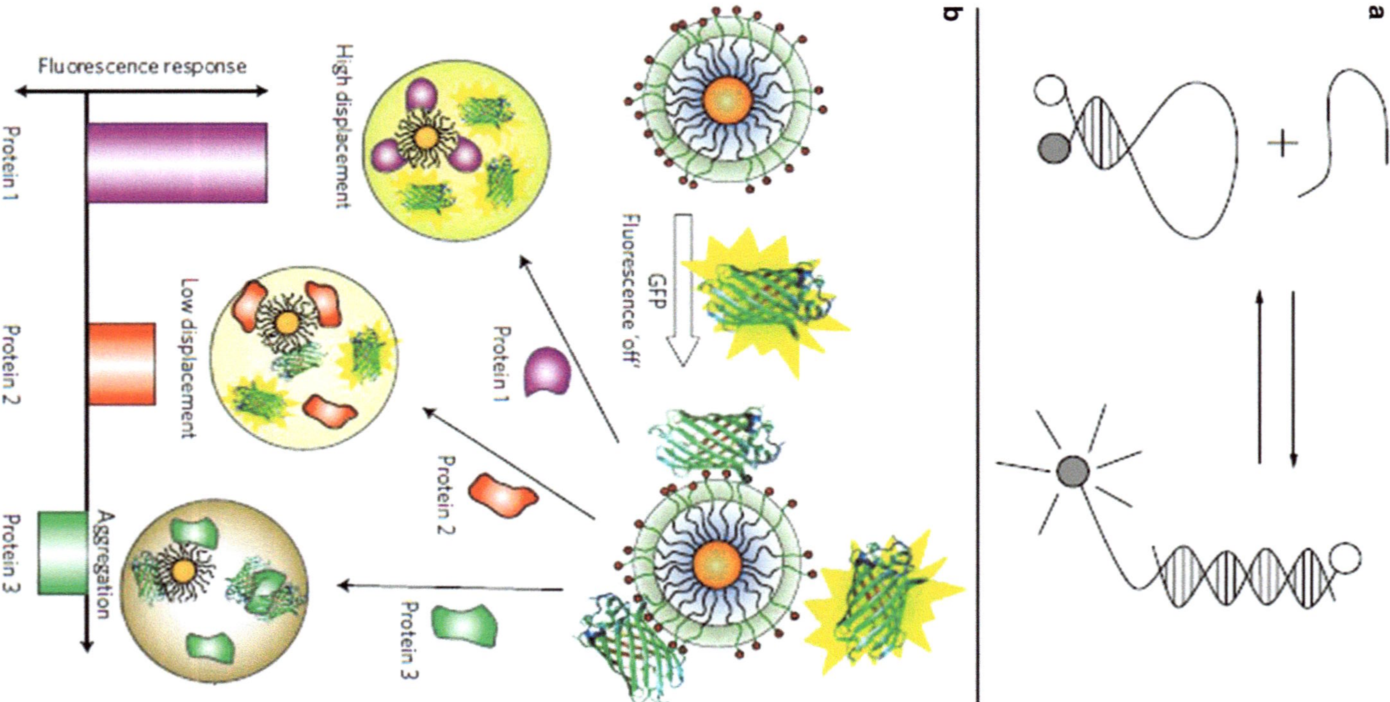

Fig. 5.23 Fluorescence-based detection: **a** schematic illustration of a turn-on fluorescence sensor using an RNA hairpin to keep the fluorophore in close proximity to the particle—binding of the complement strand results in a rigid double helix, separating the fluorophore from the particle and recovering fluorescence; **b** schematic illustration of GFP conjugated AuNP—protein binding displaces GFP in a quantifiable way that can be used to identify proteins in a sample based on the change in fluorescence response. (Modified from reference (**a**) [223] and (**b**) [226])

the integration of therapeutics and diagnostics with multimodal methodologies such as multimodal imaging, therapy, and imaging-guided drug delivery. *Ex vivo*, optical-based detection for of chemicals and biomolecules using plasmonic nanostructures is achieved using colorimetric, fluorescence, and Raman scattering techniques. The goal is to develop simple, robust, highly sensitive and highly selective techniques for analyte detection. More formulations and technologies are anticipated to soon be translated from bench top to clinical setting. Special attention ought to be paid to translational aspects such as biocompatibility, portability, and cost during the product and technology development.

Acknowledgment The author acknowledges the support from NIH/NCI R03 CA182052–01, NIH P30 GM103450, Arkansas Biosciences Institute, and startup from the University of Arkansas.

References

1. National Nanotechnology Initiative Website, in http://www.nano.gov
2. A.S. Thakor, J. Jokerst, C. Zavaleta, T.F. Massoud, S.S. Gambhir, Nano Lett **11**, 4029–4036, (2011)
3. M.W. Whitehouse, Inflammopharmacology **16**, 107–109, (2008)
4. C.L. Brown, G. Bushell, M.W. Whitehouse, D. Agrawal, S. Tupe, K. Paknikar, E.R. Tiekink, Gold Bull **40**, 245–250, (2007)
5. J.W. Alexander, Surgical infections **10**, 289–292, (2009)
6. M. Faraday, Philos Tr R Soc **147**, 145–181, (1857)
7. K.A. Wilets, R.P.V. Duyne, Annu Rev Phys Chem **58**, 267–297, (2007)
8. K. Aslan, J.R. Lakowicz, C.D. Geddes, Curr Opin Chem Biol **9**, 538–544, (2005)
9. G. Mie, Ann Phys-Berlin **330**, 377–445, (1908)
10. S. Link, Z.L. Wang, M.A. El-Sayed, J Phys Chem B **103**, 3529–3533, (1999)
11. D.B. Pedersen, S. Wang, J Phys Chem C **111**, 17493–17499, (2007)
12. W. Haiss, N.T.K. Thanh, J. Aveyard, D.G. Fernig, Anal Chem **79**, 4215–4221, (2007)
13. P.N. Njoki, I.I.S. Lim, D. Mott, H.-Y. Park, B. Khan, S. Mishra, R. Sujakumar, J. Luo, C.-J. Zhong, J Phys Chem C **111**, 14664–14669, (2007)
14. R. Weissleder, Nat Biotechnol **19**, 316–317, (2001)
15. A.M. Smith, M.C. Mancini, S. Nie, Nature nanotechnology **4**, 710–711, (2009)
16. T. Pham, J.B. Jackson, N.J. Halas, T.R. Lee, Langmuir **18**, 4915–4920, (2002)
17. N.R. Jana, L. Gearheart, C.J. Murphy, J Phys Chem B **105**, 4065–4067, (2001)
18. B. Nikoobakht, M.A. El-Sayed, Chem Mater **15**, 1957–1962, (2003)
19. S.E. Lohse, C.J. Murphy, Chem Mater **25**, 1250–1261, (2013)
20. Y. Sun, Y. Xia, J Am Chem Soc **126**, 3892–3901, (2004)
21. E.C. Dreaden, A.M. Alkilany, X. Huang, C.J. Murphy, M.A. El-Sayed, Chem Soc Rev **41**, 2740–2779, (2012)
22. S.E. Skrabalak, L. Au, X. Li, Y. Xia, Nat Protoc **2**, 2182–2190, (2007)
23 J. Zhao, A.O. Pinchuk, J.M. McMahon, S. Li, L.K. Ausman, A.L. Atkinson, G.C. Schatz, Acc Chem Res **41**, 1710–1720, (2008)
24. A.O. Govorov, H.H. Richardson, Nano Today **2**, 30–38, (2007)
25. H. Chen, L. Shao, T. Ming, Z. Sun, C. Zhao, B. Yang, J. Wang, Small **6**, 2272–2280, (2010)
26. H.H. Richardson, M.T. Carlson, P.J. Tandler, P. Hernandez, A.O. Govorov, Nano Lett **9**, 1139–1146, (2009)
27. M. Hu, G.V. Hartland, J Phys Chem B **106**, 7029–7033, (2002)
28. J. Chen, C. Glaus, R. Laforest, Q. Zhang, M. Yang, M. Gidding, M.J. Welch, Y. Xia, Small **6**, 811–817, (2010)

29. A. Mooradian, Phys Rev Lett **22**, 185–187, (1969)
30. G.T. Boyd, Z.H. Yu, Y.R. Shen, Phys Rev B **33**, 7923–7936, (1986)
31. R.A. Farrer, F.L. Butterfield, V.W. Chen, J.T. Fourkas, Nano Lett **5**, 1139–1142, (2005)
32. M.B. Mohamed, V. Volkov, S. Link, M.A. El-Sayed, Chem Phys Lett **317**, 517–523, (2000)
33. E. Dulkeith, T. Niedereichholz, T.A. Klar, J. Feldmann, G. von Plessen, D.I. Gittins, K.S. Mayya, F. Caruso, Phys Rev B **70**, 205424, (2004)
34. J. Zheng, P.R. Nicovich, R.M. Dickson, Annu Rev Phys Chem **58**, 409, (2007)
35. J.W. Strutt, Philos Mag **41**, 447–454, (1871)
36. J. Tyndall, Philos Mag **25**, 200–206, (1863)
37. K. Aslan, J.R. Lakowicz, C.D. Geddes, Anal Biochem **330**, 145–155, (2004)
38. J.J. Mock, M. Barbic, D.R. Smith, D.A. Schultz, S. Schultz, J Chem Phys **116**, 6755–6759, (2002)
39. C.J. Orendorff, T.K. Sau, C.J. Murphy, Small **2**, 636–639, (2006)
40. G.J. Nusz, S.M. Marinakos, A.C. Curry, A. Dahlin, F. Höök, A. Wax, A. Chilkoti, Anal Chem **80**, 984–989, (2008)
41. M.A. Mahmoud, M.A. El-Sayed, J Am Chem Soc **132**, 12704–12710, (2010)
42. E. Hao, G.C. Schatz, J Chem Phys **120**, 357–366, (2004)
43. L.J. Sherry, S.-H. Chang, G.C. Schatz, R.P. Van Duyne, B.J. Wiley, Y. Xia, Nano Lett **5**, 2034–2038, (2005)
44. H. Guo, F. Ruan, L. Lu, J. Hu, J. Pan, Z. Yang, B. Ren, J Phys Chem C **113**, 10459–10464, (2009)
45. L. Qin, S. Zou, C. Xue, A. Atkinson, G.C. Schatz, C.A. Mirkin, P Natl Acad Sci USA **103**, 13300–13303, (2006)
46. S.S. Aćimović, M.P. Kreuzer, M.U. González, R. Quidant, ACS Nano **3**, 1231–1237, (2009)
47. M.A. Mahmoud, J Phys Chem C **118**, 10321–10328, (2014)
48. C.V. Raman, K.S. Krishnan, Nature **121**, 501–502, (1928)
49. C.V. Raman, Indian J Phys **2**, 387–398, (1928)
50. D.A. Skoog, D.M. West, T.A. Holler, *Principles of Instrumental Analysis*, 5th edn. ed., Harcourt Brace & Co, Philadelphia, 1998
51. R.L. Garrell, Anal Chem **61**, 401A–411A, (1989)
52. M. Fleischmann, P.J. Hendra, A.J. McQuillan, Chem Phys Lett **26**, 163–166, (1974)
53. J. Kneipp, H. Kneipp, K. Kneippac, Chem Soc Rev **37**, 1052–1060, (2008)
54. N. Valley, N. Greeneltch, R.P. Van Duyne, G.C. Schatz, J Phys Chem Lett **4**, 2599–2604, (2013)
55. H. Xu, J. Aizpurua, M. Kall, P. Apell, Physical review E, Statistical physics, plasmas, fluids, and related interdisciplinary topics **62**, 4318–4324, (2000)
56. S.K. Saikin, Y. Chu, D. Rappoport, K.B. Crozier, A. Aspuru-Guzik, J Phys Chem Lett **1**, 2740–2746, (2010)
57. B.J. Kennedy, S. Spaeth, M. Dickey, K.T. Carron, J Phys Chem B **103**, 3640–3646, (1999)
58. X. Qian, J. Li, S. Nie, J Am Chem Soc **131**, 7540–7541, (2009)
59. G.J. Kovacs, R.O. Loutfy, P.S. Vincett, C. Jennings, R. Aroca, Langmuir 2, 689–694, (1986)
60. J.A. Creighton, Surf Sci **124**, 209–219, (1983)
61. H. Wei, H. Xu, Nanoscale **5**, 10794–10805, (2013)
62. K.L. Wustholz, A.-I. Henry, J.M. McMahon, R.G. Freeman, N. Valley, M.E. Piotti, M.J. Natan, G.C. Schatz, R.P.V. Duyne, J Am Chem Soc **132**, 10903–10910, (2010)
63. J. Hu, B. Zhao, W. Xu, Y. Fan, B. Li, Y. Ozaki, Langmuir **18**, 6839–6844, (2002)
64. S. Nie, S.R. Emory, Science **275**, 1102–1106, (1997)
65. E.C. Le Ru, P.G. Etchegoin, M. Meyer, J Chem Phys **125**, 204701, (2006)
66. D.A. Weitz, S. Garoff, J.I. Gersten, A. Nitzan, J Chem Phys **78**, 5324–5338, (1983)
67. P. Anger, P. Bharadwaj, L. Novotny, Phys Rev Lett (2005)
68. T.H. Tamican, F.D. Stefani, F.B. Segerink, N.F.v. Hulst, Nat Photonics **2**, 234–237, (2008)
69. H. Metiu, P. Das, Annu Rev Phys Chem **35**, 507–536, (1984)
70. F. Tam, G.P. Goodrich, B.R. Johnson, N.J. Halas, Nano Lett **7**, 469–501, (2007)
71. R. Bardhan, N.K. Grady, J.R. Cole, A. Joshi, N.J. Halas, ACS Nano **3**, 744–752, (2009)

72. S. Vukovic, S. Corni, B. Mennucci, J Phys Chem C **113**, 121–133, (2009)
73. S. Pan, Z. Wang, L.J. Rothberg, Mater Res Soc Symp Proc **818**, null-null, (2004)
74. R. Toftegaard, J. Arnbjerg, H. Cong, H. Agheli, D.S. Sutherland, P.R. Ogilby, Pure Appl Chem **83**, 885–898, (2011)
75. C. Schweitzer, R. Schmidt, Chem Rev **103**, 1685–1758, (2003)
76. Y. Zhang, K. Aslan, M.J.R. Previte, C.D. Geddes, J Fluoresc **17**, 345–349, (2007)
77. A. Srivatsan, S.V. Jenkins, M. Jeon, Z. Wu, C. Kim, J. Chen, R. Randey, Theranostics **4**, 163–174, (2014)
78. J.C. Ostrowski, A. Mikhailovsky, D.A. Bussian, M.A. Summers, S.K. Buratto, G.C. Bazan, Adv Funct Mater **16**, 1221–1227, (2006)
79. H. Mishra, B.L. Mali, J. Karolin, A.I. Dragan, C.D. Geddes, Phys Chem Chem Phys **15**, 19538–19544, (2013)
80. Y. Zhang, K. Aslan, M.J.R. Previte, S.N. Malyn, C.D. Geddes, J Phys Chem B **110**, 25108–25114, (2006)
81. M.J.R. Previte, K. Aslan, Y. Zhang, C.D. Geddes, J Phys Chem C **111**, 6051–6059, (2007)
82. R.M. Clegg, Curr Opin Biotech **6**, 103–110, (1995)
83. R. Chance, A. Prock, R. Silbey, Adv Chem Phys **37**, 65, (1978)
84. C.S. Yun, A. Javier, T. Jennings, M. Fisher, S. Hira, S. Peterson, B. Hopkins, N.O. Reich, G.F. Strouse, J Am Chem Soc **127**, 3115–3119, (2005)
85. T.L. Jennings, M.P. Singh, G.F. Strouse, J Am Chem Soc **128**, 5462–5467, (2006)
86. C.J. Breshike, R.A. Riskowski, G.F. Strouse, J Phys Chem C **117**, 23942–23949, (2013)
87. M. Li, S.K. Cushing, Q. Wang, X. Shi, L.A. Hornak, Z. Hong, N. Wu, J Phys Chem Lett **2**, 2125–2129, (2011)
88. M.P. Singh, T.L. Jennings, G.F. Strouse, J Phys Chem B **113**, 552–558, (2009)
89. H. Grönbeck, A. Curioni, W. Andreoni, J Am Chem Soc **122**, 3839–3842, (2000)
90. C.D. Bain, H.A. Biebuyck, G.M. Whitesides, Langmuir **5**, 723–727, (1989)
91. J.C. Love, L.A. Estroff, J.K. Kriebel, R.G. Nuzzo, G.M. Whitesides, Chem Rev **105**, 1103–1169, (2005)
92. C.D. Bain, G.M. Whitesides, J Am Chem Soc **111**, 7164–7175, (1989)
93. C.D. Bain, J. Evall, G.M. Whitesides, J Am Chem Soc **111**, 7155–7164, (1989)
94. J. Lipka, M. Semmler-Behnke, R.A. Sperling, A. Wenk, S. Takenaka, C. Schleh, T. Kissel, W.J. Parak, W.G. Kreyling, Biomaterials **31**, 6574–6581, (2010)
95. S.D. Perrault, C. Walkey, T. Jennings, H.C. Fischer, W.C.W. Chan, Nano Lett **9**, 1909–1915, (2009)
96. S. Nie, Nanomedicine **5**, 523–528, (2010)
97. A.Z. Wang, R. Langer, O.C. Farokhzad, Annu Rev Med **63**, 185–198, (2012)
98. S. Rana, A. Bajaj, R. Mout, V.M. Rotello, Adv Drug Deliver Rev **64**, 200–216, (2012)
99. A. Verma, F. Stellacci, Small **6**, 12–21, (2010)
100. C.H.J. Choi, C.A. Alabi, P. Webster, M.E. Davis, P Natl Acad Sci USA **107**, 1235–1240, (2010)
101. B.D. Chithrani, W.C.W. Chan, Nano Lett **7**, 1542–1550, (2007)
102. T.-G. Iversen, T. Skotland, K. Sandvig, Nano Today **6**, 176–185, (2011)
103. S.E. McNeil, J Leukocyte Biol **78**, 585–594, (2005)
104. A.E. Nel, L. Mädler, D. Velegol, T. Xia, E.M. Hoek, P. Somasundaran, F. Klaessig, V. Castranova, M. Thompson, Nat Mater **8**, 543–557, (2009)
105. X. Chen, S.T.C. Wong, *Introduction*, in: X. Chen, S. Wong (Eds.) Cancer Theranostics, Elsevier Science, Oxford, 2014, pp. 327–346
106. J. Chen, *Noble Metal Nanoparticle Platform*, in: X. Chen, S. Wong (Eds.) Cancer Theranostics, Elsevier Science, Oxford, 2014, pp. 327–346
107. V.P. Zharov, D.O. Lapotko, IEEE J Sel Topics Quantum Electron **11**, 733–751, (2005)
108. L.R. Hirsch, R.J. Stafford, J.A. Bankson, S.R. Sershen, B. Rivera, R.E. Price, J.D. Hazle, N.J. Halas, J.L. West, P Natl Acad Sci USA **100**, 13549–13554, (2003)
109. B.V. Harmon, A.M. Corder, R.J. Collins, G.C. Gobé, J. Allen, D.J. Allan, J.F.R. Kerr, Int J Radiat Biol **58**, 845–858, (1990)

110. J.M. Stern, J. Stanfield, W. Kabbani, J.-T. Hsieh, J.A. Cadeddu, J Urology **179**, 748–753, (2008)
111. W. Zhang, Z. Guo, D. Huang, Z. Liu, X. Guo, H. Zhong, Biomaterials **32**, 8555–8561, (2011)
112. L.R. Hirsch, R. Stafford, J. Bankson, S. Sershen, B. Rivera, R. Price, J. Hazle, N. Halas, J. West, P Natl Acad Sci USA **100**, 13549–13554, (2003)
113. S. Lal, S.E. Clare, N.J. Halas, Acc Chem Res **41**, 1842–1851, (2008)
114. X. Huang, I.H. El-Sayed, W. Qian, M.A. El-Sayed, J Am Chem Soc **128**, 2115–2120, (2006)
115. G. von Maltzahn, J.-H. Park, A. Agrawal, N.K. Bandaru, S.K. Das, M.J. Sailor, S.N. Bhatia, Cancer Res **69**, 3892–3900, (2009)
116. J. Chen, D. Wang, J. Xi, L. Au, A. Siekkinen, A. Warsen, Z.-Y. Li, H. Zhang, Y. Xia, X. Li, Nano Lett **7**, 1318–1322, (2007)
117. J. Chen, C. Glaus, R. Laforest, Q. Zhang, M. Yang, M. Gidding, M.J. Welch, Y. Xia, Small **6**, 811–817, (2010)
118. H. Yuan, A.M. Fales, T. Vo-Dinh, J Am Chem Soc **134**, 11358–11361, (2012)
119. J. Shao, R.J. Griffin, E.I. Galanzha, J.-W. Kim, N. Koonce, J. Webber, T. Mustafa, A.S. Biris, D.A. Nedosekin, V.P. Zharov, Sci Rep **3**, (2013)
120. V.P. Zharov, K.E. Mercer, E.N. Galitovskaya, M.S. Smeltzer, Biophysical journal **90**, 619–627, (2006)
121. R.S. Norman, J.W. Stone, A. Gole, C.J. Murphy, T.L. Sabo-Attwood, Nano Lett **8**, 302–306, (2008)
122. G.T. Hermanson, Bioconjugate Techniques, Elsevier Science, Oxford, 1996
123. H.C. Kolb, M.G. Finn, K.B. Sharpless, Angewandte Chemie (International ed in English) **40**, 2004–2021, (2001)
124. S.S. Agasti, A. Chompoosor, C.-C. You, P. Ghosh, C.K. Kim, V.M. Rotello, J Am Chem Soc **131**, 5728–5729, (2009)
125. W. Gao, J.M. Chan, O.C. Farokhzad, Mol Pharmaceut **7**, 1913–1920, (2010)
126. R. Liu, Y. Zhang, X. Zhao, A. Agarwal, L.J. Mueller, P. Feng, J Am Chem Soc **132**, 1500–1501, (2010)
127. X. Xia, M. Yang, L.K. Oetjen, Y. Zhang, Q. Li, J. Chen, Y. Xia, Nanoscale **3**, 950–953, (2011)
128. K. Knop, R. Hoogenboom, D. Fischer, U.S. Schubert, Angew Chem Int Ed **49**, 6288–6308, (2010)
129. P. Ghosh, G. Han, M. De, C.K. Kim, V.M. Rotello, Adv Drug Deliver Rev **60**, 1307–1315, (2008)
130. S. Sershen, S. Westcott, N. Halas, J. West, J Bio Mater Res A **51**, 293–298, (2000)
131. M.S. Yavuz, Y. Cheng, J. Chen, C.M. Cobley, Q. Zhang, M. Rycenga, J. Xie, C. Kim, K.H. Song, A.G. Schwartz, L.V. Wang, Y. Xia, Nat Mater **8**, 935–939, (2009)
132. G.D. Moon, S.-W. Choi, X. Cai, W. Li, E.C. Cho, U. Jeong, L.V. Wang, Y. Xia, J Am Chem Soc **133**, 4762–4765, (2011)
133. L. Tian, N. Gandra, S. Singamaneni, ACS Nano **7**, 4252–4260, (2013)
134. A.S. Angelatos, B. Radt, F. Caruso, J Phys Chem B **109**, 3071–3076, (2005)
135. J. Huang, K.S. Jackson, C.J. Murphy, Nano Lett **12**, 2982–2987, (2012)
136. G. Wu, A. Mikhailovsky, H.A. Khant, C. Fu, W. Chiu, J.A. Zasadzinski, J Am Chem Soc **130**, 8175–8177, (2008)
137. N. Lozano, W.T. Al-Jamal, A. Taruttis, N. Beziere, N.C. Burton, J. Van den Bossche, M. Mazza, E. Herzog, V. Ntziachristos, K. Kostarelos, J Am Chem Soc **134**, 13256–13258, (2012)
138. K. Niikura, N. Iyo, Y. Matsuo, H. Mitomo, K. Ijiro, ACS Appl Mater Inter **5**, 3900–3907, (2013)
139. S. Yamashita, H. Fukushima, Y. Niidome, T. Mori, Y. Katayama, T. Niidome, Langmuir **27**, 14621–14626, (2011)

140. N. Narband, S. Tubby, I.P. Parkin, J. Gil-Tomas, D. Ready, S.P. Nair, M. Wilson, Curr Nanosci **4**, 409–414, (2008)
141. L. Vigderman, E.R. Zubarev, Adv Drug Deliver Rev **65**, 663–676, (2013)
142. H. Gu, P.L. Ho, E. Tong, L. Wang, B. Xu, Nano Lett **3**, 1261–1263, (2003)
143. A.N. Grace, K. Pandian, J Bionanoscience **1**, 96–105, (2007)
144. A. Nirmala Grace, K. Pandian, Colloid Surface A **297**, 63–70, (2007)
145. A.N. Brown, K. Smith, T.A. Samuels, J. Lu, S.O. Obare, M.E. Scott, Appl Environ Microbiol **78**, 2768–2774, (2012)
146. Y. Zhao, Y. Tian, Y. Cui, W. Liu, W. Ma, X. Jiang, J Am Chem Soc **132**, 12349–12356, (2010)
147. A. Rai, A. Prabhune, C.C. Perry, J Mater Chem **20**, 6789–6798, (2010)
148. A. Elbakry, A. Zaky, R. Liebl, R. Rachel, A. Goepferich, M. Breunig, Nano Lett **9**, 2059–2064, (2009)
149. A. Barhoumi, R. Huschka, R. Bardhan, M.W. Knight, N.J. Halas, Chem Phys Lett **482**, 171–179, (2009)
150. N.L. Rosi, D.A. Giljohann, C.S. Thaxton, A.K.R. Lytton-Jean, M.S. Han, C.A. Mirkin, Science **312**, 1027–1030, (2006)
151. P.S. Ghosh, C.-K. Kim, G. Han, N.S. Forbes, V.M. Rotello, ACS Nano **2**, 2213–2218, (2008)
152. Y. Ding, Z. Jiang, K. Saha, C.S. Kim, S.T. Kim, R.F. Landis, V.M. Rotello, Mol Ther **22**, 1075–1083, (2014)
153. A. Wijaya, S.B. Schaffer, I.G. Pallares, K. Hamad-Schifferli, ACS Nano **3**, 80–86, (2008)
154. A.C. Bonoiu, S.D. Mahajan, H. Ding, I. Roy, K.-T. Yong, R. Kumar, R. Hu, E.J. Bergey, S.A. Schwartz, P.N. Prasad, P Natl Acad Sci USA (2009)
155. G.B. Braun, A. Pallaoro, G. Wu, D. Missirlis, J.A. Zasadzinski, M. Tirrell, N.O. Reich, ACS Nano 3, 2007–2015, (2009)
156. L.V. Wang, S. Hu, Science **335**, 1458–1462, (2012)
157. Y. Wang, X. Xie, X. Wang, G. Ku, K.L. Gill, D.P. O'Neal, G. Stoica, L.V. Wang, Nano Lett **4**, 1689–1692, (2004)
158. X. Yang, S.E. Skrabalak, Z.-Y. Li, Y. Xia, L.V. Wang, Nano Lett **7**, 3798–3802, (2007)
159. E.C. Cho, C. Kim, F. Zhou, C.M. Cobley, K.H. Song, J. Chen, Z.-Y. Li, L.V. Wang, Y. Xia, J Phys Chem C **113**, 9023–9028, (2009)
160. K.H. Song, C. Kim, C.M. Cobley, Y. Xia, L.V. Wang, Nano Lett **9**, 183–188, (2008)
161. C. Kim, E.C. Cho, J. Chen, K.H. Song, L. Au, C. Favazza, Q. Zhang, C.M. Cobley, F. Gao, Y. Xia, L.V. Wang, ACS Nano **4**, 4559–4564, (2010)
162. B. Wang, E. Yantsen, T. Larson, A.B. Karpiouk, S. Sethuraman, J.L. Su, K. Sokolov, S.Y. Emelianov, Nano Lett **9**, 2212–2217, (2008)
163. K. Jansen, G. van Soest, A.F. van der Steen, Ultrasound Med Biol **40**, 1037–1048, (2014)
164. M. Jeon, S. Jenkins, J. Oh, J. Kim, T. Peterson, J. Chen, C. Kim, Nonionizing photoacoustic cystography with near-infrared absorbing gold nanostructures as optical-opaque tracers, in: Nanomedicine, in press, (2013)
165. C. Kim, C. Favazza, L.V. Wang, Chem Rev **110**, 2756–2782, (2010)
166. M. Tokeshi, M. Uchida, A. Hibara, T. Sawada, T. Kitamori, Anal Chem **73**, 2112–2116, (2001)
167. D. Boyer, P. Tamarat, A. Maali, B. Lounis, M. Orrit, Science **297**, 1160–1163, (2002)
168. V.P. Zharov, E.N. Galitovskaya, C. Johnson, T. Kelly, Laser Surg Med **37**, 219–226, (2005)
169. V.P. Zharov, E.I. Galanzha, V.V. Tuchin, Photothermal imaging of moving cells in lymph and blood flow in vivo, in: A.A. Oraevsky, L.V. Wang (Eds.) Proc SPIE 5320, Photons Plus Ultrasound: Imaging and Sensing, San Jose, 2004, pp. 185–195
170. H. Jans, X. Liu, L. Austin, G. Maes, Q. Huo, Anal Chem **81**, 9425–9432, (2009)
171. W. Eck, G. Craig, A. Sigdel, G. Ritter, L.J. Old, L. Tang, M.F. Brennan, P.J. Allen, M.D. Mason, ACS Nano **2**, 2263–2272, (2008)
172. A. Wax, K. Sokolov, Laser Photonics Rev **3**, 146–158, (2009)

173. J.C. Kah, K.W. Kho, C.G. Lee, C. James, R. Sheppard, Z.X. Shen, K.C. Soo, M.C. Olivo, Int J Nanomedicine **2**, 785–798, (2007)
174. A.C. Curry, M. Crow, A. Wax, BIOMEDO **13**, 014022-014022-014027, (2008)
175. M.J. Crow, G. Grant, J.M. Provenzale, A. Wax, AJR Am J Roentgenology **192**, 1021–1028, (2009)
176. P.H. Tomlins, R. Wang, J Phys D Appl Phys **38**, 2519, (2005)
177. A.M. Zysk, F.T. Nguyen, A.L. Oldenburg, D.L. Marks, S.A. Boppart, Bio Med **12**, 051403, (2007)
178. K. Sokolov, M. Follen, J. Aaron, I. Pavlova, A. Malpica, R. Lotan, R. Richards-Kortum, Cancer Res **63**, 1999–2004, (2003)
179. J. Chen, F. Saeki, B.J. Wiley, H. Cang, M.J. Cobb, Z.-Y. Li, L. Au, H. Zhang, M.B. Kimmey, Li, Y. Xia, Nano Lett **5**, 473–477, (2005)
180. C. Loo, A. Lin, L. Hirsch, M.-H. Lee, J. Barton, N. Halas, J. West, R. Drezek, Technol Cancer Res Treatment **3**, 33–40, (2004)
181. A.L. Oldenburg, M.N. Hansen, A. Wei, S.A. Boppart, Plasmon-resonant gold nanorods provide spectroscopic OCT contrast in excised human breast tumors, in: S. Achilefu, D.J. Bornhop, R. Raghavachari (Eds.) Proc SPIE 6867, Molecular Probes for Biomedical Applications II, San Jose, 2008, pp. 68670E-68670E-68610
182. C.A. Puliafito, M.R. Hee, C.P. Lin, E. Reichel, J.S. Schuman, J.S. Duker, J.A. Izatt, E.A. Swanson, J.G. Fujimoto, Ophthalmology **102**, 217–229, (1995)
183. A. de la Zerda, S. Prabhulkar, V.L. Perez, M. Ruggeri, A.S. Paranjape, F. Habte, S.S. Gambhir, R.M. Awdeh, Optical coherence contrast imaging using gold nanorods in living mice eyes, in: Clin Exp Ophthamol, accepted, (2014)
184. C.L. Haynes, A.D. McFarland, R.P.V. Duyne, Anal Chem **77**, 338 A-346 A, (2005)
185. J. Xie, Q. Zhang, J.Y. Lee, D.I.C. Wang, ACS Nano **2**, 2473–2480, (2008)
186. X. Qian, X.H. Peng, D.O. Ansari, Q. Yin-Goen, G.Z. Chen, D.M. Shin, L. Yang, A.N. Young, M.D. Wang, S. Nie, Nat Biotechnol **26**, 83–90, (2008)
187. C.L. Zavaleta, B.R. Smith, I. Walton, W. Doering, G. Davis, B. Shojaei, M.J. Natan, S.S. Gambhir, P Natl Acad Sci USA **106**, 13511–13516, (2009)
188. J. Ando, K. Fujita, N.I. Smith, S. Kawata, Nano Lett **11**, 5344–5348, (2011)
189. R. Carriles, D.N. Schafer, K.E. Sheetz, J.J. Field, R. Cisek, V. Barzda, A.W. Sylvester, J.A. Squier, Rev Sci Instrum **80**, 081101, (2009)
190. K. Imura, T. Nagahara, H. Okamoto, J Phys Chem B **109**, 13214–13220, (2005)
191. H. Wang, T.B. Huff, D.A. Zweifel, W. He, P.S. Low, A. Wei, J.-X. Cheng, P Natl Acad Sci USA **102**, 15752–15756, (2005)
192. N.J. Durr, T. Larson, D.K. Smith, B.A. Korgel, K. Sokolov, A. Ben-Yakar, Nano Lett **7**, 941–945, (2007)
193. L. Au, Q. Zhang, C.M. Cobley, M. Gidding, A.G. Schwartz, J. Chen, Y. Xia, ACS Nano **4**, 35–42, (2009)
194. L. Tong, C.M. Cobley, J. Chen, Y. Xia, J.-X. Cheng, Angew Chem Int Ed **49**, 3485–3488, (2010)
195. P. Biagioni, D. Brida, J.-S. Huang, J. Kern, L. Duò, B. Hecht, M. Finazzi, G. Cerullo, Nano Lett **12**, 2941–2947, (2012)
196. R. Elghanian, J.J. Storhoff, R.C. Mucic, R.L. Letsinger, C.A. Mirkin, Science **277**, 1078–1081, (1997)
197. S.Y. Lin, S.W. Liu, C.M. Lin, C.H. Chen, Anal Chem **74**, 330–335, (2002)
198. Y. Kim, R.C. Johnson, J.T. Hupp, Nano Lett **1**, 165–167, (2001)
199. P.D. Beer, P.A. Gale, Angew Chem Int Ed **40**, 486–516, (2001)
200. A. Laromaine, L. Koh, M. Murugesan, R.V. Ulijn, M.M. Stevens, J Am Chem Soc **129**, 4156–4157, (2007)
201. J.S. Lee, M.S. Han, C.A. Mirkin, Angew Chem Int Ed **46**, 4093–4096, (2007)
202. J. Liu, Y. Lu, J Am Chem Soc **125**, 6642–6643, (2003)
203. A. Ogawa, Bioorg Med Chem Lett **21**, 155–159, (2011)
204. H. Yang, H. Liu, H. Kang, W. Tan, J Am Chem Soc **130**, 6320–6321, (2008)

205. C.L. Schofield, A.H. Haines, R.A. Field, D.A. Russell, Langmuir **22**, 6707–6711, (2006)
206. C.-C. You, R.R. Arvizo, V.M. Rotello, Chem Commun 2905-2907, (2006)
207. S. Watanabe, M. Sonobe, M. Arai, Y. Tazume, T. Matsuo, T. Nakamura, K. Yoshida, Chem Commun 2866-2867, (2002)
208. P. Englebienne, A. Van Hoonacker, M. Verhas, Analyst **126**, 1645–1651, (2001)
209. E. Fu, S.A. Ramsey, P. Yager, Anal Chim Acta **599**, 118–123, (2007)
210. L.A. Lyon, M.D. Musick, M.J. Natan, Anal Chem **70**, 5177–5183, (1998)
211. L. He, M.D. Musick, S.R. Nicewarner, F.G. Salinas, S.J. Benkovic, M.J. Natan, C.D. Keating, J Am Chem Soc **122**, 9071–9077, (2000)
212. H. Li, X. Ma, J. Dong, W. Qian, Anal Chem **81**, 8916–8922, (2009)
213. L. Rodriguez-Lorenzo, R. de la Rica, R.A. Alvarez-Puebla, L.M. Liz-Marzan, M.M. Stevens, Nat Mater **11**, 604–607, (2012)
214. L. Shang, S. Dong, Nanotechnology **19**, 095502, (2008)
215. J. Kneipp, H. Kneipp, B. Wittig, K. Kneipp, J Phys Chem C **114**, 7421–7426, (2010)
216. P.C. Lee, D. Meisel, J Phys Chem **86**, 3391–3395, (1982)
217. K. Kneipp, H. Kneipp, I. Itzkan, R.R. Dasari, M.S. Feld, Chem Rev **99**, 2957–2976, (1999)
218. D. Graham, B.J. Mallinder, W.E. Smith, Angew Chem Int Ed **39**, 1061–1063, (2000)
219. Y.C. Cao, R. Jin, C.A. Mirkin, Science **297**, 1536–1540, (2002)
220. L. Sun, C. Yu, J. Irudayaraj, Anal Chem **79**, 3981–3988, (2007)
221. X.X. Han, L. Chen, W. Ji, Y. Xie, B. Zhao, Y. Ozaki, Small **7**, 316–320, (2011)
222. D.S. Grubisha, R.J. Lipert, H.Y. Park, J. Driskell, M.D. Porter, Anal Chem **75**, 5936–5943, (2003)
223. B. Dubertret, M. Calame, A.J. Libchaber, Nat Biotechnol 2001, (2001)
224. E. Oh, M.-Y. Hong, D. Lee, S.-H. Nam, H.C. Yoon, H.-S. Kim, J Am Chem Soc **127**, 3270–3271, (2005)
225. D.S. Seferos, D.A. Giljohann, H.D. Hill, A.E. Prigodich, C.A. Mirkin, J Am Chem Soc **129**, 15477–15479, (2007)
226. M. De, S. Rana, H. Akpinar, O.R. Miranda, R.R. Arvizo, U.H.F. Bunz, V.M. Rotello, Nat Chem **1**, 461–465, (2009)
227. M. De, C.-C. You, S. Srivastava, V.M. Rotello, J Am Chem Soc **129**, 10747–10753, (2007)

Chapter 6
Magnetic-Metallic Nanostructures for Biological Applications

Yanglong Hou, Jing Yu and Wenlong Yang

Abstract In this chapter, we first introduce some established synthetic protocols toward magnetic-metallic nanocrystals (NCs) composed of single component metals (e.g., iron, cobalt, and nickel) and alloys (e.g., FePt, CoPt, FeCo, FeNi, and magnetic carbides). We mainly focus on colloidal chemical approaches such as thermal decomposition of organometallic precursors, chemical reduction of metal salts, sol–gel methods, reverse micelles, polyol process, and synergetic synthetic process. Current methods to protect magnetic-metallic NCs from oxidation and chemically etching, as well as their diverse magnetic properties are also discussed. Some promising biological applications of magnetic-metallic NCs are illustrated, especially in cancer diagnostic and therapy including magnetic resonance imaging (MRI), magnetic hyperthermia (MH), and ion releasing for selectively killing cells.

6.1 Introduction

Magnetic-metallic nanocrystals (NCs) have attracted tremendous attention owing to their unique chemical and physical properties. Among various synthetic protocols, colloidal chemical synthesis is efficient in controlling the morphology and composition of expected magnetic nanostructures.

For colloidal magnetic nanoparticles with sizes < 100 nm stabilized by various surfactants, the particle size distribution (also denoted as uniformity), size control, shape control, self-assembly, and phase related crystallinity are the key points in their synthesis. Chemical methods, with their intrinsic characters to be able to produce final products in large quantities and reliable qualities, have long been a straightforward pathway toward the generation of monodispersed nanoparticles. There have been several preeminent review articles concerning the chemical processes toward magnetic NCs [1–6].

As discussed in Chap. 2, generally a burst of nucleation in a solution followed by a controlled supply of original materials is critical to produce uniform nanopar-

Y. Hou (✉) · J. Yu · W. Yang
College of Engineering, Peking University, 100871 Beijing, China
e-mail: hou@pku.edu.cn

Y. Xiong, X. Lu (eds.), *Metallic Nanostructures,* DOI 10.1007/978-3-319-11304-3_6

ticles. Therefore, the synthetic protocols to generate magnetic NCs often include the formation of abundant nuclei and the growth of particles. In such a process, organo-metallic compounds (usually assisted by reducing agents) which act as precursors are injected into a special solution containing surfactants to form abundant nuclei. In addition, reagents can also be mixed at low temperature and slowly heated to a stated temperature to generate nuclei. When providing reactive species unremittingly to the system in a mild manner, particle growth occurred (also, the particle growth could be mastered by Ostwald ripening, in which growth occurred in a manner that smaller particles dissolved and deposited on the larger ones). After a size-selective precipitation process often involving the addition of a polar solvent to the reaction system, various mono-dispersed magnetic NCs can be obtained. By adjusting the synthetic parameters, such as reaction time, temperature, the concentration of reactants and surfactants or co-surfactants, and reducing conditions, the particle sizes and morphologies could be systematically tuned.

In this chapter, we first introduce various chemical synthesis methods of magnetic-metallic nanostructures, including chemical synthesis of single component magnetic nanocrystals of iron, cobalt, and nickel with tunable sizes, shapes and compositions, and subsequently their alloys. Then we discuss promising biological applications of magnetic-metallic NCs, especially in cancer diagnostics and therapy based on magnetic resonance imaging (MRI), magnetic hyperthermia (MH), and ion releasing for killing cancer cells selectively.

6.2 Chemical Synthesis of Magnetic-Metallic Nanostructures

6.2.1 Nanocrytals of Iron, Cobalt, and Nickel

6.2.1.1 Iron Nanocrystals

Metallic iron has a high magnetization at room temperature [M_s(bulk) = 212 emu/g]. For iron NCs with sizes ranging from 1 to 12 nm, a superparamagnetic behavior is expected. From a physical point of view, these properties make iron nanoparticles an ideal choice for biomedical applications such as magnetic fluid hyperthermia and magnetic resonance imaging [7–10]. Unfortunately, iron NCs do not possess good stability, especially when exposed to water and oxygen. This weakness for its reactivity is multiplied if in nanoscale, where the iron rapidly and finally oxidized. However, as catalysts, the extreme reactivity of iron NCs could be beneficial in a nonoxidizing environment, such as the Fischer-Tropsch synthesis reaction, which converts syngas (a mixture of CO and H_2 produced by steam-reforming coal) to hydrocarbons at high temperature and pressure over an appropriate catalyst, where the use of iron primarily involves promoting the formation and cleavage of the C–C bonds [11].

Mentioning the preparation of well-defined iron NCs, enormous efforts have been made early from 1940s and 1950s, which were relied on dispersing iron in mercury. Luborsky investigated the forming of iron NCs via electrodeposition iron salts in mercury in detail [12]. The methods were critical and useful, which made the testing of properties of single-domain particles come true. Taking into account the substantial solubility of iron in mercury which is toxic in vapors, and relative ease of extraction of organic solvents, organic-solvents-based methodologies have been introduced. Recently, researchers are endeavoring to keep the iron NCs more stabilized against fast oxidating [13]. In most common cases, iron NCs are synthesized via thermal decomposition of iron pentacabonyl ([$Fe(CO)_5$], also denoted as carbonyl), sonochemical decomposition of iron carbonyl, reduction of iron salts and oxides, and thermal reduction of iron(II) bis(trimethylsilyl)amide precursor, etc.

The process leading to fine monodispersity of nanoparticles or to the formation of self-organized supercrystals in an organic or aqueous solutions has long been a conundrum for researchers to resolve [14]. As to iron NCs, a good understanding of the formation is a crucial point before the fascinating applications. Besides, how to avoid the high reactivity with dioxygen and water to make iron NCs more stable if unprotected remains untoward. Works have already been done to prepare iron nanoparticles with good quality, but the magnetizations of the iron NCs were much lower than that of the bulk especially at smaller sizes [2, 15–20].

Hyeon developed a general ultra-large scale solution phase synthesis of magnetic nanoparticles. Iron-oleate complex precursors were produced via reaction of $Fe(CO)_5$ and oleic acid surfactant at a mild temperature. As UV-visible absorption spectrum and FT–IR spectrum showed, assisted by Mössbauer spectroscopic results, that iron pentacarbonyl decomposed at 120 °C and demonstrated the iron(II) complex. After aging the iron complex at 300 °C, iron nanoparticles were generated. Besides, the as-synthesized particles revealed a *bcc* α-iron structure after being heated under argon atmosphere at 500 °C, and were transformed to γ-Fe_2O_3 nanocrsytallities via oxidizing at 300 °C with a mild oxidant trimethylamine oxide ([$(CH_3)_3NO$]). While Iron-oleate complex precursors were produced via reaction of $FeCl_3$ and oleic acid surfactant, uniform 20 nm Fe nanocubes passivated by thin FeO shell were obtained [17, 21].

Dumestre et al. prepared 7 nm nanocubes organized into superlattices using the amido precursor $Fe[N(SiMe_3)_2]_2(THF)$ (Me = CH_3, THF = tetrahydrofuran) in the presence of oleic acid and hexadecylamine (HDA). How to tune the growth process to control iron particle sizes and shapes remains a problem yet [22]. Chaudret et al. reported the synthetic steps toward unoxidized, small iron NCs of homogeneous size near 1.5 nm that exhibit bulk magnetization via the reduction of $\{Fe(N[Si(CH_3)_3]_2)_2\}_2$ under H_2 atmosphere [23]. While using a mixture of long-chain acid and amines as reaction solution, well-defined and larger iron NCs with bulk magnetization were successfully synthesized. Tunable size (from 1.5 to 27 nm) and shape (spheres, cubes, or stars) of Fe NCs were successfully synthesized via tailoring the reaction parameters including the reaction time, the temperature, and the surfactant concentrations [24]. On the basis of tuned synthetic parameters, the authors promoted the environment-dependent growth mechanisms schematically

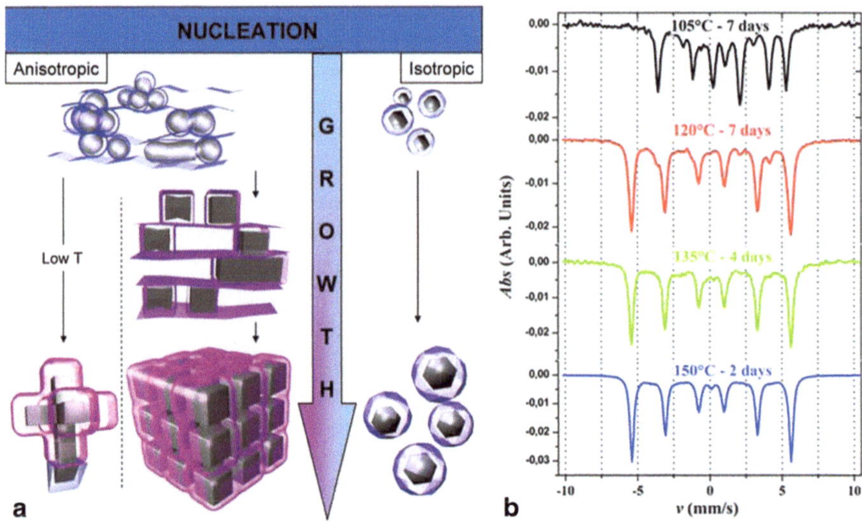

Fig. 6.1 **a** Schematic illustration of the two environment-dependent growth routes (nucleation in two different environments). The *blue* color indicated an amine-rich environment, and it represents an organic shell composition that preferentially stabilizes the {100} facets of the iron NCs. Reproduced with permission from [24]. Copyright 2008 American Chemical Society. **b** Mössbauer spectra of the iron NPs prepared after different reaction time and temperature. Reproduced with permission from [13]. (Copyright 2011 Royal Society of Chemistry)

represented in Fig. 6.1a, which involved the influence of the local acid concentration on each step of the NCs synthesis, especially on their faceting. The proposed growth protocols included nucleation, isotropic growth and coalescence, and oriented and repair mechanisms inside organic superstructures. Concerning the point of controlling the size and shape of iron NCs on a larger range, Chaudret et al. produced either ultra fine iron clusters of *ca.* 1.5 nm, polycrystalline spherical NCs with a mean diameter between 5 and 10 nm, or single-crystalline nanocubes more than 13 nm which were subsequently self-assembled into superlattices in micrometric scale successfully. The Mössbauer study also confirmed that complete Fe reduction into metallic Fe NPs can be reached at lower temperature when increasing the reaction time, as Fig. 6.1b shows. Interestingly, the magnetization of the final iron NCs was higher than 210 A m^2 kg^{-1} at 2 K (approximately equal to bulk iron). Besides, the authors reported that a dihydrogen free synthetic atmosphere may allow tuning the NCs size within quite large range and keeping high magnetization value associated with metallic iron meanwhile [13].

Iron pentacarbonyl ([Fe(CO)$_5$]), also denoted as iron carbonyl, is a metastable organometallic compound with the standard enthalpy of formation of *ca.* −185 kcal mol^{-1}, whose facile decomposition has made it an vital reagent for various iron-based NCs [25,26]. As been reported, iron pentacarbonyl has long been used extensively as a gaseous precursor to produce iron catalyst nanoparticles, specially in the flame synthesis, in the HiPco (High-pressure CO conversion) process,

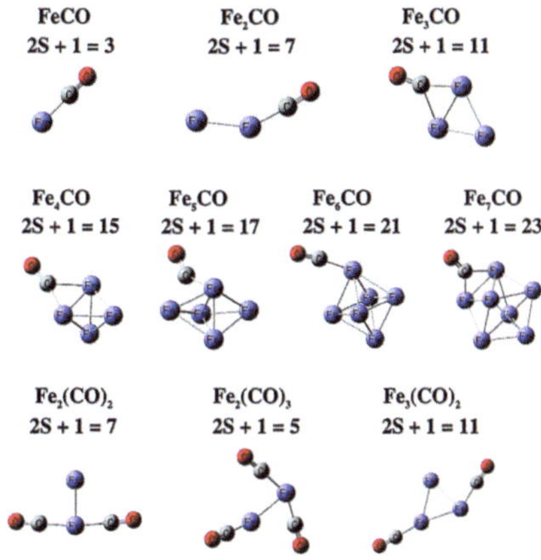

Fig. 6.2 Molecular structures of intermediates $Fe_n(CO)_m$ and their spin states. Reproduced with permission from [31]. (Copyright 2007 American Chemical Society)

and in a laminar flow reactor toward carbon nanotubes (CNTs) [27, 28]. The complete understanding of the chemical kinetics of $Fe(CO)_5$ decompositon and iron nanoparticles formation is very important, yet complicated, which requires reliable kinetic measurements for the thermal decomposition process and known thermochemical properties of related species. As pioneers, Giesen et al. used a global reaction $(Fe(CO)_5 \rightarrow Fe + 5CO)$ behind a shock wave to describe the thermal decomposition of $Fe(CO)_5$, and suggested that FeCO was a minor species, without accounting for any intermediates or byproducts for the purpose of a simplified particle growth mechanism [29]. For other studies, Rumminger et al. introduced a five-step $Fe(CO)_5$ decomposing model, which demonstrated the dissociation of five iron carbonyls $(Fe(CO)_n \rightarrow Fe(CO)_{n-1} + CO, n = 1-5)$ and presumed that the iron generate Fe atoms before forming particles when modeling the inhibition effects of $Fe(CO)_5$ on atmospheric flames [30]. Rumminger et al. have done a reasonable first approximated modeling of the iron carbonyl chemistry. However, as shown by Green et al., the conversion of $Fe(CO)_5$ into iron nanoparticles should be included in a detailed model that involved several other important intermediates and reactions [31]. The thermal decomposition of iron carbonyl and subsequent iron nanoparticles formation included iron clusters $Fe_n (n = 1, 2, 3,...)$, iron carobonyls $(Fe(CO)_n, n = 1-5)$, and other intermediates, $Fe_m(CO)_n$, as shown in Fig. 6.2.

Despite all the complications, iron carbonyl is frequently used in the formation of iron nanoparticles with/without a catalyst whose decomposition process requires seemly energy in the form of heat or sonication, an appropriate surfactant/solvent system, and means to remove the byproduct, carbon monoxide. According to the Sun group's report, monodisperse transition metal (Co, Ni, Fe, and alloy) nanoparticles with uniform size, composition, and shape were successfully synthesized via high temperature, solution phase synthesis under the control of parameters such as precursor, surfactant, and reducing rate. These nanoparticles were ideal building blocks to be assembled into close-packed superlattice for magnetic devices [32].

For further reported works, the fine tuning decomposition process of $Fe(CO)_5$ in octadecene (ODE) at $180\,^{\circ}C$ gave monodispersed Fe nanoparticles (<10 nm in

Fig. 6.3 a TEM images of Fe/Fe$_3$O$_4$ nanoparticles. (*Inset*) HRTEM image of the Fe/Fe$_3$O$_4$ nanoparticles. **b** Self-assembled Fe/Fe$_3$O$_4$ nanoparticle superlattice. Reproduced with permission from [33]. (Copyright 2006 American Chemical Society)

radius) [33]. While exposed to air, the as-synthesized particles were quickly oxidized/or agglomerated within a short time in hexane dispersion. Structural characterization of the nanoparticles showed that both metallic core and oxide shell were amorphous. With a crystalline Fe$_3$O$_4$ shell, the iron nanoparticles could get more robust protection, as shown in Fig. 6.3.

As it is known to all, stable magnetic NCs are the essential building blocks for fabricating next generation exchange-coupled nanocomposites magnets that are capable of storing of high-density magnetic energy. Among the iron NCs studies thus far generated via the thermal decomposition of iron carbonyls, the reported works showed a low magnetization value (90.6 A m^2 kg^{-1}/$_g$Fe), which inhibited the wide applications of iron NCs. Taking advances of the previous work, Sun et al. demonstrated an exciting process toward body centered cubic (*bcc*-Fe) iron nanoparticles via the decomposition of Fe(CO)$_5$ in the presence of oleylamine (OAm), oleic acid (OA), and hexadecylammonium chloride (HAD·HCl), which could be further converted to crystalline Fe$_3$O$_4$ when exposed to ambient environments or just with a thin crystalline Fe$_3$O$_4$ shell to stimulate the robust protection effects of the Fe@Fe$_3$O$_4$ architecture, as shown in Fig. 6.4 [34]. At room temperature, these NPs were ferromagnetic and exhibited a high magnetization at 2 T with a saturation magnetization value M$_s$ = 164 A m^2 kg^{-1}$_{Fe}$ and coercive field H$_c$ = 14 mT. As reported, the crystalline Fe$_3$O$_4$ protected Fe NPs, showed little magnetization change over 24 h, and were stable even after a month of air exposure period.

To further investigate their works, the authors monitored the Fe NCs growth in the reaction to demonstrate the role halide ions played in the synthetic steps. Contrast to the previous works showing that halide ions could tune different crystalline facets growth of Pd based NCs, the Cl$^-$/Br$^-$ ions did not show crystal-facet dependent growth. As the authors declared, the strong binding between Fe–Cl$^-$/Br$^-$, retarded the growth of Fe on the seeding nanoparticle surfaces, which favored a thermodynamic growth of Fe over the existing Fe nucleis and seeds into the *bcc* structure [35].

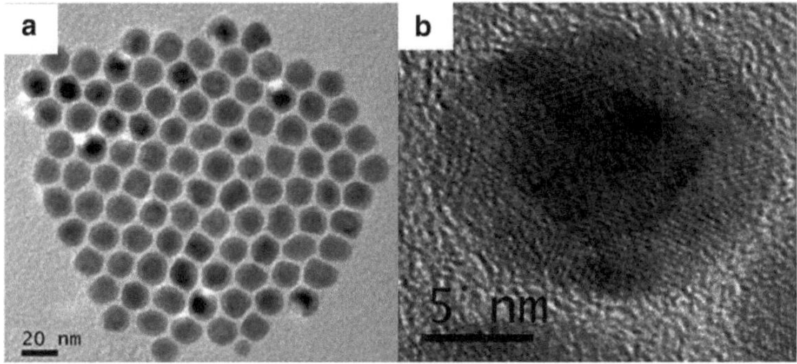

Fig. 6.4 **a** TEM image of the *bcc*-Fe NCs obtained from the re-dispersion of the plate assembly in hexanes. **b** HRTEM image of a single *bcc*-Fe NC revealing the metallic Fe core and Fe$_3$O$_4$ shell with Fe {110} and Fe$_3$O$_4$ {222} planes indicated. Reproduced with permission from [34]. (Copyright 2011 American Chemical Society)

6.2.1.2 Nanocrystals of Cobalt and Nickel

Similarly to the synthetic steps toward iron NCs, the fabrication of cobalt NCs is commonly via the thermal decomposition or chemical reduction of organometallic precursors, and cobalt salts.

Murray and Sun et al. have investigated the monodispersed metallic magnetic NCs of cobalt intensively using different synthetic procedures, which were organized into 2 and 3-D (2 and 3-Dimensional) superlattices for further fabrication of magnetic devices in potential. Via the injection of superhydride (LiBEt$_3$H) solution in dioctyl ether into a hot cobalt chloride solution at 200 °C assisted by oleic acid and trialkylphosphine as surfactants, the formation of small metal clusters were induced simultaneously as nucleis for the growth of nanoparticles. Finally, the metal clusters grew to magnetic NCs via heating consistently at 200 °C. Controlled by the steric bulkiness of the stabilizing effects of oleic acid and co-surfactants, particles size could be finely modulated. Interestingly, the authors reported that short-chain alkylphosphines may allow faster growth which resulted in the bigger products, while bulkier surfactants hindered the particle growth process and favored the generation of smaller nanoparticles [32, 36, 37].

Cobalt atoms could be tuned to generate three types of phases including hexagonal close packed (*hcp*), face-centered cubic (*fcc*), and ε-phase with distinct magnetic properties. Furthermore, heating the ε-Co nanoparticles at 300 °C could generate *hcp*-Co NCs directly.

Numerous work has been done concerning the thermal decomposition of dicobalt octacarbonyl Co$_2$(CO)$_8$ in the presence of a surfactant mixture composed of oleic acid (OA), lauric acid and trioctylphosphine (TOP) or trioctylphosphine oxide (TOPO) by Alivistos and Bawendi et al. Through controlling the reaction parameters, such as the precursor/surfactant ratio, the reaction temperature and the injection time, particles in a wide diameter scale from 3 to 17 nm were generated successfully [38, 39].

For the purpose of increasing density in microelectronics and magnetic-storage industries via the production of devices that function reproducibly at every smaller dimensions, monodispersed Co-NCs (radius 5 nm, $\sigma = 5\%$) with a metastable cubic phase isomorphous with β-Mn generated via high-temperature solution-phase synthetic method were organized into ordered arrays through a controlled evaporation of the solvent. Excitingly, the authors demonstrated spin-dependent electron transport, which was reported by Pileni et al. previously [40]. Besides, at temperatures below 20K, the device magnetoresistance ratios were approaching the maxium predicted for cobalt islands with randomly oriented preferred magnetic axes. And with the increasing of temperatures and bias-voltage, magnetoresistance was suppressed [41].

Generally, the *hcp*-cobalt nanoparticles were synthesized by using polyol process, where high boiling alcohol was applied as both reductant and solvent [32]. In a typical synthetic process, nanoparticles were isolated by the so-called size-selective precipitation. By controlling the steric bulkiness of the reacted phosphine stabilizers, and the relative concentration ratio of precursor and stabilizer, particle size could be tuned in a diameter scale of 3–13 nm. As it has been reported initially on the synthesis of cadmium sulfide quantum dots, colloidal assemblies such as reverse micelles are good candidates to reduce the size of crystallities to form nanoscale particles [42, 43]. Reverse micelles, which could be defined as water-in-oil droplets stabilized by a monolayer of surfactants, have then been induced for the generation of various nanoparticles including semiconducting, metallic, and/ or magnetic nanoparticle as ideal instruments to investigate the quantum size effect on optical, catalytic, or magnetic properties. In a typical process toward cobalt NCs via reverse micelle synthetic steps, collisions followed by exchange between micellar droplets induced chemical reaction of cobalt ions and sodium borohydride $NaBH_4$ to form cobalt nanoparticles. The cobalt containing reactants could be other $CoCl_2$, $Co(AOT)_2$, and other cobalt salts, while the solution often included didodecyldimethylammonium bromide (DDAB), sodium bis(2-ethylhexyl)sulfosuccinate (NaAOT) and related organics to form micelles [44, 45].

As one of the transition metals that exhibit characterized magnetism and other interesting properties, nickel reveals various applications such as in hydrogen storage and catalysis in potential. Numerous works have been explored for the synthetic methods toward nickel NCs, which involves high temperature organometallic decomposition [46–48], electrochemical reduction [49], and chemical reduction [50].

Based on the previous methods described by Murry and Son et al., Zanchet et al. reported a systematic chemical route toward 4–16 nm nickel NCs on the chemical reduction of nickel salts ($Ni(CH_3COO)_2$ or $Ni(acac)_2$ (acac = acetylacetonate) [32, 51, 52]. Considering the growing demand of nickel NCs, T. Ohta et al. improved the TOPO-oleic acid by introducing HAD into the metal colloidal synthetic system. By modification of surfactants, the solution reduction of $Ni(acac)_2$ (acac = acetylacetonate) by sodium borohydride or superhydride in dichlorobenzene was employed to produce nickel nanoparticles in a diameter scale of 3–11 nm [50]. In a standard anaerobic procedure investigated by Guo et al., the Bis(1,5-cyclooctadiene)nickel ($Ni(COD)_2$) was introduced as the nickel precursor. Following the protocol of

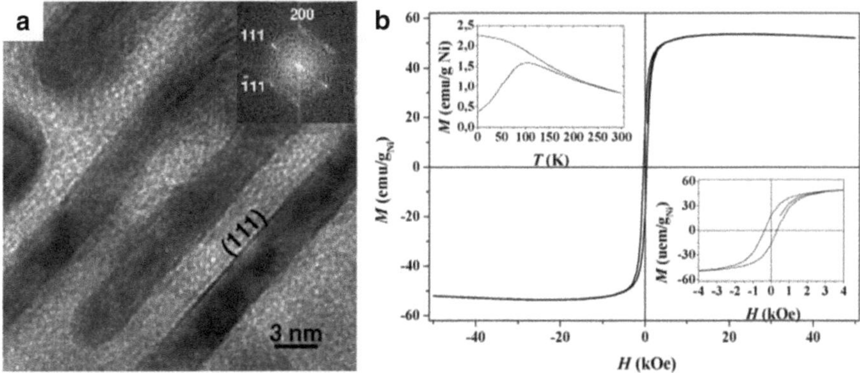

Fig. 6.5 **a** High-resolution TEM of a nanorods. **b** Hysteresis loop for Ni nanorods at 2 K. Insets: (*bottom right*) enlargement at low field; (*top left*) ZFC/FC curve (10 Oe). Reproduced with permission from [55]. (Copyright 2001 American Chemical Society)

preparing Co NCs, the authors used toluene, 1,2-dichlorobenzene (DCB), or octyl ether as solvents, trioctylamine (TOA), OA, and TOPO and their combinations as surfactants to generate nickel NCs [47]. As evidenced by Chaudret et al., via the hydrogenation of the olefinic ligands into the corresbonding, coordinatively inert alkanes, nanoparticles could possess noncontaminated surfaces, which display magnetic properties analogous to that of clusters prepared in ultrahigh vacuum [53, 54].

As has been reported in the indium synthesis, through stabilizing polymers in the surface, the shape could be tuned from spherical nanoparticles to nanowires depending on the ligands present in the reacting environment mediated by trioctylphosphineoxide (TOPO) or hexadecylamie (HDA), respectively [54]. Inspired by these results, the authors thus investigated the effect of HDA on the shape and magnetic properties of nickel nanoparticles, using the complex $Ni(COD)_2$ (COD = cycloocta-1, 5-diene) as precursor. As the results showed, in this typical organometallic based synthetic method, increasing the concentration of amines favored the formation of nickel nanorods nearly monodispersed in diameter, which most probably induced through the specific coordination of the amines, Fig. 6.5 shows the TEM images of the nickel nanorods and the corresponding hysteresis loops [55].

6.2.2 Nanocrystals of Alloys of Iron, Cobalt, and Nickel

6.2.2.1 M–Pt (M = Fe, Co) Alloys Nanocrystals

Aside from single composition magnetic NCs, nanostructured alloys such as FePt and CoPt have received intensive investigations concerning their magneto-anisotropy and chemical stability [56]. FePt nanoparticles are a vital class of magnetic nanoparticles which contains a near-equal atomic percentage of Fe and Pt known to possess a chemically disordered face-centered cubic (*fcc*) structure or chemically disordered face-centered tetragonal (*fct*, also denoted as L_{10}) structure. Compared to the commonly used high-moment nanoparticles of Co and Fe, along with the large coercive materials such as $SmCo_5$ and $Nd_2Fe_{14}B$, Fe–Pt interactions dramatically render the FePt nanoparticles chemically much more stable, thus making them especially useful for practical applications in solid-state devices and biomedicine [57].

Concerning the deficiency in physical deposition process (such as vacuum-decompostion [58–60], gas-phase evaporation [61]) toward FePt NCs that include magnetically soft structure, aggregation, various insulator matrices, and barriers to make them into regular arrays, solution-phase synthesis offers a unique step toward monodispersed FePt NCs and nanoparticles superlattices. Generally, chemical synthesis of FePt nanoparticles included thermal decomposition of $Fe(CO)_5$ and reduction of $Pt(acac)_2$, and co-reduction of metal salts. Sun et al. reported the monodispersed FePt nanoparticles using a combination of oleic acid and oleylamine to stabilize the FePt colloids and prevent oxidation. Based on the reduction of $Pt(acac)_2$ (acacm = acetylacetonate, $CH_3COCHCOCH_3$) by a diol and the decomposition of $Fe(CO)_5$ in high-temperature solutions, the authors modified the synthetic procedure by using a long-chain 1,2-hexadecanediol to reduce the $Pt(acac)_2$ to Pt metal. The size and composition of FePt NCs could be fine tuned by modulation of precursors and the solvent [62]. Further studies revealed that exclusion of additional reducing agent such as 1,2-alkanediol in the reaction mixture promoted better size control of the FePt NCs [63, 64]. Notably, Howard et al. investigated the detailed co-reduction process toward FePt NCs, with the diameter scale of 5–10 nm by the reduction of $Pt(acac)_2$ and iron acetylacetonates or $Na_2Fe(CO)_4$ [65].

6.2.2.2 M–Fe (M = Co, Ni) Alloys Nanocrystals

Concerning the unique intrinsic magnetic properties with large permeability and high saturation magnetization, FeCo alloys nanostructures attracted intensive deliberations for biomedical applications and self-organizing into superlattices for further devices in potential.

Chaubey et al. successfully obtained bimetallic FeCo NCs with well-modulated particle size and size distribution via the reductive decomposition of $Fe(acac)_3$ and $Co(acac)_2$ in a solution including OA, OAm, and 1,2-hexadecanediol under $Ar + H_2$ atmosphere [66]. Bimetallic nanoparticles (CoRu, CoRh, CoPt or RuPt) have been obtained in Chaudret's group by co-decomposition under H_2 of appropriate organometallic precursors [67]. Inspired by this, the authors reported a similar route that leads to 20-nm FeCo nanoparticles using $Fe(CO)_5$, $Co(\eta^3-C_8H_{13})(\eta^4-C_8H_{12})$, and $Co(N(SiMe_3)_2)_2$ as precursors and a mixture of HAD, OA, and stearic acid (SA) as stabilizing agents at 150 °C under 3 bar of H_2. Furthermore, the FeCo NCs could be organized into supercrystals from a macroscopic to a nanometric scale [68].

Recently, Soulantica et al. synthesized a topologically L structured Fe-Co dumbbell, the growth steps were studied by advanced electron microscopy techniques denoted as electro tomography, as Fig. 6.6 shows [69, 70]. In this special topological dumbbell structure, the soft anisotropy of the Fe nanocubes and the dipolar

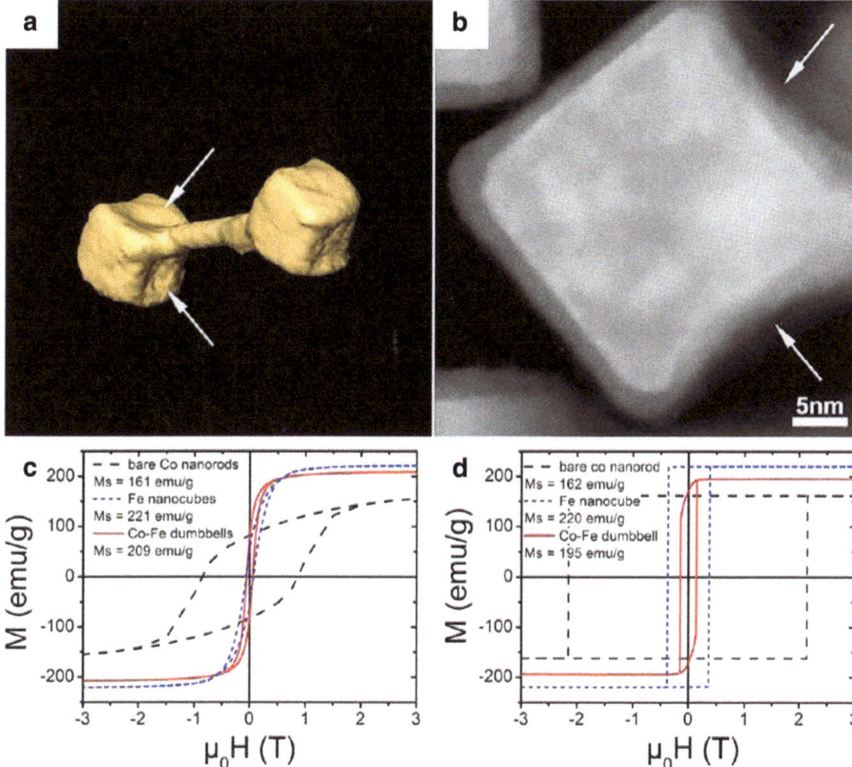

Fig. 6.6 **a** 3-D representation of the reconstructed volume along different viewing directions of the Fe–Co dumbbell. **b** HAADF-STEM image of the particle focusing on the Fe cube. **c** M-H hysteresis loops at 2 K for Co nanorods (*black*), Fe nanocubes (*blue*), and Fe–Co dumbbell (*red*). **d** Simulated M-H hysteresis loops along the nanorod axis for a single Co nanorod (*black*), a single Fe nanocube along the <110> direction (*blue*), and a single Fe–Co dumbbell along the long axis (*red*) using the OOMMF micromagnetic code at 0 K. Reproduced with permission from [69]. (Copyright 2014 American Chemical Society)

magnetic interactions induced a drastic decrease of the magnetic anisotropy of the nanohybrids, the coercive field of which became only slightly higher than the one of interacting Fe nanocubes.

As a unique magnetic alloy, nanostructured FeNi alloys possess various applications from recording heads and transformers to magnetic shielding materials. Up to now, researchers have developed a wide range of synthetic steps toward nanostructured FeNi alloys, such as hydrogen reduction of metal salts, hydrothermal reduction methods, and electroleaching processes [71–74]. Li et al. reported the synthesis of face centered cubic $FeNi_3$ nanoplatelets with controlled size through a solution reduction method. The size of the $FeNi_3$ nanoplatelets could be controlled by varying the concentrations of the reactants with the lattice constants of 92 nm in diameter, which was larger than that of the bulk [74]. K. T. Leung prepared bimetallic FeNi concave nanocubes with high Miller index planes through controlled

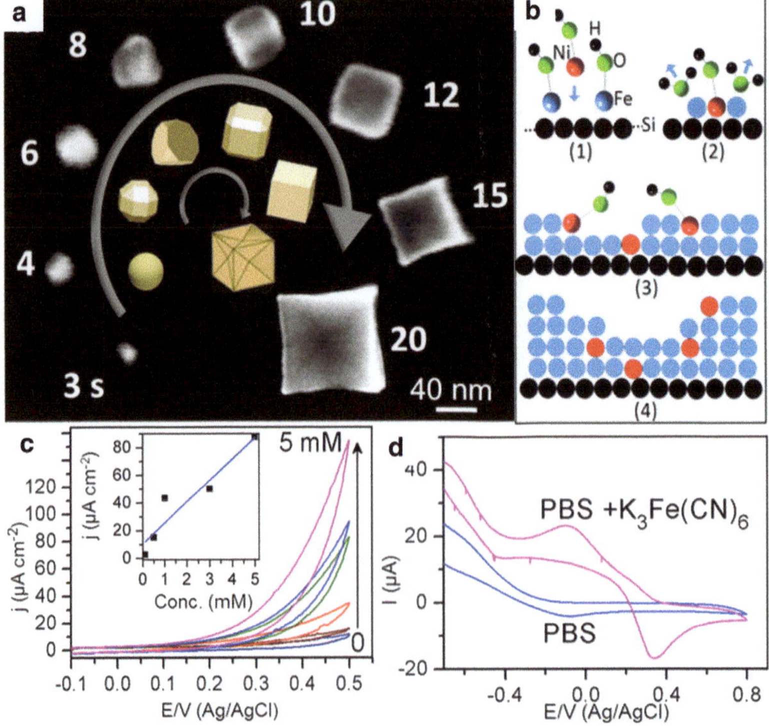

Fig. 6.7 **a** SEM images of the typical nanocubes with increasing deposition time from 3 to 20 s. **b** Schematic illustration of the plausible mechanism for the formation of higher-index facets of a concave nanocube: (1) selective adsorption of $FeOH^+$ over $NiOH^+$; (2) faster reduction of $FeOH^+$ than $NiOH^+$; (3, 4) formation of step or kink defects and ultimately higher-index planes. **c** CVs of concave nanocubes at different concentrations of 4-aminophenol (scan rate = 50 mV s^{-1}). The inset shows the linear relationship between current density and concentration. **d** CVs of concave FeNi nanocubes in 10 mM PBS with and without 1 mM $K_3Fe(CN)_6$. Reproduced with permission from [49]. (Copyright 2013 American Chemical Society)

triggering of the different growth kinetics of iron and nickel via a material-independent electroleaching process. With the high-index facets exposed, these concave nanocubes 10-fold more active toward electrodetection of 4-aminophenol than cuboctahedrons, providing a label-free sensing approach for monitoring toxins in water and pharmaceutical wastes, as shown in Fig. 6.7 [49].

6.2.2.3 M–C (M = Fe, Co,Ni) Carbide Alloys Nanocrystals

Magnetic carbide nanostructures such as iron carbides, cobalt carbides, and nickel carbides have long attracted intensive attention due to their chemical characteristics that make them as valid and sustainable alternatives to noble metals in catalysis, biomedicine as contrast agents and drug delivers, and to air-sensitive metals or oxides for applications under harsh conditions [75].

As methods that possessed more defined and chemically pure structures, physical methods such as physical vapor decomposition, plasma, and laser acted as active synthetic pathways toward magnetic carbide nanostructures, yet, physical methods made the variety of as-synthesized products restricted [76–78]. Commonly, classical chemical procedures toward transition metal carbides involved reactions of parental metals, corresponding metal oxides or other metal precursors with special carbon sources such as coal and gaseous hydrocarbons, which are always performed at high temperature higher than $1000\,^\circ C$ [75, 79].

6.2.2.4   Iron Carbides Nanosrystals

Giordano et al. extended the use of a so-called urea-glass route to the preparation of iron carbide, which could be denoted as a sol–gel chemical pathway. By heating treatment of the gel phase obtained by mixing an iron precursor with urea, a black powder constituted of Fe_3C with superparamagnetic properties could be obtained. As the authors declared, the urea partially replaced the water with oxygen atoms presumably in the first coordination sphere around the metal ion [75, 80].

Inspired by the Fischer–Tropsch process, Chaudret et al. reported a tunable organometallic synthesis of monodispersed iron carbides and core-shell iron–iron carbide nanoparticles with a diameter of ca. 13.1 nm and 13.6 nm, respectively, which was based on the decomposition of $Fe(CO)_5$ on $Fe(0)$ seeds that allow the modulation of carbon atoms diffusion and tune the magnetic anisotropy of the final products, as shown in Fig 6.8. Interestingly, the iron/iron carbide core/shell NCs possessed the anisotropy axes aligned along the magnetic field direction in the ferromagnetic regime, which were the typical behaviors of magnetic nanoparticles [81].

Hou et al. reported a facile chemie douce route for the synthesis of single phased Fe_5C_2 NPs that involved the reaction of iron carbonyl ($[Fe(CO)_5]$) with octadecylamine in the presence of CTAB (Hexadecyl trimethyl ammonium Bromide) under mild temperatures (up to $350\,^\circ C$). The size of the iron carbide NPs can be tuned by tailoring the concentration of $Fe(CO)_5$. In this typical synthetic pathway, bromide ions acted partially as an agent that favor thermodynamic growth of iron to *bcc*-Fe nanoparticles, which were then promoting the cleavage of C–C bonds of the amines decomposing process to form unsaturated hydrocarbons and cyanides as carbon sources, as shown in Fig 6.8c [82].

Ionic Liquids (ILs), which is a class of organic solvents possessing specific physico-chemical properties such as high fluidity, thermal stability, and ionic conductivity, low melting temperatures, and extended temperature range in the liquid state, etc [83 84]. Recently, Guari et al. reported the thermal decomposition of $Fe_x(CO)_y$ precursors for the synthesis of nanoparticles of iron carbides and their superstructures with sizes ranging from 2.8 to 15.1 nm using imidazolium-based ionic liquids as solvents, stabilizers, and carbon source, as shown in Fig. 6.9 [85].

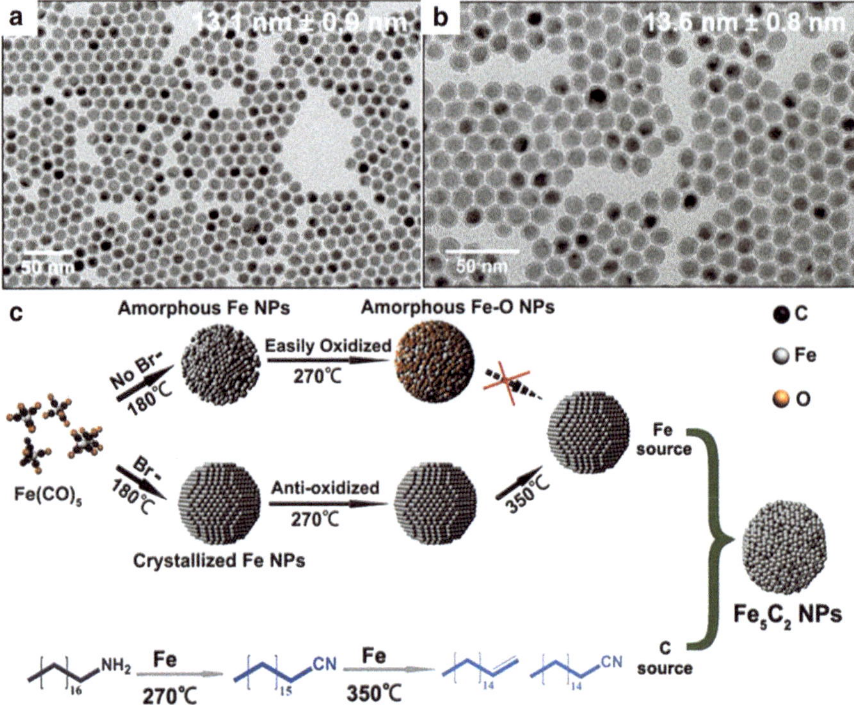

Fig. 6.8 a TEM analysis of iron carbide NCs of about 13.1 nm. **b** TEM analysis of iron/iron carbide nanoparticles of about 13.6 nm. Reproduced with permission from [81]. Copyright 2012 American Chemical Society. **c** Schematic Illustration of the Formation Mechanism of Fe_5C_2 NPs. Reproduced with permission from [82]. (Copyright 2012 American Chemical Society)

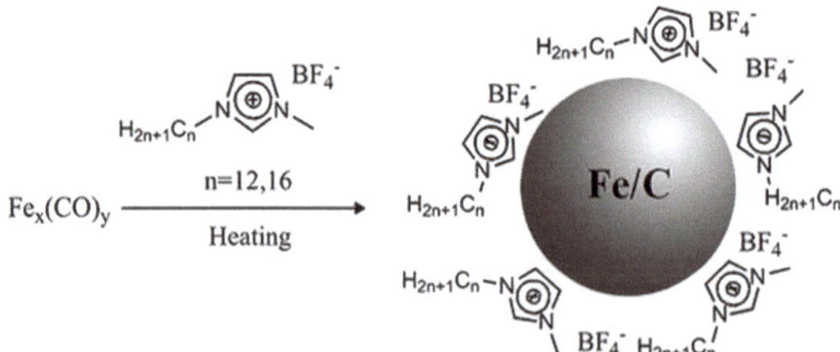

Fig. 6.9 Schematic representation of the procedure used for the synthesis of Fe/C NPs by thermal decomposition of $Fe_x(CO)_y$ complexes in ILs. Reproduced with permission from [85]. (Copyright 2013 Springer-Verlag)

6.2.2.5 Cobalt Carbides Nanocrystals

As a new kind of permanent magnet, cobalt carbides nanostructures have received dramatical attention for its interesting structural and physical properties. Doping carbon into metallic cobalt structure increased the magnetocrystalline anisotropy and observed the coercivity, which was caused by the crystalline reconstruction. Via applying an extensively investigated pathways toward metal and oxides nanostructures denoted as the polyol process, whereby metal salts were dissolved in a polyhydric alcohol (polyol) and heated to elevated temperatures, cobalt carbides have been successfully obtained [86, 87].

Zhang et al. reported the controlled synthesis of cobalt carbide (Co_2C and Co_3C) nanoparticels and nanowires through a facile polyol reduction procedure assisted with surfactant. By adjusting the category and concentration of the surfactant, shape, and size of the cobalt carbide NCs could be controlled, thus accordingly tuning the magnetic properties of the products [88]. As reported, $[OH^-]$ ions has shown to intensively affect the shape and metallic phase during the polyol process toward cobalt nanoparticles [89–92]. Besides, halide ions could also be applied to control the size and shape of special nanoparticles by oxidatively etching the surface and preventing polycrystalline agglomeration [93, 94]. Based on the reported works, Carpenter et al. investigated the effects of various $[OH^-]$ and $[Cl^-]$ ions toward the final products of cobalt carbides nanoparticles. It revealed that, via designed control of $[OH^-]$ and $[Cl^-]$ ions in the solution, single phased Co_2C forming agglomerations of ellipsoidal grains could be obtained. Apart from the phase control, ions could also tune the particle shapes from nanoflower-like to sea-urchin-like morphologies [95].

6.2.2.6 Nickel Carbide Nanocrystals

Compared to pure *hcp*-Ni, which was predicted to be highly magnetic with saturation magnetization values approximate to *fcc*-Ni, Ni_3C is nominaly nonmagnetic, despite a very small magnetic moment caused by special regions with substoichiometric interstitial carbon atoms [96–99]. Besides, nickel carbides show great activities in various catalytic reactions, such as methane reforming [100], conversion of cellulose [101], steam reforming, and other energy related reactions [102].

Omata et al. demonstrated the thermolysis of nickel acetylacetonate [Ni(II)(acac)$_2$] in oleylamine, hard X-ray photoemission spectroscopy (HX-PES) showed that hexagonal nickel carbide was synthesized, Fig. 6.10 shows the crystal structure of Ni and Ni_3C and their corresponding magnetism related characterizations of the as-synthesized samples [103]. Similarly, Raymond E. Schaak et al. proposed a synthetic protocol using nickel 2,4-pentanedionate as nickel salts that was closely related to existing colloidal routes, which yielded particles attributed to both *hcp*-Ni and Ni_3C. Reaction time could be used to tune average carbon content via a continuous Ni_3C_{1-x} solid solution [104].

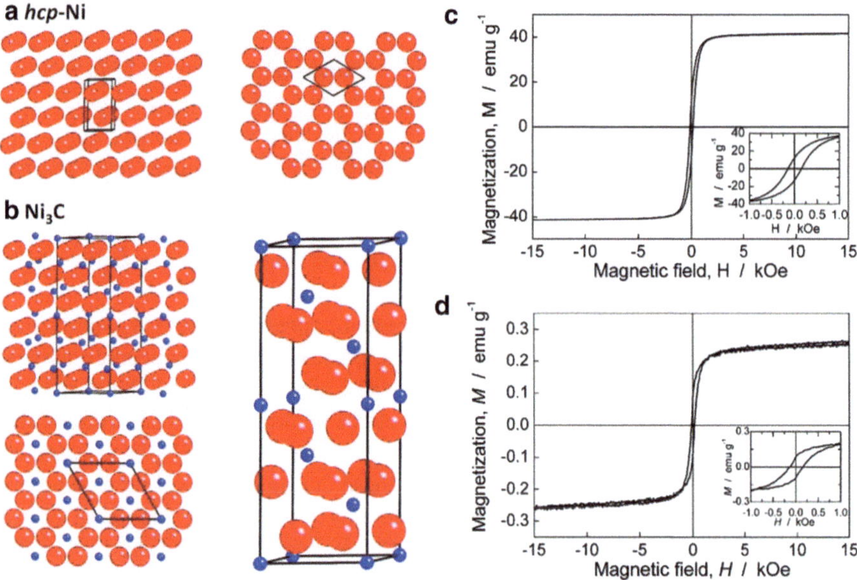

Fig. 6.10 **a** Crystal structures of hexagonal *hcp*-Ni. **b** Crystal structures of rhombohedral Ni_3C, highlighting the similarities between the structures and the location of the interstitial carbon in Ni_3C. Nickel atoms are represented by large *red* spheres, and carbon atoms are represented by small *blue* spheres. The unit cell of Ni_3C in **b** shows the superlattice structure. Reproduced with permission from [104]. Copyright 2011 American Chemical Society. **c** M-H hysteresis loops of *fcc*-Ni NCs. **d** M-H hysteresis loops of the hexagonal nickel carbide NCs. Insets show the coercivity. Reproduced with permission from [103]. (Copyright 2008 American Chemical Society)

6.3 Biological Applications of Magnetic-Metallic Nanocrystals

6.3.1 Magnetic-Metallic Nanocrystals for Magnetic Hyperthermia

Magnetic hyperthermia (MH) is a promising application of magnetic nanoparticles (MNPs) in biomedicine. Usually, after injecting MNPs into the tumor, the patient is immersed in an alternating magnetic field (AMF) with an appropriate frequency and amplitude, which can raise the temperature of the tumor site. This temperature increase can improve the efficiency of chemotherapy and even directly kill the tumor cells by apoptosis or necrosis.

Various mechanisms for the heat generation by MNPs under the AMF are reported. Among them, hysteretic properties and relaxation behavior are the two dominating ones. Hysteretic property, which is originated from the internal energy of magnetic particles, is the primitive mechanism, especially for the situation of ferromagnetic nanoparticles. While the relaxation behavior, which includes Brownian

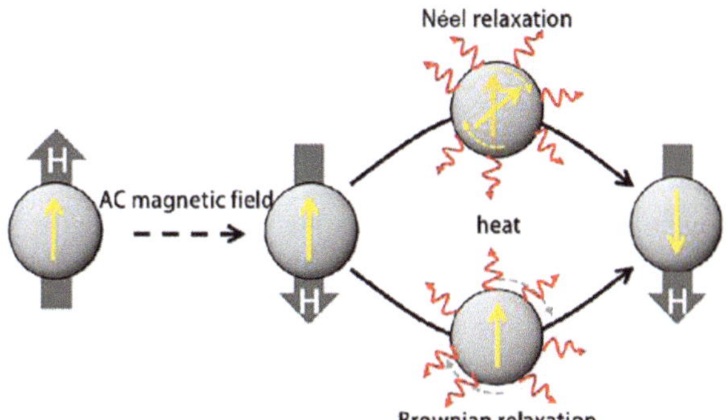

Fig. 6.11 Néel and Brownian relaxation processes. Reproduced with permission from [107]. (Copyright 2011 American Chemical Society)

and Néel relaxation pathway, is another mechanism for magnetic hyperthermia, particularly in the situation of superparamagnetic nanoparticles (NPs) [105]. Brownian relaxation is caused by the entire magnetic particle rotating, while Néel relaxation is due to the magnetic moment rotating within the magnetic core (Fig. 6.11) [106, 107]. The competition of these two relaxation processes is controlled by the faster one. In the case of intracellular MNPs, the intracellular components usually hinder the movement of NPs, which results in heat contribution mostly from the Néel relaxation [108, 109].

According to the heat generation mechanisms, several factors have been found to influence the heating power, including the AMF and the structures and magnetic properties of NPs. When applying a magnetic field to MNPs, hysteresis loop is generated, whose area A corresponds to a dissipated energy. This dissipated energy is the origin for the heat. Under the excitation of AMF, A represents heat released from one cycle. Then, power generated by MNPs is finally determined by specific absorption rate (SAR or SLP) = Af, where f is the frequency of the magnetic field. According to this equation, an increase in f, magnetic field applied (H), and saturation magnetization (M_s) is beneficial. However, there is a physiological limitation in clinical application due to the tolerance of human body on magnetic field, with $H_{max} f < 5 \times 10^9$ Am^{-1} s^{-1} [110]. Typical values used in medical treatments so far are 100 kHz and 20 mT [111]. There are several experimentally used ways to improve SAR in clinical. The first point is to optimize the MNPs used. For example, high M_s can increase the area of the hysteresis loop which eventually leads to the temperature increase. For a given MNPs, their size, morphology, and mono-dispersion are also major parameters. For instance, poly-dispersed NPs are proved to exhibit considerable low heat generation rate compared with mono-dispersed ones, i.e., mono-dispersed MNPs are preferred. Intrinsically, magnetic materials possess magnetically disordered spin glass like layers near the surface due to the reduced spin–spin exchange coupling energy at the surface. Upon reducing the size of magnetic materials to nanoscale, the surface canting effects are dramatically reduced in M_s, which is described as $M_s = M_{s0}[(r-d)/r]^3$, where r is the size, M_{s0} is the saturation magnetization of bulk materials, and d is the thickness of disordered surface layer. This phenomenon is especially noticeable when the diameter of NPs is smaller than

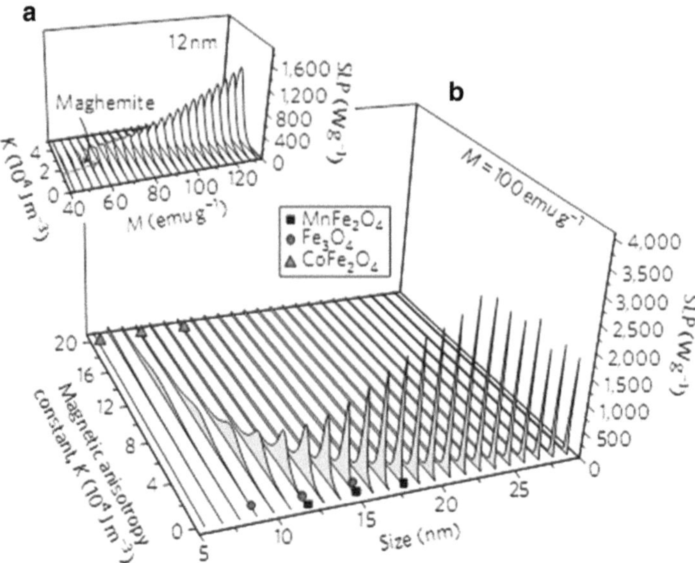

Fig. 6.12 Simulated plot of SLP based on NPs' **a** magnetic field and magnetic anisotropy constant at size of 12 nm, **b** size and magnetic anisotropy constant at a magnetization value M of 100 emu/g. Reproduced with permission from [114]. (Copyright 2011 Nature Publishing Group)

5 nm. Therefore, size of NPs is another critical factor for MH, and the larger size is favored [112]. However, MNPs with large size tend to aggregate under physiological environment, which is an adverse factor in biomedical application. Therefore, size of the MNPs should be in an appropriate range. Optimizing the anisotropy (K) is another important protocol to improve heating power. An optimum anisotropy increases the SAR and reduces the detrimental effects of the size distribution of MNPs. It is reported that, K depends on the magnetic field used in the hyperthermia experiments and the MNP magnetization [113]. By varying these parameters, Cheon et al. proved all of them have great effects on SPL based on ferrite magnetic NPs, and optimizing them would dramatically increase SAR (Fig. 6.12) [114].

Regarding the ease of fabrication and their approved clinical usage, iron oxide based NPs (IONPs) with controlled sizes and surface chemistry have been studied extensively for MH applications. However, their further applications have been limited due to their relatively low M_s, with SAR values commonly less than 100 W/g [115]. One of the possible ways to solve this problem is to use metallic NPs with higher M_s, e.g., cobalt or iron NPs. Take Co NPs for example, SAR values increased to 500 W/g and 720 W/g at 400 kHz, 13 kA/m and 410 kHz, 10 kA/m respectively [9, 115], offering the opportunity for higher heating. However, issues of biocompatibility are the main barrier for the application of Co NPs in diseases therapy. Fe NPs on the contrary, consist of essential element Fe only, and in higher M_s than Co NPs, being better candidate for MH. Hadjipanayis et al. reported that for achieving

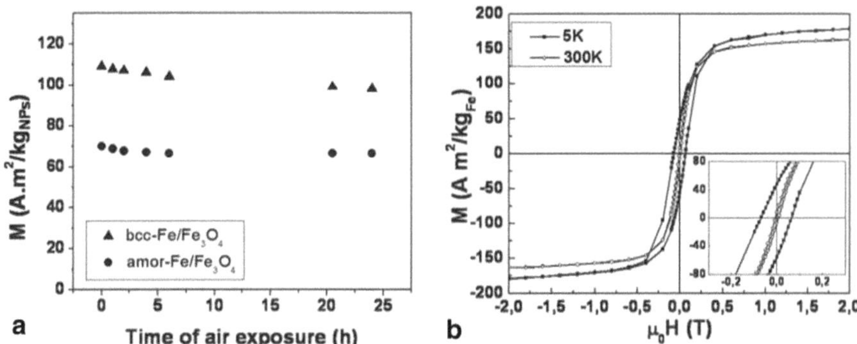

Fig. 6.13 Magnetic characterization of the Fe/Fe$_3$O$_4$ NPs. **a** The change of the magnetization values as a function of air exposure time of the *bcc*-Fe/Fe$_3$O$_4$ and the amor-Fe/Fe$_3$O$_4$ NPs of similar size at 1.5 T and room temperature. **b** Magnetic hysteresis loops of the *bcc*-Fe/Fe$_3$O$_4$ NPs recorded at 5 K (*solid cubes*) and 300 K (*open circles*). *Inset*: a close look of the hysteresis loops between 0.3 and 0.3 T. Reproduced with permission from [34]. (Copyright 2011 American Chemical Society)

the same temperature change, the local hyperthermia, Fe NPs require a dose of only 1/20 that of IONPs [116]. However, Fe NPs used in this work is synthesized by directly reducing the metal salt solution with a sodium borohydride solution, which has the M$_s$ of only about 70 emu/g. Sun et al. prepared body centered cubic (*bcc*-) Fe NPs via a thermal decomposition of iron pentacarbonyl in the presence of hexadecylammonium chloride. These NPs have a magnetization close to the bulk value. But they would be quickly oxidized once exposed to air. They further controlled, oxidized the surface layers of the bcc-Fe into crystalline Fe$_3$O$_4$ by exposing to ambient environments, which dramatically improved their stability, with little M$_s$ changed over 24 h and only about 20 % dropped after a month by exposing to air (Fig. 6.13a). At room temperature, these NPs are ferromagnetic and exhibit a high magnetization at 2 T with M$_s$ = 164 Am^2 kg^{-1} Fe and coercive field H$_c$ = 14 mT (Fig. 6.13b). Due to their high M$_s$ and small H$_c$, these NPs are with a SAR value of 140 W g^{-1} Fe, which is much more efficient heater than both the amor-Fe/Fe$_3$O$_4$ and Fe$_3$O$_4$ NPs, being one of the highest reported NPs thus far under similar conditions [34]. Iron carbide NPs, with similar properties with Fe NPs and much higher stability, showing great potential in MH. Chaudret et al. tuned the magnetic property of iron carbide NPs by adjusting their Fe content, and found that the optimized NPs possess SAR = 415 W g^{-1}, which is three times larger than the values reported on the best chemically synthesized MNPs under the situation of f = 96 kHz and H$_{max}$ = 20 mT (Fig. 6.14) [81].

Alloys such as FePt, CoPt, CoFe et al. are more stable than Fe NPs, and their M$_s$ are relatively higher, being good candidates for HM. Maenosono et al. estimated the comparative heating rates for aqueous monodispersions of the various MNPs including magnetite, maghemite, FeCo, and L1$_0$-phase FePt, finding out that FeCo NPs have the largest heating rate [117]. Based on this, Lacroix et al. reported the application of single-domain FeCo in MH with a moderate saturation field to

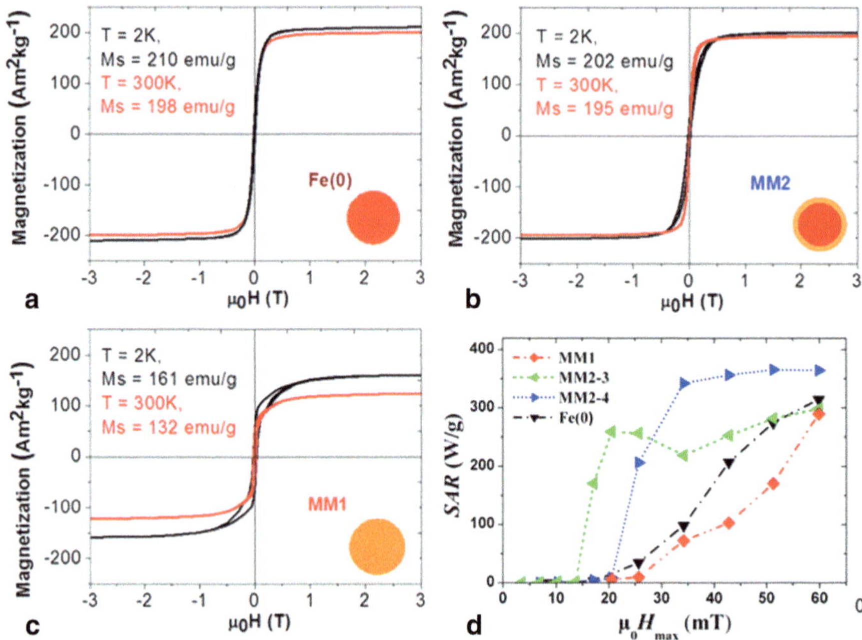

Fig. 6.14 Magnetization data of **a** ~9.6 nm iron NPs (Fe(0)), **b** ~13.1 nm/iron carbides core shell MM2 NPs and **c** ~14.2 nm iron carbide MM1 NPs at 2 and 300 K. **d** SAR measurements of MNPs at $f=54$ kHz. Reproduced with permission from [81]. (Copyright 2012 American Chemical Society)

maximize the losses, whose value is comparable to the highest of the literature [118]. However, the size range of FeCo where the heating is possible is larger than 30 nm, which is much larger than the usual size range of standard ferrofluids (D = 8–10 nm) [117]. Moreover, the heat release rate is far too rapid to be safe in human beings [119]. In addition, due to the spontaneous magnetization, the stability of magnetic colloid becomes impaired when D > 20 nm. However, *fcc*- and L1$_0$-phase FePt NPs yield the largest heating rates in the size range of D < 20 nm, and the heat rate is controllable and suitable for MH [117]. They are therefore regarded as promising candidates in MH.

6.3.2 Magnetic-Metallic Nanocrystals for Magnetic Resonance Imaging

Magnetic resonance imaging (MRI) is another important application of MNPs in biological field. Generally, there are two modes in MRI, which are termed as T_1-weighted MRI and T_2-weighted MRI. The former one provides positive signal, which is bright in images, while the latter one shows a darken figure. On the other hand, MNPs used in MRI can be divided into paramagnetic, ferromagnetic, and superparamagnetic NPs, with the latter two having almost the equal impact. Interestingly, paramagnetic NPs, mainly chelates of paramagnetic ions are usually for T_1-weighted MRI, and the ferromagnetic or superparamagnetic NPs are good candidates for T_2-weighted MRI. Metallic NPs, usually ferromagnetic or superparamagnetic, are therefore commonly applied in the T_2-weighted MRI.

The T_2-weighted contrast ability based on MNPs is usually estimated by the spin–spin relaxivity r_2 and it is dependent on their M_s, size and distance between the MNPs and the surrounding protons (d). The higher M_s is beneficial, as the value of r_2 is reported to be proportional to M_s^2 [120]. In the situation of MNPs, M_s is in positive correlation with the diameter of NPs, and therefore, larger size is favorable. However, cellular uptake and bio-circulation are also dependent on NPs' size, with the smaller ones easily been engulfed into cells and having longer circulation time, medium diameter are therefore optimal. In addition, r_2 has a relationship with d^{-6}, i.e., reducing the distance between MNPs and surrounding protons, which can dramatically increase the r_2 value [120].

According to these principles, various metallic NPs have been developed for the contrast agent in T_2-weighted MRI. As discussed previously, Fe NPs have greater M_s and coercivity than clinically used MRI contrast agent IONPs based on hysteresis loops, and therefore, can induce significantly higher r_2 value at a comparable particle size [116]. By further improving the M_s using bcc-Fe NPs, Sun et al. make the bcc-Fe/Fe$_3$O$_4$ NPs with r_2 value comparable with optimized ferrite NPs ($r_2 = 220$ mM^{-1} s^{-1}) [34]. Recently, iron carbides are also applied in MRI based on their high M_s and stability. Hou et al. prepared Fe$_5$C$_2$ NPs by a facile and mild wet-chemistry route. These NPs displayed an r_2 relaxivity of 464.02 mM^{-1} s^{-1}, which is among the highest of all MRI probes reported. By further coupling to a tumor-targeting ligand, these NPs were able to home into tumors and induce more prominent signal change than Fe$_3$O$_4$ NPs (Fig. 6.15) [121]. Gao et al. confirmed this result, by reporting, that Fe$_5$C$_2$ NPs exhibit twice as high as r_2 value of spherical Fe$_3$O$_4$ NPs, and are able to produce much more significant MRI contrast enhancement than conventional Fe$_3$O$_4$ NPs in living subjects [122]. The Hou's group further made full use of their high absorption in near-infrared (NIR) which lies on the carbon coating on the Fe$_5$C$_2$ NPs, developing a multifunctional probe for cancer diagnosis and therapy based on MRI, photoacoustic tomography, and photothermal therapy (PTT). The carbon shell on the other hand, can protect the Fe$_5$C$_2$ NPs from oxidizing, which makes the NPs with high M_s for at least 6 months. With the conjugation of targeted affibody, it can target tumor cells with low cytotoxicity, and selectively kill them through laser radiation (Fig. 6.16) [123].

FePt NPs is another important metallic material for enhancing the T_2-shortening effect and improving in molecular recognition ability as MRI contrast agents due to their superior magnetic properties. It is suggested that chemically disordered fcc-FePt NPs with a mean diameter of 9 nm displayed a higher T_2-shortening effect than conventional superparamagnetic IONPs, with the R_2/R_1 relaxivity ratio being 3–14 times improved [124]. Further coating a shell layer such as Fe$_3$O$_4$ and Fe$_2$O$_3$ can improve the contrast ability. Xu et al. revealed that FePt@Fe$_2$O$_3$ core-shell NPs exhibited stronger MR signal attenuation effect compared with FePt@Fe$_3$O$_4$ core-

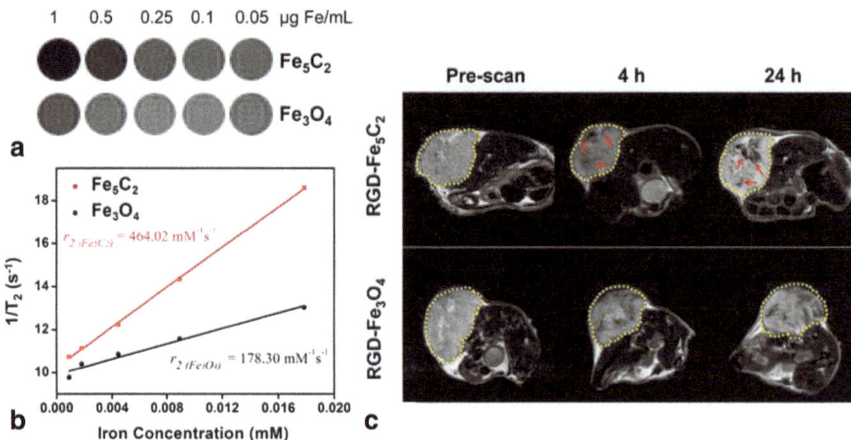

Fig. 6.15 **a** Phantom studies with Fe_5C_2 and Fe_3O_4 NPs dispersed in 1 % agarose gel at different concentrations on a 7 T magnet. **b** r_2 relaxivities derived from the imaging results in **a**. **c** MR imaging results. U87MG tumor-bearing mice were intravenously injected with RGD–Fe_5C_2 or RGD–Fe_3O_4 NPs (10 mg Fe/kg). Scans were performed before, and 4 and 24 h after the administration. In both the RGD–Fe_5C_2 and RGD–Fe_3O_4 groups, black areas (in the RGD–Fe_5C_2 group highlighted by *red arrows*) were found distributed in the tumors (circled by *yellow dotted lines*). Reproduced with permission from [121]. (Copyright 2014 John Wiley & Sons)

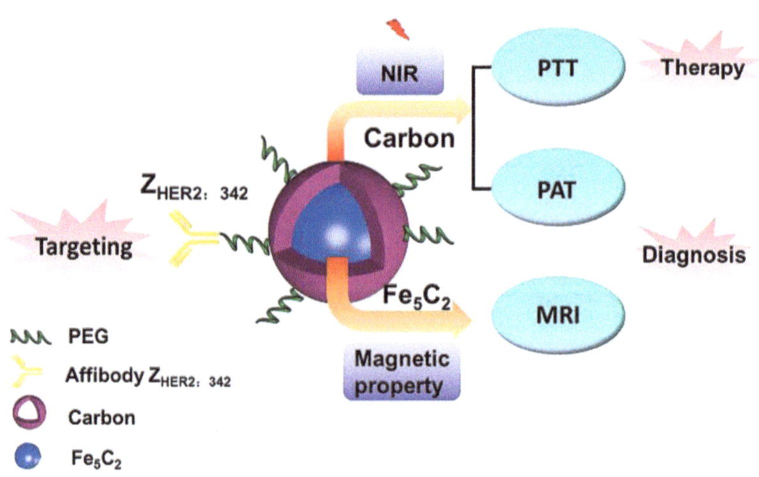

Fig. 6.16 Schematic illustration for the design of Fe_5C_2 NPs as a targeted theranostic platform. Reproduced with permission from [123]. (Copyright 2014 John Wiley & Sons)

shell NPs and bare FePt alloy NPs (Fig. 6.17) [125]. By conjugating folic acid on the surface, the NPs showed obvious tumor MRI contrasts after injecting intravenously [126]. However, these FePt@Fe_2O_3 yolk-shell NPs exhibit relatively low IC_{50} value originating from the fact that the slow oxidation and releasing of FePt

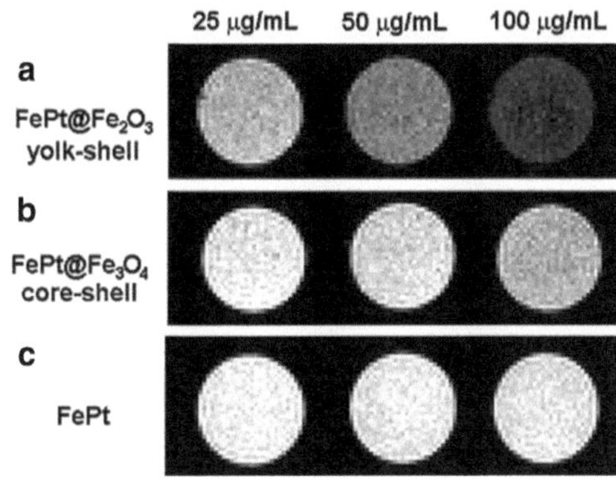

Fig. 6.17 T_2*-weighted MR images of **a** FePt@Fe$_2$O$_3$ yolk-shell NPs, **b** FePt@ Fe$_3$O$_4$ core-shell NPs, and **c** FePt NPs from a 7.0 T MRI system at 25, 50, and 100 µg/ mL in water (containing 1 % agarose gel). Reproduced with permission from [125]. (Copyright 2008 American Chemical Society)

yolks [125]. Silica coating, on the contrary, can enhance the biocompatibility. Lai et al. developed silica-coated FePt with both significant T_1 and T_2 MRI contrast abilities and without obvious cytotoxicity. Moreover, the silica shell enables the incorporation of fluorescein isothiocyanate inside the NPs, which shows the potential in fluorescent imaging [127]. In addition, FePt NPs have a high X-ray absorption, being a candidate for contrast agent in computed tomography (CT). Chen et al. synthesized 3, 6, and 12 nm FePt NPs for the application as a dual modality contrast agent for CT/MRI molecular imaging. The bio-distribution analysis revealed the highest serum concentration and circulation half-life for 12 nm-FePt, followed by 6 nm-FePt then 3 nm-FePt, and the 3 nm-FePt showed higher brain concentrations. After conjugating anti-Her2 antibody on FePt NPs, selective contrast enhancement of Her2/neu over expression cancer lesions in both CT and MRI was found in tumor-bearing animal after tail vein injection [128].

Other alloys such as FeCo and CoPt NPs are with high potential in improving contrast in MRI [129, 130]. Take FeCo NPs for example, Dai et al. synthesized FeCo/graphitic-shell NPs through a chemical vapor deposition method, whose M_s is close to bulk FeCo, being the highest M_s obtained for magnetic NPs (Fig. 6.18a). The NPs therefore exhibit both ultra-high r_1 and r_2 relaxivities, with the value of 70 mM^{-1} s^{-1} and 644 mM^{-1} s^{-1} respectively (Fig. 6.18b and c). Mesenchymal stem cells are able to internalize these NPs, showing high T_2-contrast enhancement in MRI, and long-lasting T_1-contrast enhancement for vascular MRI in rabbits was also achieved. In addition, the graphitic shell, on the one hand, makes NPs with a superior chemical stability against HF etching and oxidation resistance, without any degradation in M_s over a period of one month by exposing in air [131]; on the other hand, endows the NPs high absorbance in NIR for PTT, as well as the high drug loading due to the π-π stacking on the graphitic shell [132].

6.3.3 Ion Releasing for Selectively Killing Cancer Cell

Taking the advantage of the instability of metallic NPs, such as *fcc*-FePt NPs under acidic environment, which can lead to a leakage of iron ions, that may be developed

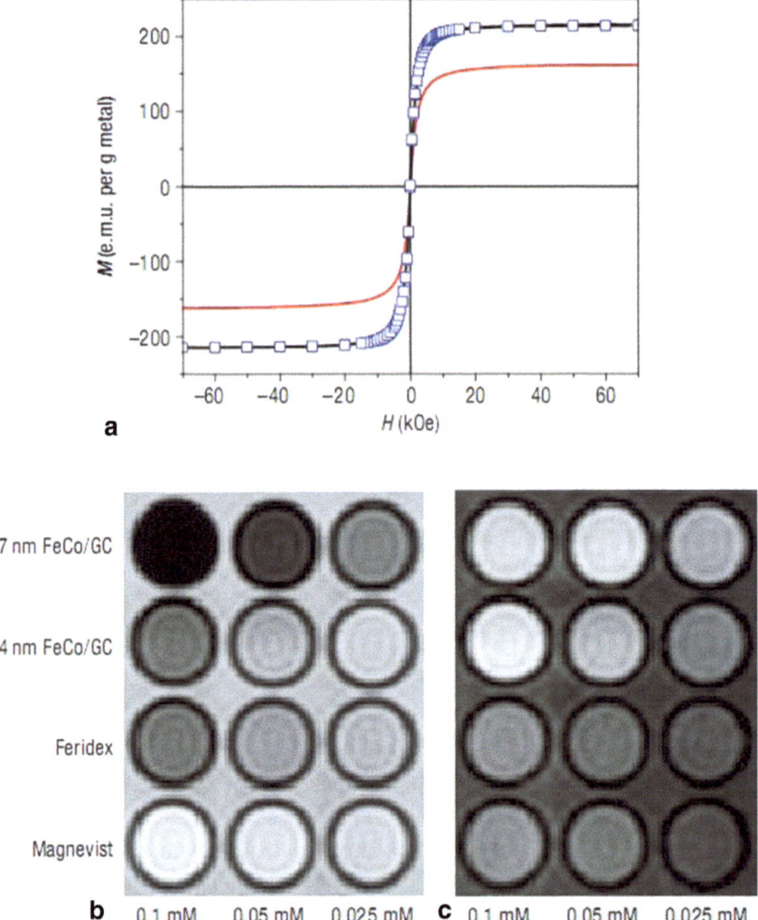

Fig. 6.18 **a** Room-temperature magnetization versus field data for ~7 nm (*black line*) and ~4 nm (*red line*) NPs measured shortly after synthesis, and after 1 month exposure in ambient air for the ~7 nm NCs (*blue* symbols; no degradation from the *solid black line*). **b** MR images of various contrast agents at three metal concentrations generated on a T_2-weighted spin-echo sequence with an echo time (TE) of 60 ms and pulse repetition time (TR) of 3000 ms. **c** MR images with a T_1-weighted spin-echo sequence with TE of 12 ms and TR of 300 ms. Reproduced with permission from [131]. (Copyright 2006 Nature Publishing Group)

as a new class of candidates as anticancer drugs. By incubating *fcc*-FePt NPs in dialysis tubing at pH 7.4 and 4.8, the Sun's group proved that NPs are stable at neutral pH conditions (pH = 7.4) but releasing Fe at low pH environments (pH = 4.8) (Fig. 6.19a). Within a cell, the released Fe could catalyze H_2O_2 decomposition into ROS, which lead to a fast membrane lipid oxidation and cell death. By incubating NPs with A2780 cells which are loaded with DCFH-DA, a ROS indicator that would be fluorescent in the presence of ROS, they found that *fcc*-FePt NPs exhibited

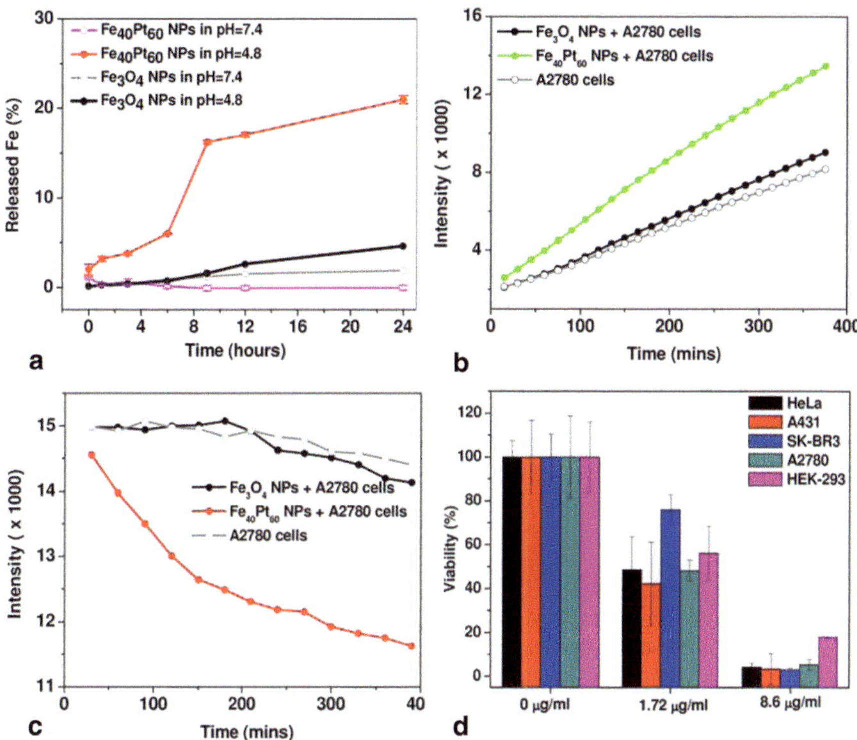

Fig. 6.19 **a** Fe release from the *fcc*-FePt and Fe_3O_4 NPs in PBS at different pH values. **b** Time-dependent fluorescent intensity from DCFH-DA labelled A2780 cells with *fcc*-FePt and Fe_3O_4 NPs (Fe concentration: 1.5 µg/ml) in HBSS. **c** Fluorescent emission intensity detected at 595 nm from C11-BODIPY labelled A2780 cells and those treated with FePt or Fe_3O_4 NPs. **d** MTT cellular viability of several tumour cell lines incubated with *fcc*-FePt NPs at different iron concentrations at 37 °C for 24 h. Reproduced with permission from [133]. (Copyright 2009 American Chemical Society)

a 50 % higher fluorescence after 6 h compared to cells incubated with the Fe_3O_4 NPs and the control group, proving that the *fcc*-FePt NPs induce the formation of excess ROS in A2780 cells (Fig. 6.19b). By further assessing the membrane lipid damage via the oxidation of a fluorescent dye C11-BODIPY, whose fluorescence would be decreased during this process, it showed a much weaker fluorescence after treating A2789 cells with *fcc*-FePt NPs compared with that treated with Fe_3O_4 NPs and the control group (Fig. 6.19c), indicating a strong membrane oxidation was induced by *fcc*-FePt NPs. Interestingly, as shown in the MTT study, the FePt-initiated catalytic formation of ROS results in serious toxicity to various cancer cells, including A2780, HeLa, A431, Sk-Br3, and HEK-293 cell (Fig. 6.19d), revealing that this iron release process may be a general protocol as a cancer therapy [133].

6.4 Conclusions and Perspectives

Over the past decades, considerable progress has been made in the synthetic methods toward magnetic-metallic NCs in a large scale in diameter and various morphologies, which possess important applications including ultrahigh-density magnetic storage, electrochemical catalysis, and biological systems. Colloidal chemical synthesis, which includes chemical reduction, decomposition, and precipitation of corresponding precursors, are still fascinating areas worth pursuing. The generalized synthetic protocols, intensively modulated morphologies and compositions of the final products and large-scale synthesis of magnetic-metallic NCs with specialities into organized alignments are needed. There is also a need to intensively investigate the toxicity of MNPs before they are applied clinically, and to find novel metallic materials to be screened for maximum absorption of magnetic field lines for higher MH and MRI. The field will open for broader and in depth investigations that may bring magnetic-metallic NCs into clinical trials in the near future, with interdisciplinary collaborations from physics, chemistry, biology, pharmacy, and clinical medicine.

Acknowledgment This work was partially supported by National Natural Science Foundation of China (NSFC) (nos. 51125001 and 51172005), the National Basic Research Program of China (no. 2010CB934601), the Research Fellowship for International Young Scientists of the National Natural Science Foundation of China (grant no. 51250110078), the Doctoral Program (no. 20090001120010), the Natural Science Foundation of Beijing (2122022), and PKU COE-Health Science Center Seed Fund.

Reference

1. T. Hyeon, Chem. Commun., 927 (2003).
2. D. L. Huber, Small, **1**, 482 (2005).
3. A.-H. Lu, E. L. Salabas and F. Schüth, Angew. Chem. Int. Ed., **46**, 1222 (2007).
4. M. Willard, L. Kurihara, E. Carpenter, et al., Int. Mater. Rev., **49**, 125 (2004).
5. F. Liu, J.-H. Zhu, Y.-L. Hou, et al., Chin. Phys. B, **22**, 107503 (2013).
6. R. Hao, R. Xing, Z. Xu, et al., Adv. Mater., **22**, 2729 (2010).
7. M. C. Bautista, O. Bomati-Miguel, X. Zhao, et al., Nanotechnology, **15**, S154 (2004).
8. A. H. Habib, C. L. Ondeck, P. Chaudhary, et al., J. Appl. Phys., **103**, 07A307 (2008).
9. R. Hergt, S. Dutz, R. Müller, et al., J. Phys.: Condens. Matter, **18**, S2919 (2006).
10. S.-J. Cho, B. R. Jarrett, A. Y. Louie, et al., Nanotechnology, **17**, 640 (2006).
11. G. P. Van Der Laan and A. A. C. M. Beenackers, Catalysis Reviews, **41**, 255 (1999).
12. F. Luborsky, J. Phys. Chem., **61**, 1336 (1957).
13. A. Meffre, S. Lachaize, C. Gatel, et al., J. Mater. Chem., **21**, 13464 (2011).
14. Y. Yin and A. P. Alivisatos, Nature, **437**, 664 (2005).
15. S. Sun and H. Zeng, J. Am. Chem. Soc., **124**, 8204 (2002).
16. D. L. Huber, E. L. Venturini, J. E. Martin, et al., J. Magn. Magn. Mater., **278**, 311 (2004).
17. T. Hyeon, S. S. Lee, J. Park, et al., J. Am. Chem. Soc., **123**, 12798 (2001).
18. D. Kim, J. Park, K. An, et al., J. Am. Chem. Soc., **129**, 5812 (2007).
19. H. Yang, T. Ogawa, D. Hasegawa, et al., Phys. Status solidi (a), **204**, 4013 (2007).
20. A. Shavel, B. Rodríguez-González, M. Spasova, et al., Adv. Funct. Mater., **17**, 3870 (2007).

21. J. Park, K. An, Y. Hwang, et al., Nat. Mater., **3**, 891 (2004).
22. F. Dumestre, B. Chaudret, C. Amiens, et al., Science, **303**, 821 (2004).
23. L.-M. Lacroix, S. Lachaize, A. Falqui, et al., J. Appl. Phys., **103**, 07D521 (2008).
24. L.-M. Lacroix, S. b. Lachaize, A. Falqui, et al., J. Am. Chem. Soc., **131**, 549 (2008).
25. H. Honda, T. Noro and E. Miyoshi, Theor. Chem. Acc., **104**, 140 (2000).
26. R. C. Weast, M. J. Astle and W. H. Beyer, *CRC handbook of chemistry and physics*, (CRC press Boca Raton, FL, 1988).
27. M. J. Height, J. B. Howard, J. W. Tester, et al., Carbon, **42**, 2295 (2004).
28. P. Nikolaev, M. J. Bronikowski, R. K. Bradley, et al., Chem. Phys. Lett., **313**, 91 (1999).
29. A. Giesen, J. Herzler and P. Roth, The Journal of Physical Chemistry A, **107**, 5202 (2003).
30. M. Rumminger, D. Reinelt, V. Babushok, et al., Combust. Flame, **116**, 207 (1999).
31. J. Z. Wen, C. F. Goldsmith, R. W. Ashcraft, et al., J. Phys. Chem. C, **111**, 5677 (2007).
32. C. Murray, S. Sun, H. Doyle, et al., MRS Bull., **26**, 985 (2001).
33. S. Peng, C. Wang, J. Xie, et al., J. Am. Chem. Soc., **128**, 10676 (2006).
34. L. M. Lacroix, N. Frey Huls, D. Ho, et al., Nano Lett., **11**, 1641 (2011).
35. S. Zhang, G. Jiang, G. Filsinger, et al., Nanoscale, **6**, 4852 (2014).
36. S. Sun and C. B. Murray, J. Appl. Phys., **85**, 4325 (1999).
37. Z. L. Wang, Z. Dai and S. Sun, Adv. Mater., **12**, 1944 (2000).
38. D. P. Dinega and M. Bawendi, Angew. Chem. Int. Ed., **38**, 1788 (1999).
39. V. F. Puntes, K. M. Krishnan and A. P. Alivisatos, Science, **291**, 2115 (2001).
40. C. Petit, T. Cren, D. Roditchev, et al., Adv. Mater., **11**, 1198 (1999).
41. C. Black, C. Murray, R. Sandstrom, et al., Science, **290**, 1131 (2000).
42. P. Lianos and J. K. Thomas, Chem. Phys. Lett., **125**, 299 (1986).
43. C. Petit and M. Pileni, J. Phys. Chem., **92**, 2282 (1988).
44. C. Petit and M. P. Pileni, J. Phys. Chem. B, **103**, 1805 (1999).
45. J. P. Chen, C. M. Sorensen, K. J. Klabunde, et al., J. Appl. Phys., **76**, 6316 (1994).
46. M. Green, Chem. Commun., 3002 (2005).
47. G. Cheng, V. F. Puntes and T. Guo, J. Colloid Interface Sci., **293**, 430 (2006).
48. N. Cordente, C. Amiens, B. Chaudret, et al., J. Appl. Phys., **94**, 6358 (2003).
49. N. Moghimi, M. Abdellah, J. P. Thomas, et al., J. Am. Chem. Soc., **135**, 10958 (2013).
50. Y. Hou, H. Kondoh, T. Ohta, et al., Appl. Surf. Sci., **241**, 218 (2005).
51. M. R. Diehl, J.-Y. Yu, J. R. Heath, et al., J. Phys. Chem. B, **105**, 7913 (2001).
52. S. U. Son, Y. Jang, J. Park, et al., J. Am. Chem. Soc., **126**, 5026 (2004).
53. M. Respaud, J. Broto, H. Rakoto, et al., Phys. Rev. B, **57**, 2925 (1998).
54. K. Soulantica, A. Maisonnat, M.-C. Fromen, et al., Angew. Chem. Int. Ed., **40**, 448 (2001).
55. N. Cordente, M. Respaud, F. Senocq, et al., Nano Lett., **1**, 565 (2001).
56. J.-P. Wang, Proc. IEEE, **96**, 1847 (2008).
57. S. Sun, Adv. Mater., **18**, 393 (2006).
58. K. R. Coffey, M. A. Parker and J. K. Howard, IEEE Trans. Magn., **31**, 2737 (1995).
59. J. Li, M. Mirzamaani, X. Bian, et al., J. Appl. Phys., **85**, 4286 (1999).
60. D. Weller and A. Moser, Magnetics, IEEE Transactions on, **35**, 4423 (1999).
61. S. Stappert, B. Rellinghaus, M. Acet, et al., J. Cryst. Growth, **252**, 440 (2003).
62. S. Sun, C. B. Murray, D. Weller, et al., Science, **287**, 1989 (2000).
63. M. Chen, J. Kim, J. Liu, et al., J. Am. Chem. Soc., **128**, 7132 (2006).
64. J. Chen, T. Herricks, M. Geissler, et al., J. Am. Chem. Soc., **126**, 10854 (2004).
65. L. E. M. Howard, H. L. Nguyen, S. R. Giblin, et al., J. Am. Chem. Soc., **127**, 10140 (2005).
66. G. S. Chaubey, C. Barcena, N. Poudyal, et al., J. Am. Chem. Soc., **129**, 7214 (2007).
67. D. Zitoun, C. Amiens, B. Chaudret, et al., J. Phys. Chem. B, **107**, 6997 (2003).
68. C. Desvaux, C. Amiens, P. Fejes, et al., Nat. Mater., **4**, 750 (2005).
69. N. Liakakos, C. Gatel, T. Blon, et al., Nano Lett., **14**, 2747 (2014).
70. B. Zhang and D. S. Su, Angew. Chem. Int. Ed., **52**, 8504 (2013).
71. P. Bose, S. Bid, S. Pradhan, et al., J. Alloys Compd., **343**, 192 (2002).
72. Y. Chen, X. Luo, G.-H. Yue, et al., Mater. Chem. Phys., **113**, 412 (2009).
73. X. G. Li, A. Chiba and S. Takahashi, J. Magn. Magn. Mater., **170**, 339 (1997).

74. H. Wang, J. Li, X. Kou, et al., J. Cryst. Growth, **310**, 3072 (2008).
75. C. Giordano and M. Antonietti, Nano Today, **6**, 366 (2011).
76. R. Andrievski, J Mater Sci, **32**, 4463 (1997).
77. V. B. Raut, T. Seth, A. U. Pawar, et al., Int. J. Nanotechnol., **7**, 1098 (2010).
78. P. Schaaf, M. Kahle and E. Carpene, Appl. Surf. Sci., **247**, 607 (2005).
79. J. A. Nelson and M. J. Wagner, Chem. Mater., **14**, 1639 (2002).
80. C. Giordano, A. Kraupner, S. C. Wimbush, et al., Small, **6**, 1859 (2010).
81. A. Meffre, B. Mehdaoui, V. Kelsen, et al., Nano Lett., **12**, 4722 (2012).
82. C. Yang, H. Zhao, Y. Hou, et al., J. Am. Chem. Soc., **134**, 15814 (2012).
83. T. Welton, Chem. Rev., **99**, 2071 (1999).
84. K. Marsh, J. Boxall and R. Lichtenthaler, Fluid Phase Equilib., **219**, 93 (2004).
85. L. Lartigue, J. Long, X. Dumail, et al., J. Nanopart. Res., **15**, 1 (2013).
86. Y. Sun and Y. Xia, Adv. Mater., **353**, 833 (1991).
87. K. TekaiaáElhsissen, J. Mater. Chem., **6**, 573 (1996).
88. Y. Zhang, G. S. Chaubey, C. Rong, et al., J. Magn. Magn. Mater., **323**, 1495 (2011).
89. T. Hinotsu, B. Jeyadevan, C. N. Chinnasamy, et al., J. Appl. Phys., **95**, 7477 (2004).
90. C. Osorio-Cantillo and O. Perales-Perez, J. Appl. Phys., **105**, 07A332 (2009).
91. Y. Soumare, C. Garcia, T. Maurer, et al., Adv. Funct. Mater., **19**, 1971 (2009).
92. G. Viau, C. Garcıa, T. Maurer, et al., Phys. Status solidi (a), **206**, 663 (2009).
93. J. An, B. Tang, X. Zheng, et al., J. Phys. Chem. C, **112**, 15176 (2008).
94. B. Wiley, T. Herricks, Y. Sun, et al., Nano Lett., **4**, 2057 (2004).
95. Z. J. Huba and E. E. Carpenter, J. Appl. Phys., **111**, 07B529 (2012).
96. X. He, L. T. Kong and B. X. Liu, J. Appl. Phys., **97**, 106107 (2005).
97. D. A. Papaconstantopoulos, J. L. Fry and N. E. Brener, Phys. Rev. B, **39**, 2526 (1989).
98. M. Podgórny and J. Goniakowski, Phys. Rev. B, **42**, 6683 (1990).
99. L. Yue, R. Sabiryanov, E. M. Kirkpatrick, et al., Phys. Rev. B, **62**, 8969 (2000).
100. A. M. Gadalla and B. Bower, Chem. Eng. Sci., **43**, 3049 (1988).
101. N. Ji, T. Zhang, M. Zheng, et al., Angew. Chem., **120**, 8638 (2008).
102. D. Trimm, Catal. Today, **49**, 3 (1999).
103. Y. Goto, K. Taniguchi, T. Omata, et al., Chem. Mater., **20**, 4156 (2008).
104. Z. L. Schaefer, K. M. Weeber, R. Misra, et al., Chem. Mater., **23**, 2475 (2011).
105. J. L. Dormann, D. Fiorani and E. Tronc, *in Advances in Chemical Physics*, Vol 98, ed. by. I. Prigogine and S. A. Rice, (John Wiley & Sons, New York, 1997), p. 283.
106. S. L. McGill, C. L. Cuylear, N. L. Adolphi, et al., IEEE Trans. Nanobiosci., **8**, 33 (2009).
107. D. Yoo, J.-H. Lee, T.-H. Shin, et al., Acc. Chem. Res., **44**, 863 (2011).
108. C. Riviere, C. Wilhelm, F. Cousin, et al., Eur. Phys. J. E, **22**, 1 (2007).
109. C. S. S. R. Kumar and F. Mohammad, Adv. Drug Deliv. Rev., **63**, 789 (2011).
110. A. Jordan, R. Scholz, P. Wust, et al., J. Magn. Magn. Mater., **201**, 413 (1999).
111. P. Wust, U. Gneveckow, M. Johannsen, et al., Int. J. Hyperthermia, **22**, 673 (2006).
112. M. P. Morales, S. Veintemillas-Verdaguer, M. I. Montero, et al., Chem. Mater., **11**, 3058 (1999).
113. J. Carrey, B. Mehdaoui and M. Respaud, J. Appl. Phys., **109**, 083921 (2011).
114. J.-H. Lee, J.-t. Jang, J.-s. Choi, et al., Nat. Nanotechnol., **6**, 418 (2011).
115. M. Zeisberger, S. Dutz, R. Müller, et al., J. Magn. Magn. Mater., **311**, 224 (2007).
116. C. G. Hacliipanayis, M. J. Bonder, S. Balakrishanan, et al., Small, **4**, 1925 (2008).
117. S. Maenosono and S. Saita, IEEE Trans. Magn., **42**, 1638 (2006).
118. L. M. Lacroix, R. B. Malaki, J. Carrey, et al., J. Appl. Phys., **105**, 023911 (2009).
119. R. Kappiyoor, M. Liangruksa, R. Ganguly, et al., J. Appl. Phys., **108**, 094702 (2010).
120. S. H. Koenig and K. E. Kellar, Magnetic Resonance in Medicine, **34**, 227 (1995).
121. W. Tang, Z. Zhen, C. Yang, et al., Small, **10**, 1245 (2014).
122. G. Huang, J. Hu, H. Zhang, et al., Nanoscale, **6**, 726 (2014).
123. J. Yu, C. Yang, J. Li, et al., Adv. Mater., (2014), DOI: 10.1002/adma.201305811.
124. S. Maenosono, T. Suzuki and S. Saita, J. Magn. Magn. Mater., **320**, L79 (2008).
125. J. Gao, G. Liang, J. S. Cheung, et al., J. Am. Chem. Soc., **130**, 11828 (2008).

126. Y. Liu, K. Yang, L. Cheng, et al., Nanomedicine: NBM, **9**, 1077 (2013).
127. S.-W. Chen, J.-J. Lai, C.-L. Chiang, et al., Rev. Sci. Instrum., **83**, 064701 (2012).
128. S.-W. Chou, Y.-H. Shau, P.-C. Wu, et al., J. Am. Chem. Soc., **132**, 13270 (2010).
129. I. A. Choi, Y. Li, D. J. Kim, et al., Chem. Asian. J., **8**, 290 (2013).
130. X. Meng, H. C. Seton, L. T. Lu, et al., Nanoscale, **3**, 977 (2011).
131. W. S. Seo, J. H. Lee, X. Sun, et al., Nat. Mater., **5**, 971 (2006).
132. S. P. Sherlock, S. M. Tabakman, L. Xie, et al., ACS Nano, **5**, 1505 (2011).
133. C. Xu, Z. Yuan, N. Kohler, et al., J. Am. Chem. Soc., **131**, 15346 (2009).

Chapter 7
Metallic Nanostructures for Electrocatalysis

Zhenmeng Peng

Abstract The metal electrocatalyst research requires molecular-level knowledge of the electrocatalytic reactions and the controlling factors that determine the reaction kinetics. Several important factors, including electronic structure and geometric structure of metal surfaces, third-body effect, and bifunctional effect, are discussed for their roles in electrocatalysis. Chemical stability and surface restructuring/segregation of metals, which need practical consideration in the electrocatalyst research, are introduced. The electrochemistry of oxygen, hydrogen, carbon-containing compounds, and nitrogen-containing compounds is reviewed. Discussions are focused on recent advances in preparing metallic nanostructures for these reactions, and on understanding the relationship between the physical parameters of metallic nanostructures and the electrocatalytic property.

7.1 Introduction

Metallic nanostructures for electrocatalysis have been an area of active research for decades and can find important applications in many research areas and industries [1–6], including energy generation, chemicals production, electrochemical sensors, environmental control, and so on. With increasing concerns over environments and depletion of fossil fuels in recent years, more research efforts have been put into utilizing alternative energy resources and developing clean energy technologies. For example, polymer electrolyte membrane fuel cells (PEMFCs), which can generate electric energy by electrochemically reacting hydrogen and oxygen, have been researched for automobiles [7–11]. Metal–air batteries have been developed to improve the power density and efficiency of portable energy devices [12–15]. Electrochemical water splitting has been recently considered as a clean and alternative technology for producing H_2 [16–18], which is both energy carrier and important

Book Chapter to "Metallic Nanostructures: form Controlled Synthesis to Applications"

Z. Peng (✉)
Department of Chemical and Biomolecular Engineering, University of Akron,
Akron, OH 44325, USA
e-mail: zpeng@uakron.edu

© Springer International Publishing Switzerland 2015
Y. Xiong, X. Lu (eds.), *Metallic Nanostructures*, DOI 10.1007/978-3-319-11304-3_7

chemical commodity. All these technologies involve electrochemical reactions and require the use of metallic nanostructures for promoting the reaction kinetics. Research of metallic nanostructures as efficient electrocatalyst thus becomes crucial to advance these energy technologies.

Metal electrocatalyst researches in the early years have once been highly empirical, primarily based on trial-and-error experiments and research experiences. It was caused by the fact that electrocatalysis is governed by multiple parameters of both reaction and catalyst and that these parameters often intervene with each other to affect the catalytic property. With rapid advances in both experimental and theoretical tools nowadays, electrocatalytical reactions can be better understood at molecular level. Many theories have also been proposed to guide metal electrocatalyst research and development. In this chapter, we will first introduce the fundamentals for metal electrocatalysis and electrocatalyst, and then discuss recent examples of metallic nanostructures for electrocatalysis.

7.2 Electrochemical Reaction

7.2.1 Thermodynamics of Electrochemical Reaction

An electrochemical reaction involves the transfer of electrons between two substances: one, a solid (electrode), and the other, a liquid (electrolyte) in most cases. In principle, every redox chemical reaction, $aA + bB \rightarrow cC + dD$, can be written into two electrochemical reactions, in which one gains the electrons being generated from the other [19]:

$$aA \rightarrow cC + ne^- \tag{7.1}$$

$$bB + ne^- \rightarrow dD \tag{7.2}$$

An electrochemical cell can be built by separating occurrence of the two electrochemical reactions at different electrodes and connecting the electrodes using electrolyte and an external circuit (Fig. 7.1). Thus, the electrochemical reactions are also called half-cell reactions. One major difference between a redox chemical reaction and an electrochemical cell made of the reaction is their energy conversion. A redox chemical reaction releases/absorbs thermal energy during the course of reaction, whereas an electrochemical cell generates/consumes electric energy. The open-circuit voltage of the cell (E_{cell}) can be correlated with the Gibbs free energy change of the reaction (ΔG) using the following equation:

$$\Delta G = -nFE_{cell} \tag{7.3}$$

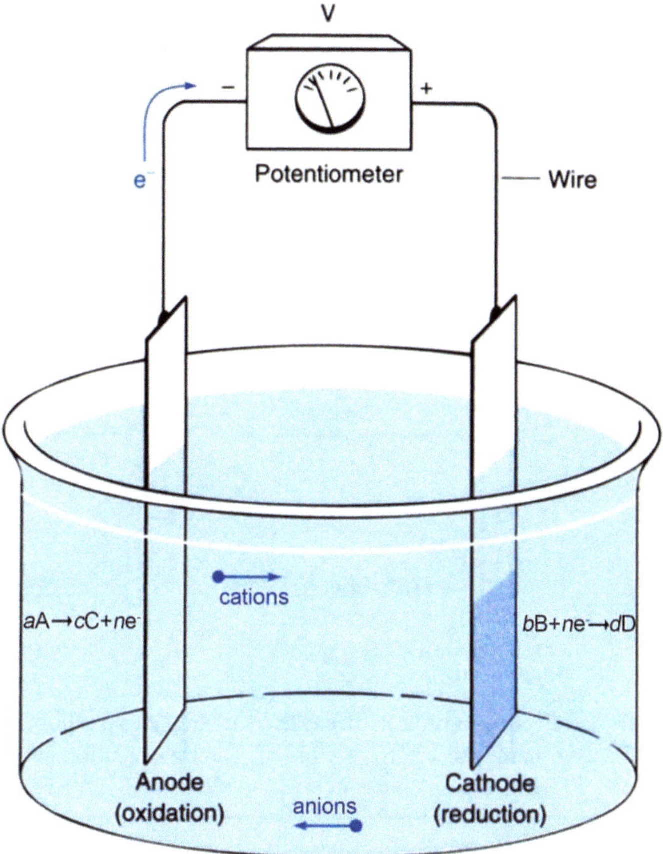

Fig. 7.1   An electrochemical cell constructed using redox chemical reaction $aA + bB \rightarrow cC + dD$

where F is the Faraday constant and has a value of 96485 C/mol. An electrochemical cell made of spontaneous redox chemical reactions (negative ΔG) has a positive E_{cell} and is often called galvanic cell. It can be utilized to convert chemical energy stored in fuel molecules into electric energy. On the other hand, one made of nonspontaneous redox reactions (positive ΔG) is called electrolytic cell and can be utilized to produce valuable chemicals by inputting electric energy into the reaction system. One example is the reaction between H_2 and O_2, which is a spontaneous redox reaction [20]:

$$2H_2 + O_2 \rightarrow 2H_2O \quad (\Delta G^0 = -475 \text{ KJ/mol}) \tag{7.4}$$

A PEMFC can be constructed by separating H_2 oxidation and O_2 reduction at anode and cathode, having a PEM in between, and connecting the electrodes with an electric circuit. The two half-cell reactions can be depicted as follows:

Table 7.1 Standard redox potentials of half-cell reactions

Half-cell reaction	$E^0_{C/A}$ (V vs. SHE)
$N_2 + 6H_2O + 6e^- \Leftrightarrow 2NH_3 + 6OH^-$	-0.77
$N_2 + 4H^+ + 4e^- \Leftrightarrow N_2H_4$	-0.39
$CO_2 + 2H^+ + 2e^- \Leftrightarrow HCOOH$	-0.25
$CO_2 + 2H^+ + 2e^- \Leftrightarrow CO + H_2O$	-0.10
$2H^+ + 2e^- \Leftrightarrow H_2$	0.00
$CO_2 + 6H^+ + 6e^- \Leftrightarrow CH_3OH + H_2O$	0.04
$CO_2 + 12H^+ + 12e^- \Leftrightarrow CH_3CH_2OH + 3H_2O$	0.08
$O_2 + 2H_2O + 4e^- \Leftrightarrow 4OH^-$	0.401
$O_2 + 2H^+ + 2e^- \Leftrightarrow H_2O_2$	0.695
$O_2 + 4H^+ + 4e^- \Leftrightarrow 2H_2O$	1.229

$$2H_2 \rightarrow 4H^+ + 4e^- \tag{7.5}$$

$$O_2 + 4H^+ + 4e^- \rightarrow 2H_2O \tag{7.6}$$

H_2 is electrochemically oxidized into protons because of its higher susceptibility to lose electrons than O_2. O_2 gains the electrons from H_2 electro-oxidation and is reduced into H_2O. Such a PEMFC has an open-circuit cell voltage of $E^0_{cell} = 1.229$ V if being operated under the standard condition.

Redox potentials are defined for half-cell reactions to represent their ability to gain or lose electrons electrochemically. The voltage of cells constructed using any two half-cell reactions can thus be described using the following equation:

$$E_{cell} = E_{cathode} - E_{anode} \tag{7.7}$$

The redox potentials are not absolute values, but rather relative ones comparing to reference electrode reactions. One most often used reference is the standard hydrogen electrode (SHE), which operates under the standard condition and has a defined redox potential of $E^0_{H^+/H_2} = 0V$. By constructing an electrochemical cell using any given half-cell reaction and the SHE and measuring the cell voltage, its redox potential vs. SHE can be determined using the following equation:

$$E_{C/A} = E^0_{cell} + E^0_{H^+/H_2} = E^0_{cell} \tag{7.8}$$

Table 7.1 lists a few half-cell reactions of electrocatalysis importance and their equilibrium redox potentials under standard condition ($E^0_{C/A}$). The redox potential under nonstandard conditions can be calculated using the Nernst equation:

$$E_{C/A} = E_{C/A}^0 - \frac{RT}{nF} ln \frac{[A]^a}{[C]^c} \tag{7.9}$$

Correspondingly, the open-circuit voltage of any electrochemical cell operating under nonstandard conditions can also be calculated:

$$E_{cell} = E_{B/D}^0 - E_{C/A}^0 \frac{RT}{nF} ln \frac{[C]^c [D]^d}{[A]^a [B]^b} \tag{7.10}$$

7.2.2 Kinetics of Electrochemical Reaction

Many theories have been developed for understanding the kinetics of chemical reaction. In the transition state theory, reactants overcome an energy barrier to generate active intermediates, which further react to form products. The rate constant, k, can be described using the Arrhenius equation [21]:

$$k = Ae^{-E_a/RT} \tag{7.11}$$

where A is a pre-exponential constant, E_a is the activation energy, R is the gas constant, and T is the reaction temperature. For an elementary reaction, the reaction rate is simply the production of k and concentration(s) of the reactant(s). For a complex reaction that contains multiple elementary steps, the reaction kinetics is determined by the rate-limiting step. The rate law is an algebraic function of k, concentration(s) of the reactant(s), and some constant parameters.

The kinetics of electrochemical reactions can also been understood using the transition state theory. Comparing to chemical reactions, electrode potential (E) serves as one additional factor for controlling the electrochemical reaction kinetics. The electrode potential can alter the energy barrier and consequently influence the reaction rate (Fig. 7.2). If we consider an elementary half-cell reaction $O + ne \underset{k_b}{\overset{k_f}{\Leftrightarrow}} R$ and use current density (i) to represent the reaction rate, the relationship between i and E can be depicted using the Butler–Volmer equation [22]:

$$i = i_0 \left[e^{-\alpha nF \eta/RT} - e^{(1-\alpha)nF \eta/RT} \right] \tag{7.12}$$

$$i_0 = nFAk_f^{1-\alpha}k_b^\alpha [O]^{1-\alpha}[R]^\alpha \tag{7.13}$$

$$\eta = E - E_{O/R} \tag{7.14}$$

where i_0 is the exchange current density, α is the charge transfer coefficient, η is the overpotential, E and $E_{O/R}$ are the electrode potential and redox potential, and k_f and k_b are the rate constant for the forward and reverse half-cell reaction. When $\eta = 0$,

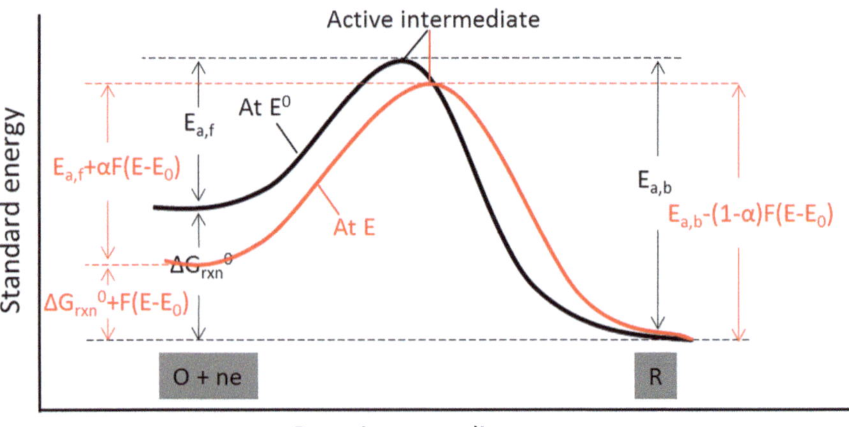

Fig. 7.2 Free energy changes during a reaction and effects of a potential change on the reaction

which represents the electrode stays at the redox potential, i is equal to zero and the half-cell reaction is at its equilibrium. The i value is positive and increases rapidly with η when $E<E_{O/R}$ (negative η), indicating a reduction process. It becomes negative and the value changes exponentially with η when $E>E_{O/R}$ (positive η), corresponding to an oxidation process (Fig. 7.3).

The half-cell reactions can be classified into reversible and irreversible systems based on their intrinsic reaction kinetics. For reversible half-cell reactions, electrons can be rapidly exchanged at the electrode and the i_0 value is relatively large. A small η can lead to a large increase in i (Fig. 7.3). The apparent reaction rate is limited by the diffusion of reacting species rather than the fast reaction kinetics. Cyclic voltammetry of such reactions shows defined distance between peak potentials, and the peak current densities can be calculated using the Randles–Sevcik equation (Fig. 7.4):

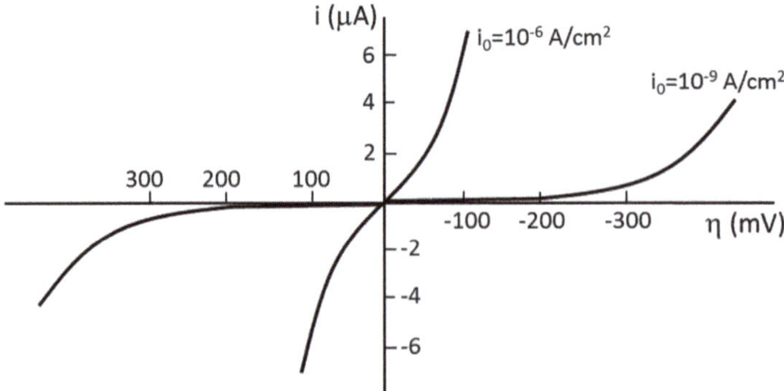

Fig. 7.3 Current density (i) vs. overpotential (μ) curves for the reaction $O+ne \underset{k_b}{\overset{k_f}{\rightleftharpoons}} R$ with different i_0

Fig. 7.4 Cyclic voltammetry curves of reversible and
irreversible electrochemical
processes

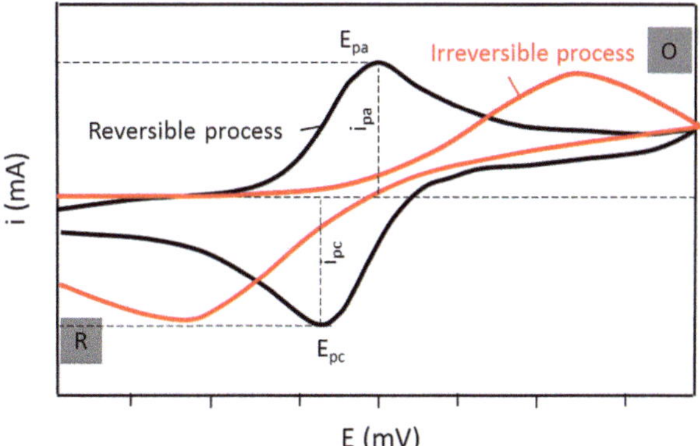

$$E_{pa} - E_{pc} = \frac{59}{n} mV \tag{7.15}$$

$$i_p = 2.69 \times 10^8 \times n^{3/2} \times EA \times D^{1/2} \times v^{1/2} \times C \tag{7.16}$$

where E_{pa} and E_{pc} are the anodic and cathodic peak potentials, i_{pc} and i_{pc} are the corresponding peak currents, EA is the electrode area, D is the diffusion coefficient of reacting species, v is the potential scan rate, and C is the concentration of reacting species ([O] for calculating i_{pc} and [R] for calculating i_{pa}).

For irreversible processes (those with sluggish electron exchange and thus small i_0), the two peaks are reduced in size and widely separated when being compared with reversible systems. Totally irreversible systems are characterized by a shift of the peak potential with the scan rate:

$$E_p = E_{O/R} - \frac{RT}{\alpha n_a F} \left[0.78 - ln \left(\frac{k}{D^{1/2}} \right) + ln \left(\frac{\alpha n_a F v}{RT} \right)^{1/2} \right] \tag{7.17}$$

where n_a is the number of electrons involved in the charge-transfer step, k is the rate constant (k_f for calculating E_{pc} and k_b for calculating E_{pa}), and v is scan rate. The peak current density can be given by:

$$i_p = 2.99 \times 10^8 \times n \times (\alpha \times n_a)^{1/2} \times EA \times D^{1/2} \times v^{1/2} \times C \tag{7.18}$$

The i_p value is still proportional to the bulk concentration, but will be lower in height (depending on the value of α).

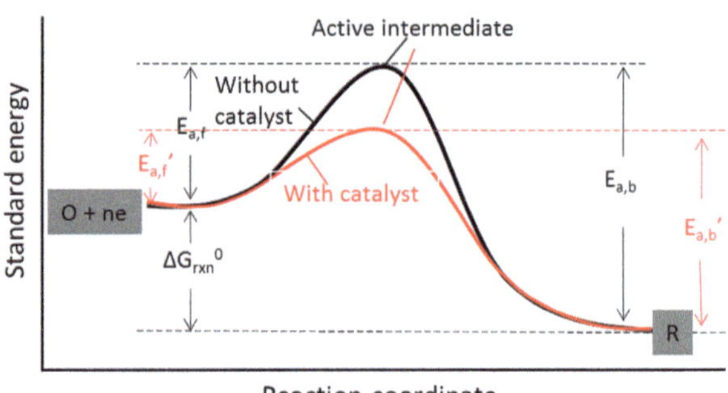

Fig. 7.5 Free energy changes during an electrochemical reaction with and without metal electrocatalysis

7.3 Fundamentals for Metal Electrocatalysis

7.3.1 Mechanism of Metal Electrocatalysis

As being discussed in Chap. 7.2.2, electrochemical reactions can be classified into reversible and irreversible processes based on their exchange current density, i_0. One characteristic of the irreversible processes is a much smaller i_0 and thus the requirement of a much larger η to achieve a same i as for the reversible processes (Eq. 7.12 and Fig. 7.3). If an electrochemical cell is constructed using irreversible half-cell reactions, the working cell voltage will be significantly smaller than the open-circuit value. Consequently, the energy conversion efficiency will be significantly lower than the theoretical value.

The irreversible half-cell reaction kinetics can be improved by metal electrocatalysis. From Eq. 7.13, i_0 is affected by several parameters, including the number of electrons being transferred (n), the pre-exponential constant (A), the charge transfer coefficient (α), the concentration of reacting species ([O] and [R]), and the rate constants (k_f and k_b). k_f and k_b depict the intrinsic kinetics of forward and reverse half-cell reactions. Similar to chemical reactions, k_f and k_b for the half-cell reactions also follow the Arrhenius law (Eq. 7.11) and are functions of their activation energy ($E_{a,f}$ and $E_{a,b}$). The irreversible half-cell reactions have large $E_{a,f}$ and $E_{a,b}$ values, which cause small k_f and k_b. In consequence, a small i_0 is resulted and a large η is required for the half-cell reaction to occur effectively.

Metals can change the reaction rate by promoting a different molecular path for half-cell reactions, which is named metal electrocatalysis. Rather than directly exchanging electrons with electrode, the reacting species first interact with metal surface atoms. The interaction can lead to the generation of altered active intermediates, which further exchange electrons with electrode and react [23]. By altering the reaction path, the energy barrier can be effectively decreased, leading to improved rate constant and thus increased reaction kinetics (Fig. 7.5) [24].

7.3.2 *Kinetics of Metal Electrocatalysis*

The overall reaction rate of an electrocatalytic reaction is limited by the rate of the slowest step in the path, including diffusion of reactant(s) from bulk electrolyte to catalyst surface, electrocatalytic reaction, and diffusion of product(s) from catalyst surface back to bulk electrolyte. Here we focus on electrocatalysis and assume the diffusions are much faster than the electrocatalytic reaction rate, which means that the concentrations of reacting species on catalyst surface are equal to those in bulk electrolyte and the overall reaction is controlled by metal electrocatalysis. If we consider a simple half-cell reaction $O + ne \underset{k_b}{\overset{k_f}{\rightleftharpoons}} R$ occurring under metal electrocatalysis, the reaction path typically involves three steps including reactant adsorption, surface reaction, and product desorption:

$$O + S \underset{k_{-A}}{\overset{k_A}{\rightleftharpoons}} O \cdot S \tag{7.19}$$

$$O \cdot S + ne \underset{k_{-S}}{\overset{k_S}{\rightleftharpoons}} R \cdot S \tag{7.20}$$

$$R \cdot S \underset{k_{-D}}{\overset{k_D}{\rightleftharpoons}} R + S \tag{7.21}$$

where S stands for metal active sites, and $O \cdot S$ and $R \cdot S$ are adsorbed O and R species. The overall reaction rate law (r) is determined by the rate-limiting step, which can fall in one of the following three situations for different reactions and/or using different catalysts.

1. Adsorption-limited process:

$$r = \frac{(k_A[O] - k_{-A}K_R[R]/K_S) \cdot C_t}{1 + K_R[R] + K_R[R]/K_S} \tag{7.22}$$

2. Surface reaction-limited process:

$$r = \frac{(k_S K_O[O] - k_{-S}K_R[R]) \cdot C_t}{1 + K_O[O] + K_R[R]} \tag{7.23}$$

3. Desorption-limited process:

$$r = \frac{(k_D K_S K_O[O] - k_{-D}[R]) \cdot C_t}{1 + K_O[O] + K_S K_O[O]} \tag{7.24}$$

where k_A and k_{-A} are the rate constants for [O] adsorption, k_S and k_{-S} are the rate constants for the surface reaction, k_D and k_{-D} are the rate constants for [R] desorption, K_O and K_R are the equilibrium constants for O and R adsorption, K_S is the equilibrium constant for the surface reaction, and C_t is the concentration of metal active sites. The electrochemical reaction is at equilibrium, i.e. $r=0$, at $\eta=0$.

The exchange current density, i_0, is largely determined by the rate constants of the rate-limiting step from the aforementioned rate law expressions. Taking the surface reaction-limited process as one example, both K_O and K_R are significantly larger in values than k_S and k_{-S}. If we consider a special case where K_O and K_R have similar values and O and R have comparable concentrations, the apparent rate constants for the forward and reverse half-cell reaction under metal electrocatalysis will become $k_f' \approx k_s / 2$ and $k_b' \approx k_{-s} / 2$. In another word, the rate constants for the overall electrochemical reaction are directly determined by the rate constants of the surface reaction step for a surface reaction-limited mechanism. If we take k_f' and k_b' values into Eq. 7.13, we can describe i_0 as a function of k_S and k_{-S}:

$$i_0 \approx nFA(k_S / 2)^{1-\alpha}(k_{-S} / 2)^{\alpha}[O]^{1-\alpha}[R]^{\alpha} \tag{7.25}$$

From this equation, the electrochemical reaction kinetics of a surface reaction-limited process can be promoted by using metal catalysts that exhibit high k_S and k_{-S} values.

7.4 Fundamentals for Metal Electrocatalyst

An electrocatalyst by definition is a catalyst that participates in an electrochemical reaction and alters the reaction kinetics without being consumed in the process. Many types of materials can serve as electrocatalyst, among which metals are most often used for their outstanding property. Chapter 7.3 provides discussions on the mechanism of metal electrocatalysis and the kinetics of electrocatalytic reactions. Similar to heterogeneous catalyst, metal electrocatalyst is reaction specific. A metal electrocatalyst good for one electrochemical reaction might be completely inactive for another. For a same electrochemical reaction, some metals exhibit high activity, while some others are inert. For surface sensitive electrochemical reactions, the property of metal electrocatalyst can be significantly altered by the nature of the metal surfaces exposed. From Chaps. 7.2 and 7.3, we learn the electrochemical reaction kinetics is determined by i_0, which can be affected by the interaction between reacting species and metal active sites. In this section, we discuss several major factors that can affect the metal-reacting species interaction and thus the reaction kinetics.

7.4.1 Electronic Effect

As being discussed in Chap 7.3, metal electrocatalysis involves interaction between reacting species and metal active sites. At molecular level, the interaction is caused by the different energy levels of their outer-layer electrons, which lead to electron

Table 7.2 Shifts in d-band centers, ε_d, of surface impurities (A) and overlayers (B) relative to the clean metal values (bold).

	Fe	Co	Ni	Cu	Ru	Rh	Pd	Ag	Ir	Pt	Au
A											
Fe	**−0.92**	−0.05	−0.20	−0.13	−0.29	−0.54	−1.24	−0.83	−0.36	−1.09	−1,42
Co	0.01	**−1.17**	−0.28	−0.16	−0.24	−0.58	−1.37	−0.91	−0.36	−1.19	−1.56
Ni	0.09	0.19	**−1.29**	0.19	−0.14	−0.31	−0.97	−0.53	−0.14	−0.80	−1.13
Cu	0.56	0.60	0.27	**−2.67**	0.58	0.32	−0.64	−0.70	0.58	−0.33	−1.09
Ru	0.21	0.26	0.01	0.12	**−1.41**	−0.17	−0.82	−0.27	0.02	−0.62	−0.84
Rh	0.24	0.34	0.16	0.44	0.04	**−1.73**	−0.54	0.07	0.17	−0.35	−0.49
Pd	0.37	0.54	0.50	0.94	0.24	0.36	**−1.83**	0.59	0.53	0.19	0.17
Ag	0.72	0.84	0.67	0.47	0.84	0.86	0.14	**−4.30**	1.14	0.50	−0.15
Ir	0.21	0.27	0.05	0.21	0.09	−0.15	−0.73	−0.13	**−2.11**	−0.56	−0.74
Pt	0.33	0.48	0.40	0.72	0.14	0.23	−0.17	0.44	0.38	**−2.25**	−0.05
Au	0.63	0.77	0.63	0.55	0.70	0.75	0.17	0.21	0.98	0.46	**−3.56**
B											
Fe	**−0.92**	0.14	−0.04	−0.05	−0.73	−0.72	−1.32	−1.25	−0.95	−1.48	−2.19
Co	−0.01	**−1.17**	−0.20	−0.06	−0.70	−0.95	−1.65	−1.36	−1.09	−1.89	−2.39
Ni	0.96	0.11	**−1.29**	0.12	−0.63	−0.74	−1.32	−1.14	−0.86	−1.53	−2.10
Cu	0.25	0.38	0.18	**−2.67**	−0.22	−0.27	−1.04	−1.21	−0.32	−1.15	−1.96
Ru	0.30	0.37	0.29	0.30	**−1.41**	−0.12	−0.47	−0.40	−0.13	−0.61	−0.86
Rh	0.31	0.41	0.34	0.22	0.03	**−1.73**	−0.39	−0.08	0.03	−0.45	−0.57
Pd	0.36	0.54	0.54	0.80	−0.11	0.25	**−1.83**	0.15	0.31	0.04	−0.14
Ag	0.55	0.74	0.68	0.62	0.50	0.67	0.27	**−4.30**	0.80	0.37	−0.21
Ir	0.33	0.40	0.33	0.56	−0.01	−0.03	−0.42	−0.09	**−2.11**	−0.49	−0.59
Pt	0.35	0.53	0.54	0.78	0.12	0.24	0.02	0.19	0.29	**−2.25**	−0.08
Au	0.53	0.74	0.71	0.70	0.47	0.67	0.35	0.12	0.79	0.43	**−3.56**

The impurity/overlayer atoms are listed horizontally and the host entries are listed vertically. The surfaces considered are the closest packed, and the overlayer structure are pseudomorphic. All values are in eV, and the e_d values are relative to the Fermi level (reprinted with permission from [31], copyright 2007 Elsevier)

transfer to minimize the system energy. The process is usually accompanied by chemical bond breakage/formation, including chemisorption and intermediate generation. The d-band center of metals relative to the Fermi level, ε_d, has been identified as a good measure for describing the strength of interaction with reacting species [25–29]. Nørskov and coworkers have calculated ε_d values of different metals as well as their alloys and overlayer structures (Table 7.2) [30]. They have achieved promising successes in using the data for explaining chemisorption of molecules to metals.

From Table 7.2, we learn that the ε_d alters with metals, which helps to explain why they have different interaction with reacting species and exhibit different electrocatalytic property. For example, the platinum group metals (including Ru,

Rh, Pd, Ir, and Pt) are effective electrocatalyst for CO electro-oxidation reaction, whereas the coinage metals (including Cu, Ag, and Au) are inactive toward this reaction [32–34]. The little activity of coinages metals could be attributed to their significantly deeper d-band centers comparing to the platinum group metals, which leads to much weaker interaction with CO molecules. The arguments are in good consistence with Fourier transform infrared spectroscopy (FTIR) studies, which observe little CO adsorption to the coinage metals but strong CO adsorption to the platinum group metals.

Another important message drawn from Table 7.2 is that the ε_d of metals can be adjusted by alloying or creating heterogeneous overlayers. The shift in ε_d has been attributed to charge transfer and/or lattice change due to incorporation of a second metal. For instance, the ε_d of pure Pt is -2.25 eV, which enables Pt to interact moderately with many reacting species and makes it an active catalyst for a variety of electrochemical reactions, for instance oxygen reduction reaction (ORR), methanol oxidation reaction (MOR), and ammonia oxidation reaction (AOR) [20, 35–42]. Fundamental studies have suggested a more negative ε_d for Pt in ORR to suppress the adsorption of hydroxyl species, which compete with ORR and negatively influence the reaction kinetics. From Table 7.2, the ε_d for surface Pt can be shifted negatively when it is alloyed with selected elements (Pt-M, M = Fe, Co, Ni, Cu, etc) or is overlayered on these metals. In experiments, these alloys and overlayer structures have been demonstrated with improved ORR kinetics than pure Pt [43–47], which validates the d-band theory.

7.4.2 Geometric Effect

The physical structure of metals depends primarily on the arrangements of atoms that make up the metals. Most of metal electrocatalyst nanoparticles are crystalline in structure, i.e., the metal atoms being arranged in a pattern that repeat themselves in three dimensions. Body-centered cubic (*bcc*), face-centered cubic (*fcc*), and hexagonal close-packed (*hcp*) structures are most often observed for crystalline metals. The atom arrangement, or geometry, can vary significantly with different crystal structure and planes. Correspondingly, the geometric structure of metal surface can be greatly altered by exposing different crystal planes. Figure 7.6 shows the geometric structure of (100), (110), and (111) surface planes of an *fcc* structure, which have different atom arrangements and atom-atom distances.

It is often that multiple active sites are required simultaneously to catalyze electrochemical reactions. How these metal active sites arrange can thus dramatically alter their interaction with the reacting species, which lead to an altered reaction kinetics. These kind of electrochemical reactions are grouped as surface sensitive reactions, in which the surface geometry of metal nanoparticles can play a determining role. For instance, CO can be oxidized into CO_2 under Pt electrocatalysis, and the reaction is found sensitive to Pt surface. Mechanistic studies suggest an Eley–Rideal (E-R) mechanism in an alkaline solution [48]:

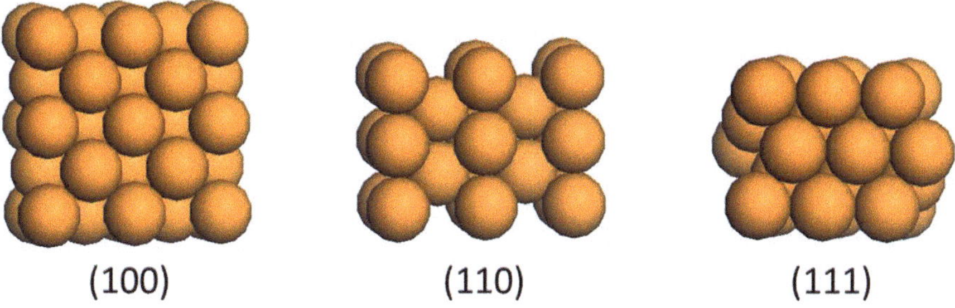

Fig. 7.6 Schematic illustration of (100), (110), and (111) surface planes of *fcc* metals

$$CO_{ads} + 2OH^- \rightarrow CO_2 + H_2O + 2e^- \tag{7.26}$$

Adsorbed CO molecules on Pt react with OH⁻ in the bulk phase, which leads to production of CO_2 and H_2O and meanwhile generation of electrons. Figure 7.7 shows the voltammetric CO oxidation on single crystalline Pt (111), (110), and (100) [48]. The main anodic peaks are resulted from E-R CO oxidation, while the small shoulder peaks in the double layer region are attributed to the reaction between CO_{ads} and OH_{ads} adsorbed on defect sites via a Langmuir–Hinshelwook (L-H) mechanism. The reaction exhibits varying onset potentials on different Pt surfaces, following the order $E_{Pt(110)} < E_{Pt(100)} < E_{Pt(111)}$. It suggests different reaction kinetics in order of $i_{0,Pt(110)} > i_{0,Pt(100)} > i_{0,Pt(111)}$. A plausible explanation is that CO molecules form multiple bonds with Pt active sites when they adsorb to the surface. The CO_{ads} on Pt (111) is primarily triple bonded while the CO_{ads} on Pt (110) and (100) are less bonded due to geometric constraint. The more stabilized CO_{ads} on Pt (111) cause a larger E_A and thus a smaller i_0 for electro-oxidation comparing to that on the other two Pt surfaces.

7.4.3   Other Effects

Besides the electronic and geometric structures of metals in controlling the intrinsic kinetics of electrochemical reactions, some other factors can also play an important role. Here, we briefly discuss two major effects which often involve in metal electrocatalysis.

7.4.3.1   Third-Body Effect

A third-body effect describes the role of second metal atoms, which by themselves are inactive and do not directly involve in electrocatalysis, in blocking the active sites for a side reaction or in blocking the adsorption of poisoning species, which require multiple adjacent active sites for adsorption. In this effect, the second metal

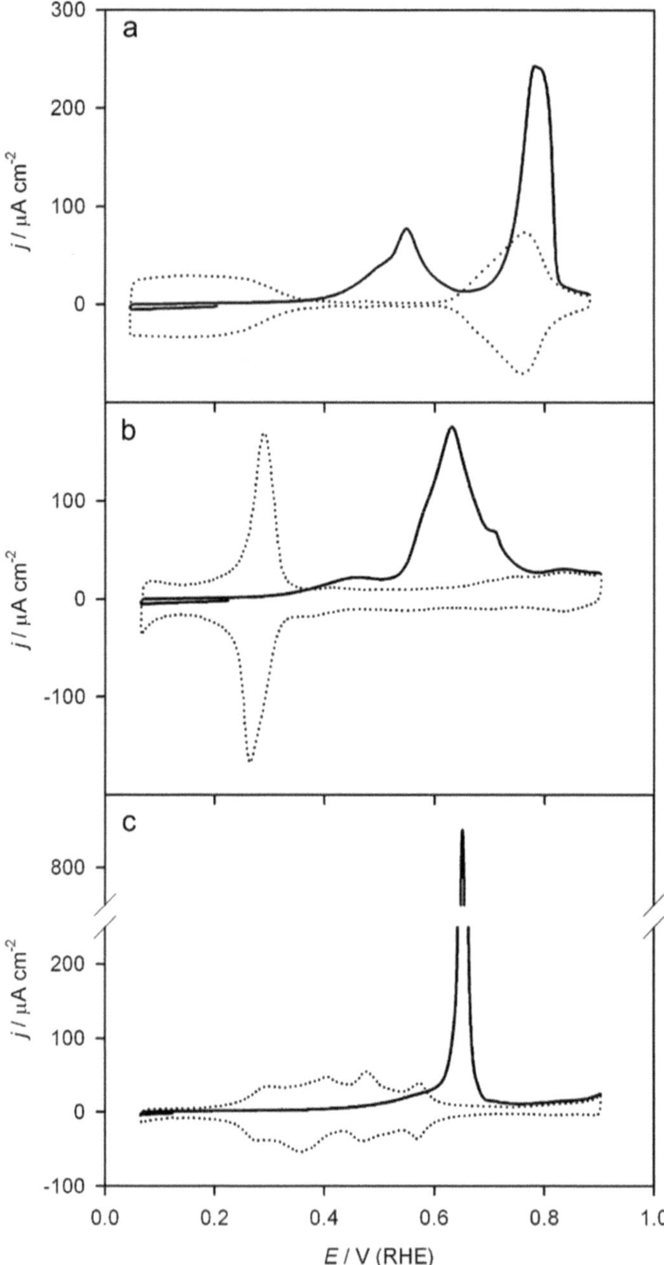

Fig. 7.7 Electrochemical oxidation of adsorbed CO on Pt (hkl) in 0.1 M NaOH solution: (a) Pt (111), (b) Pt (110), and (c) Pt (100). Scan rate = 50 mV/s, background CV (*dotted line*) collected immediately after CO stripping (reprinted with permission from [48], copyright 2004 Elsevier)

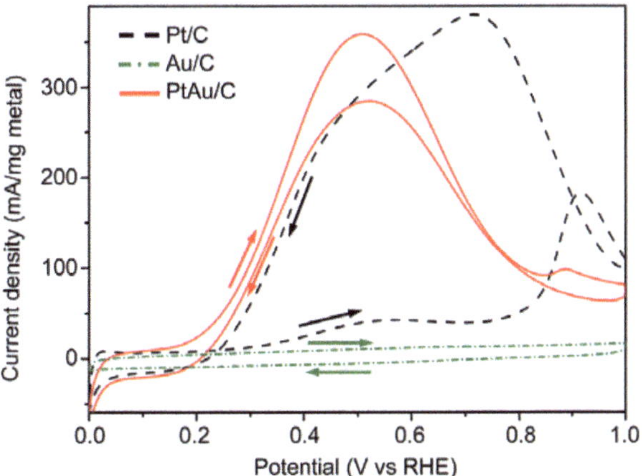

Fig. 7.8 Cyclic voltammetry curves of FAOR using Pt/C, Au/C, and PtAu/C alloy nanoparticles (reprinted with permission from [49], copyright 2009 Springer)

atoms does not improve the reaction turnover frequency (TOF) on individual active sites of the first metal, but alter the distribution of reaction pathways by breaking the active sites into smaller ensembles.

One example is the formic acid electro-oxidation reaction (FAOR), which have two major reaction pathways (dehydrogenation and dehydration) under metal electrocatalysis [49]:

$$HCOOH \xrightarrow{\text{dehydrogenation}} CO_2 + 2H^+ + 2e^- \tag{7.27}$$

$$HCOOH \xrightarrow{\text{dehydration}} CO_{ads} + H_2O \rightarrow CO_2 + 2H^+ + 2e^- \tag{7.28}$$

FA molecules are directly electro-oxidized to CO_2 in a dehydrogenation pathway, while CO_{ads}-like intermediates are first generated and then oxidized to CO_2 in a dehydration pathway. The reaction occurs via both pathways when pure Pt is used. The generated CO_{ads} through FA dehydration readily adsorb to Pt and block the active sites for further FAOR, causing decreased amount of the active sites and thus diminished activity (Fig. 7.8) [49]. PtAu alloy nanoparticles exhibit significantly higher activity than pure Pt, which can be attributed to the third-body effect. Although Au is inert for FAOR, it can modify the Pt surface by breaking the Pt atoms into smaller ensembles. As the dehydration path requires multiple adjacent Pt sites, it is effectively suppressed due to decreased availability of such site ensembles. The FAOR thus primarily undergoes the dehydrogenation pathway on PtAu and generates less CO_{ads} poisoning species than on pure Pt, leading to increased reaction rate and thus current density.

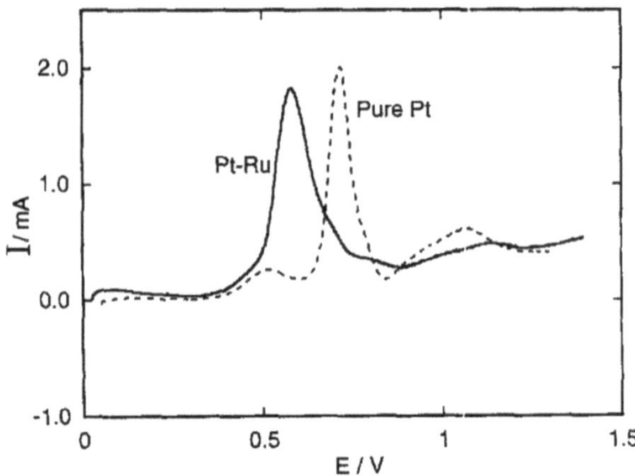

Fig. 7.9 Electrochemical oxidation of adsorbed CO on pure Pt and bimetallic Pt–Ru electrodes in 3 M H_2SO_4 at 10 mV/s (reprinted with permission from [50], copyright 1998 Elsevier)

7.4.3.2 Bifunctional Effect

In the bifunctional effect two types of active sites, which have distinct roles in the electrochemical reaction, are in presence and in close adjacency. It is often that an electrochemical step involves two reacting species, both of which need activation but impose different requirements on the active sites. A good catalyst needs to contain two types of active sites, which function together to activate both species and to ensure efficient reaction between the two species.

For example, electro-oxidation of adsorbed CO on Pt ($CO_{ads, Pt}$) in an acid electrolyte is considered following a L-H mechanism [50]:

$$H_2O + Pt \rightarrow OH_{ads,Pt} + H^+ + e^- \tag{7.29}$$

$$CO_{ads,Pt} + OH_{ads,Pt} \rightarrow CO_2 + H^+ + e^- \tag{7.30}$$

The reaction has little activity below 0.6 V vs. reversible hydrogen electrode (RHE), which is limited by Step (7.29) because pure Pt cannot effectively adsorb OH species until the potential (Fig. 7.9). Bimetallic Pt–Ru is discovered to be a better electrocatalyst for this reaction, with more negative onset potential (i.e., bigger i_0) comparing to pure Pt. One plausible explanation is that Ru adsorbs OH groups efficiently at low potentials, which react with adsorbed CO on Pt sites:

$$H_2O + Ru \rightarrow OH_{ads,Ru} + H^+ + e^- \tag{7.31}$$

$$CO_{ads,Pt} + OH_{ads,Ru} \rightarrow CO_2 + H^+ + e^- \tag{7.32}$$

The reaction mechanism is altered by using the bimetallic Pt–Ru, in which Ru promotes the OH adsorption and Pt promotes the CO adsorption. They function together to catalyze the reaction at a lower potential than pure Pt.

It needs to be noted that all the four effects are functions of the physical parameters of metal electrocatalyst, some of which are common. Adjustment in one physical parameter can lead to variations in multiple effects. In another word, it is often that multiple effects work simultaneously to determine the electrochemical reaction kinetics, although one of them might play a dominating role. For instance, the higher FAOR activity on PtAu alloy than on pure Pt can be mainly attributed to a third-body effect (Chap. 7.4.3.1), in which surface Au atoms by themselves are inert for the reaction. Their presence helps to break surface Pt atoms into discontinuous ensembles, suppresses the undesired pathway, and leads to increased reaction rate. Besides the third-body effect, the addition of Au atoms also alters the geometric structure of the particle surface. It is because that Au and Pt have different atom size and electron affinity, which cause surface restructuring and changes in the lattice parameters. From Table 7.2, the ε_d for surface Pt shifts up when being alloyed with Au, which indicates an alteration of the electronic structure. Thus, besides the dominating third-body effect, the geometric and electronic effects could also possibly play a role in the FAOR reaction. Because of the fact that multiple effects can work together and influence each other, it is sometimes difficult to disentangle their separate impact on the reaction. A synergistic effect, a more general term which describes the creation of an effect greater than the sum of individual effects due to the interaction of multiple elements in electrocatalyst, is sometimes used to include all possible effects.

7.4.4 Practical Considerations in Metal Electrocatalyst Research

Some practical issues also need consideration in the metal electrocatalyst research. It is because metals have their own physical and chemical properties, which can affect and even determine the electrochemical property.

7.4.4.1 Chemical Stability of Metals

The chemical stability of metals requires practical consideration in researching metal electrocatalyst. A good metal electrocatalyst should well maintain its structure and chemical identity under the working condition. For instance, mechanistic studies suggest a down-shift in the d-band center of Pt, which has a ε_d of -2.25 eV, for promoting the ORR kinetics [51]. If we only consider the electronic effect, Cu has a moderately deeper ε_d of -2.67 eV and could be expected to be a better electrocatalyst than Pt. However, Cu is not a suitable ORR catalyst at all. It has a low redox potential and can readily dissolve in acid electrolyte in the ORR potential range. Moreover, the low redox potential of Cu causes high coverage of OH species even at low potentials, which compete with ORR and negatively influence the

Table 7.3 Standard redox potentials for metals

Half-cell reaction	$E^0_{C/A}$ (V vs. SHE)
$Fe^{2+} + 2e^- \Leftrightarrow Fe$	−0.447
$Co^{2+} + 2e^- \Leftrightarrow Co$	−0.28
$Ni^{2+} + 2e^- \Leftrightarrow Ni$	−0.257
$Cu^{2+} + 2e^- \Leftrightarrow Cu$	0.342
$Ru^{2+} + 2e^- \Leftrightarrow Ru$	0.455
$Rh^{3+} + 3e^- \Leftrightarrow Rh$	0.758
$Pd^{2+} + 2e^- \Leftrightarrow Pd$	0.951
$Ag^+ + e^- \Leftrightarrow Ag$	0.7996
$Ir^{3+} + 3e^- \Leftrightarrow Ir$	1.156
$Pt^{2+} + 2e^- \Leftrightarrow Pt$	1.18
$Au^{3+} + 3e^- \Leftrightarrow Au$	1.498

reaction kinetics. Another example is the intermetallic PtPb electrocatalyst for both MOR and FAOR at low potential range [52, 53]. It is discovered as significantly better catalyst for the two reactions than pure Pt, which could possibly be attributed to a combination of third-body and bifunctional effects. The Pb surface atoms can effectively adsorb OH groups at low potentials, which serve as one of the reacting species, and break surface Pt atoms into smaller ensembles, which help to suppress the undesired pathway. They function synergistically with Pt active sites for promoting the reaction kinetics. However, it is observed the intermetallic PtPb rapidly loses the activity if the working electrode is ramped to potentials above the Pb redox potential. It is because Pb begins to leach out above the potential, which leaves a pure Pt surface and loses the third-body and bifunctional effects. Table 7.3 lists the equilibrium redox potential for some metals under the standard condition ($E^0_{A_{x+}/A}$), which can be used as a qualitative measure for evaluating their chemical stability under the working condition.

7.4.4.2 Metal Surface Restructuring and Segregation

Another practical consideration in the electrocatalyst research is restructuring and/or segregation of metal surfaces, which affect the catalytic property by altering the surface geometry. The real surface of working metal electrocatalyst, which is often resultant of complex interplays between many factors, including surface restructuring, surface segregation, and surface interaction with reacting species, can be quite different from the ideal bulk structure.

It is a general phenomenon that the surface atoms can restructure and have different arrangement from those in the bulk, which is driven by a positive surface energy. The surface atoms can also restructure under the interaction with reacting species. For instance, *fcc* Pt (100) surface under vacuum consists of a quasi-hexagonal structure, rather than the square symmetry (100) lattice. The surface restructuring is driven by the surface energy minimization. When the Pt (100) is exposed to CO atmosphere, CO molecules adsorb to the surface and can cause restructuring of the quasi-hexagonally arranged surface Pt atoms back into a square-like structure. This surface restructuring process is driven by the interaction between adsorbed CO and surface Pt atoms [54].

Surface aggregation often occurs in alloys, which causes a different surface composition from the bulk phase. The phenomenon is originated from the different property between elements, including metal–metal bond strength and atom size. Norskov and coworkers have calculated the segregation energies of all binary combinations of transition metals (Fig. 7.10) [28], which provide a valuable measure to estimate the tendency of metals to segregate in their alloy surfaces. For instance, Pt–Ru is among the most active electrocatalysts for MOR [55–59]. From the calculated segregation energies, we learn that Ru strongly antisegregates when being alloyed with Pt, whereas Pt strongly segregates when being alloyed with Ru. The resultant surface structure under such segregation effects consists of separate Ru sites, each of which is surrounded by several Pt atoms. Such a surface structure is favorable for MOR, with the isolated Ru sites to provide OH species and the adjacent Pt sites to activate methanol oxidation.

7.5 Current Development of Metallic Nanostructures for Electrocatalysis

As the electrochemical reactions differ from each other in many aspects, including reaction mechanism, intrinsic kinetics, reaction conditions, and so on, the imposed requirements on metal electrocatalyst are often different. As a result, different electrochemical reactions are found with different metallic nanostructures as the best performed catalyst. In general, a good metal electrocatalyst for an electrochemical reaction possesses the following characteristics: (1) it exhibits excellent activity toward the reaction; (2) it is chemically stable and durable under the reaction condition; and (3) it is cost-effective. Here we briefly introduce a few important electrochemical reactions and discuss the current catalyst development.

7.5.1 Oxygen Electrochemistry

7.5.1.1 Oxygen Reduction Reaction

Oxygen reduction reaction (ORR) has been intensively studied over the past few decades because of its fundamental complexity and practical importance for many applications, including PEMFCs and metal–air batteries. The overall reaction in an acidic electrolyte can be described as follows:

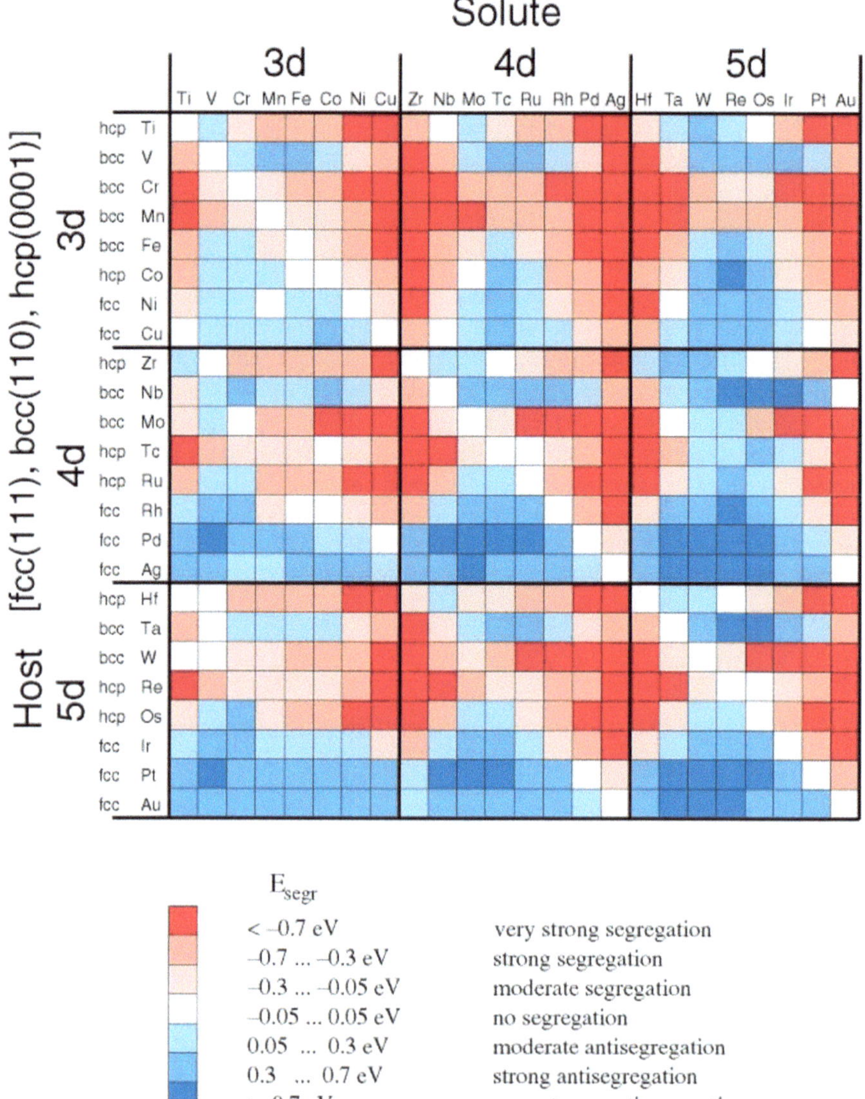

Fig. 7.10 Surface segregation energies for transition metal solutes in the close-packed surfaces of transition metal hosts (reprinted with permission from [28], copyright 2002 Annual Reviews)

$$O_2 + 4H^+ + 4e^- \rightarrow 2H_2O \quad (E_0 = 1.229\,\text{V vs. SHE}) \tag{7.33}$$

There are two difficulties in the research of ORR metal electrocatalysts. One is the low intrinsic exchange current density, i_0, which leads to very sluggish reaction kinetics and large overpotential. The other is the high working potential in acidic electrolyte, which imposes rigid requirement on the chemical stability of metals.

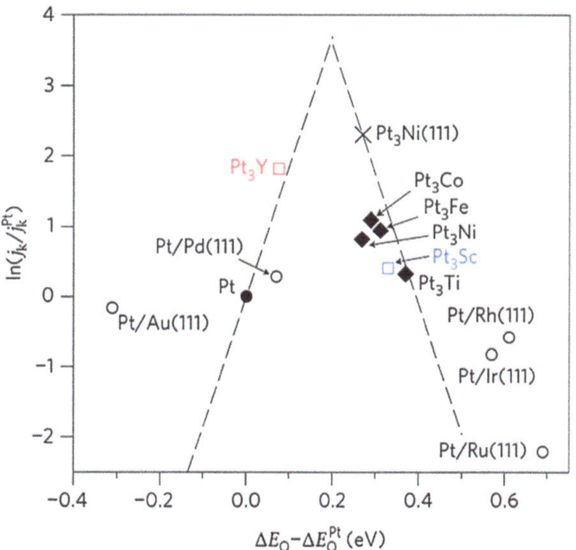

Fig. 7.11 Experimental kinetic current density of metal electrocatalysts as a function of the calculated oxygen adsoprtion energy, ΔE_O, with respect to that of Pt (111) (reprinted with permission from [51], copyright 2013 Royal Society of Chemistry)

Pt has been found the most effective metal for both reaction kinetics and stability considerations. However, even the i_0 can be increased to around 10^{-10} A/cm^2 using Pt, it is still not sufficient for the reaction [60]. The current state-of-the-art Pt/C catalyst exhibits around 0.2 mA/cm^2 Pt and 0.2 A/mg Pt at 0.9 V vs. RHE in ORR [20]. Large amount of expensive Pt metal is needed to generate usable currents with high energy conversion efficiency in both PEMFCs and metal–air batteries. To advance these technologies the Pt usage must be dramatically reduced, with targets of 0.7 mA/cm^2 Pt and 0.44 A/mg Pt at 0.9 V vs. RHE being set by the DOE [61]. Intensive research activities have been conducted in search of active Pt structure in the past years to reach the targets, which led to the development of many types of alloy [62–68], skin-layer [69–72], core-shell and thin-film electrocatalysts [73–76].

Mechanistic studies on Pt suggest that the ORR begins with adsorption of O$_2$ to the active sites (Eq. 7.34), which is a fast process. The adsorbed O$_2$, O$_2$·S, gains one electron from the electrode and reacts with one proton to generate active intermediate HOO·S (Eq. 7.35), which is further electroreduced to H$_2$O.

$$O_2 + S \Leftrightarrow O_2 \cdot S \quad \text{(fast)} \tag{7.34}$$

$$O_2 \cdot S + H^+ + e^- \Leftrightarrow HOO \cdot S \quad \text{(rate-limiting step)} \tag{7.35}$$

The HOO·S generation is discovered to be the rate-limiting step on Pt [77], which determines the i_0 value and controls the ORR kinetics. The oxygen adsorption energy to the metal surface (ΔE_O), which is a function of the metal electronic structure, has been found an effective descriptor for the reaction. A valcano-like relationship between the ORR activity and ΔE_O has been discovered (Fig. 7.11), suggesting an optimal ΔE_O for a perfect metal electrocatalyst [51].

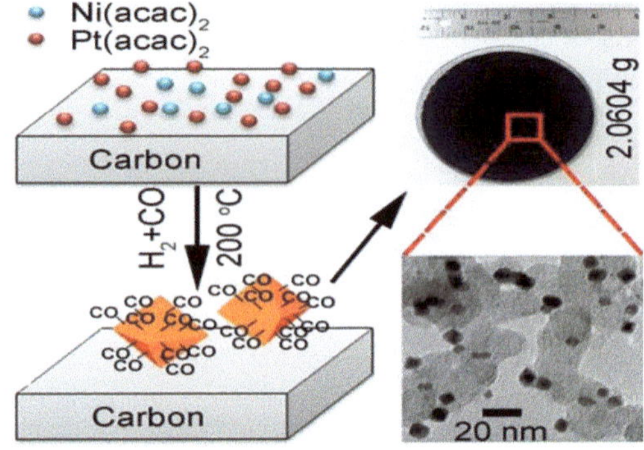

Fig. 7.12 Preparation and TEM of octahedral Pt–Ni nanoparticles on C support using solid-state chemistry (reprinted with permission from [61], copyright 2014 American Chemical Society)

Studies show that single crystalline Pt$_3$Ni (111) can be about 90 times more active than the state-of-the-art Pt/C [62]. The largely improved kinetics could be attributed to a down-shift in the ε_d of surface Pt and the (111) surface geometry. A plausible explanation is that the rate-limiting step favors a more negative ε_d (or a less negative ΔE_O) comparing to that of pure Pt. Meanwhile, the (111) surface geometry is also favored to decrease the surface coverage of adsorbed OH species, which negatively influence the reaction by blocking the active sites for Eq. 7.35. The discovery has motivated the research of octahedral Pt–Ni alloy nanoparticles, which are enclosed by the (111) planes and have large specific active area [62, 63, 78–81].

Peng and coworkers has recently developed a scalable, surfactant-free, and low-cost solid-state chemistry method for making octahedral Pt–Ni alloy nanoparticles on carbon support (Pt-Ni/C) [61]. In the method, octahedral Pt–Ni/C can be prepared by simply impregnating both platinum and nickel acetylacetonates onto C support and reducing them at 200 °C in 120/5 cm^3/min CO/H$_2$ for 1 h. Mechanistic studies suggest that the octahedral Pt–Ni production is resultant of employing both CO and H$_2$ gases, wherein H$_2$ aids transportation and reduction of the metal precursors on C support and CO is responsible for the particle morphology formation. Figure 7.12 shows the schematic illustration of the method and the produced octahedral Pt–Ni nanoparticles.

The octahedral Pt–Ni/C nanoparticles with different composition were studied for the ORR property, which exhibit significantly higher activity than the commercial Pt/C (Fig. 7.13). The octahedral Pt$_{1.5}$Ni/C shows the highest activity, with a j_{area} of 3.99 mA/cm^2 Pt being harvested at 0.90 V vs. RHE. It is about 20 times of the value for the Pt/C (0.2 mA/cm^2 Pt). The j_{mass} of the octahedral Pt$_{1.5}$Ni/C is 1.96 A/mg Pt and is 10 times as high comparing to the Pt/C (0.19 A/mg Pt). However, the long-term stability of the octahedral Pt-Ni/C catalyst remains a challenge. After 4000 cycles of linear potential sweeps between 0.60 and 1.00 V vs. RHE at a scan rate of 50 mV/s, the j_{area} and j_{mass} for the Pt$_{1.5}$Ni/C decrease to 2.17 mA/cm^2 Pt and 0.97 A/mg Pt at 0.9 V vs. RHE, although the electrochemical active surface area (ECSA) decays by only 8 %. The instability of the octahedral Pt–Ni/C could be attributed to a gradual Ni dissolution, which is chemically unstable in acid and under high potential and tends to leach out.

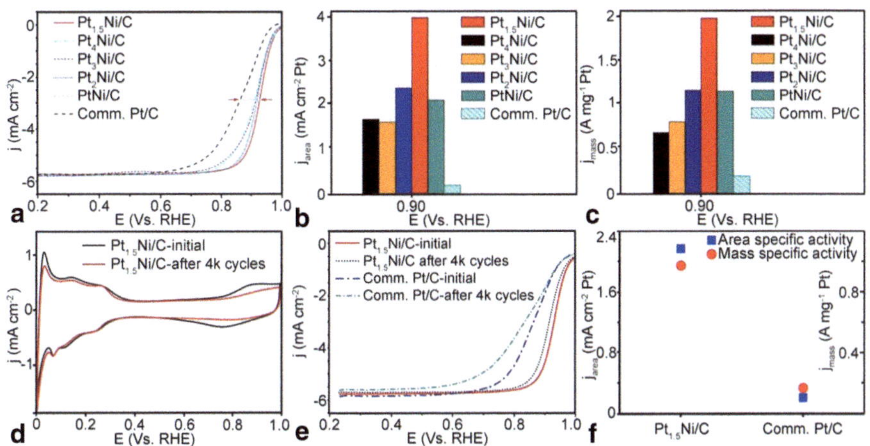

Fig. 7.13 ORR property of octahedral Pt–Ni/C. **a** ORR polarization curves and **b, c** active area and mass-specified ORR current densities (j_{area} and j_{mass}) of PtNi/C, $Pt_{1.5}$Ni/C, Pt_2Ni/C, Pt_3Ni/C, Pt_4Ni/C, and commercial Pt/C in O_2-saturated 0.1 M $HClO_4$ at room temperature, with a scan rate of 10 mV/s and an electrode rotating rate of 1600 rpm, and **d** Cyclic voltammograms, **e** ORR, and **f** j_{area} and j_{mass} of $Pt_{1.5}$Ni/C and commercial Pt/C after accelerated stability test (reprinted with permission from [61], copyright 2014 American Chemical Society)

7.5.1.2 Oxygen Evolution Reaction

Oxygen evolution reaction (OER) differs from ORR in the reaction mechanism, although it is considered as the reverse process of ORR. Thus, the characteristics of a good OER catalyst are different from that for ORR. The OER catalyst research is of particular interest for water splitting, and regenerative fuels cells and batteries. Mechanistic studies suggest the following OER pathway on metals in acidic electrolyte [60, 82]:

$$H_2O + S \Leftrightarrow HO \cdot S + H^+ + e^- \tag{7.36}$$

$$HO \cdot S \Leftrightarrow O \cdot S + H^+ + e^- \tag{7.37}$$

$$O \cdot S + H_2O \Leftrightarrow HOO \cdot S + H^+ + e^- \tag{7.38}$$

$$HOO \cdot S \Leftrightarrow S + O_2 + H^+ + e^- \tag{7.39}$$

It is proposed that Eq. 7.37 is the rate-limiting step when the active sites interacts weakly with oxygenated species, while Eq. 7.38 is the rate-limiting step if the interaction is strong. Based on the mechanism, metals are actually more in their oxidative state during electrocatalysis.

A perfect OER catalyst must have optimal interaction with oxygenated species to balance the kinetics between the two steps. A valcano plot has been obtained

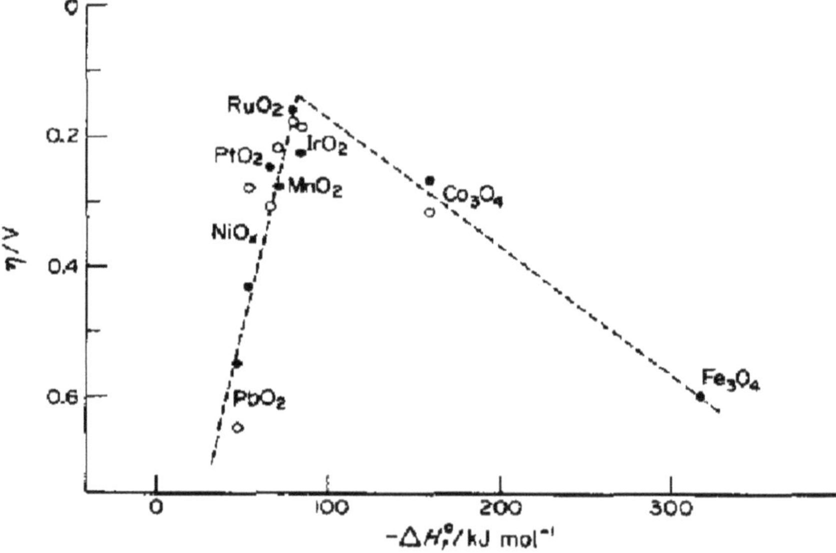

Fig. 7.14 Volcano plot of the overpotential for OER vs. the enthalpy of the lower to higher oxide transition. *Solid circles*: in acidic electrolyte, *hollow circles*: in alkaline electrolyte (reprinted with permission from [82], copyright 1984 Elsevier)

for the OER performance using different metals (Fig. 7.14) [82]. The change of enthalpy for metal oxidation has been found a good descriptor for predicting the OER property. Ru and Ir possess the most moderate values among all metals and are thus found the most effective elements for the reaction. In practice, Ru and Ir oxide nanoparticles are often used as OER electrocatalyst.

7.5.2 Hydrogen Electrochemistry

7.5.2.1 Hydrogen Evolution Reaction

Hydrogen evolution reaction (HER) receives considerable attention in recent years for H_2 production via electrochemical and photo-electrochemical water splitting. The reaction in an acidic electrolyte can be written as follows [51]:

$$2H^+ + 2e^- \rightarrow H_2 \quad (E_0 = 0.00\text{V vs. SHE}) \tag{7.40}$$

The reaction pathway on metals is proposed as follows:

$$H^+ + e^- + S \Leftrightarrow H{\cdot}S \quad \text{(rate-limiting step)} \tag{7.41}$$

$$H{\cdot}S + H{\cdot}S \Leftrightarrow H_2 + 2S \tag{7.42}$$

Fig. 7.15 The HER activity on metals as a function of the metal–hydrogen bond strength (reprinted with permission from [51], copyright 2013 Royal Society of Chemistry)

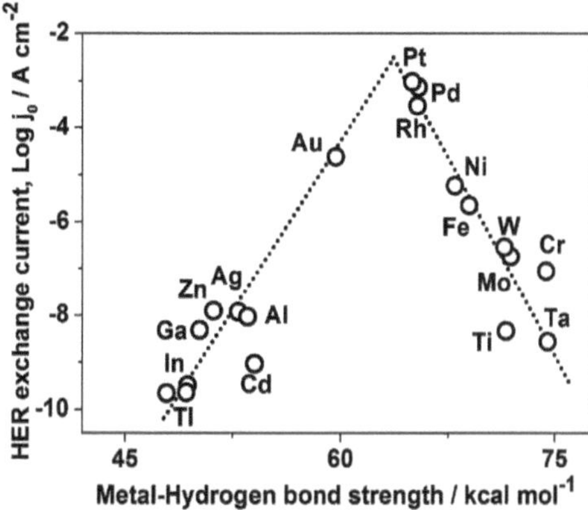

A valcano-like relationship for the HER kinetics has been discovered as function of the metal–hydrogen bond strength, which can be related to the electronic structure of metal surfaces (Fig. 7.15) [51]. Pt possesses the most optimal metal–hydrogen interaction among all the metals and thus exhibits the highest HER activity. Pt/C nanoparticles are often used as HER electrocatalyst.

7.5.2.2   Hydrogen Oxidation Reaction (HOR)

Hydrogen oxidation reaction (HOR) is the reverse process of HER and serves as the anode half-cell reaction for PEMFCs. Similar to HER, Pt is the most effective metal in HOR for its optimal electronic interaction with hydrogen, and is thus most often used as the electrocatalyst. It needs noted that the i_0 value for HOR on Pt can reach 10^{-3} A/cm^2 [83], which is around several orders of that for ORR on Pt. Much less amount of Pt is needed for HOR than for ORR. Thus, the electrocatalyst development is more focused on ORR side in the PEMFC research.

7.5.3   Electrochemistry of Carbon-Containing Compounds

7.5.3.1   Methanol Oxidation Reaction

Methanol has been considered as an alternative fuel for constructing direct methanol fuel cells (DMFCs), in which methanol oxidation reaction (MOR) serves as an anode reaction. The reaction has slow kinetics because of its complex reaction pathway. Despite the many mechanistic studies, the clear MOR mechanism is still under active debates. Figure 7.16 shows a multipathway scheme for the reaction, which is proposed based on the intermediate species identified in experiments [84]. Several routes involve the generation of CO_{ads}-like species, which can accumulate on metal surface and block the active sites. Pure Pt exhibits a low activity toward MOR, which is probably caused by a poisoning effect by the generation of CO_{ads}.

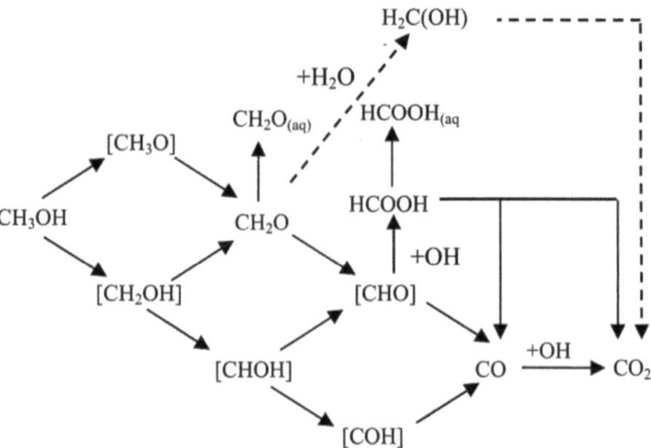

Fig. 7.16 Multipathway scheme for MOR on metal surface, with pathway in *dashed line* not being well documented (reprinted with permission from [84], copyright 2008 Springer).

Based on current understanding of the reaction, a good MOR electrocatalyst should effectively suppress CO_{ads} generation and efficiently remove generated CO_{ads}. PtRu/C alloy nanoparticles have been found a better catalyst than pure Pt for the reaction. The property improvement has been attributed to both the third-body and bifunctional effects. As being discussed in Chap. 7.4.3, the incorporation of Ru breaks surface Pt atoms into smaller ensembles, which helps to suppress CO_{ads} generation because the process requires multiple adjacent Pt sites. Meanwhile, surface Ru atoms adsorb hydroxyl species (OH_{ads}), which helps to react with and thus remove CO_{ads}. Abruna and coworkers have recently prepared PtPb intermetallic nanoparticles, which exhibit exceptionally higher MOR activity [52]. They reported one to two orders of increase in the current density and no detectable CO_{ads} formation comparing to pure Pt. They concluded that the CO_{ads} generation pathway was completely suppressed for MOR on the intermetallic PtPb (Fig. 7.17).

7.5.3.2 Formic Acid Oxidation Reaction

As being discussed in Chap. 7.4.3, formic acid oxidation reaction (FAOR) have major dehydrogenation and dehydration pathways in parallel (Eqs. 7.27 and 7.28) [49]. The dehydration pathway generates CO_{ads}, which can poison metal electrocatalyst by blocking the active sites. Pure Pt is a poor catalyst for the reaction, with the low activity caused by CO_{ads} accumulation on the surface. Pure Pd exhibits significantly higher FAOR activity than Pt [85–89]. Spectroscopic studies suggest that FAOR favors the dehydrogenation pathway on Pd comparing to Pt, which could be attributed to the electronic effect. The dehydration pathway and thus CO_{ads} generation are suppressed on Pd, leading to improved FAOR rate.

The FAOR kinetics can also be improved by adding a second metal for modifying the Pt surface. A range of PtM (M=Au, Bi, Pb, etc) alloy nanoparticles exhibit greatly improved FAOR kinetics comparing to pure Pt [90–97]. Similar to the alloy electrocatalysts for MOR, the improvement can be attributed to the third-body and/or bifunctional effect. Peng et al. has recently demonstrated the importance of Pt

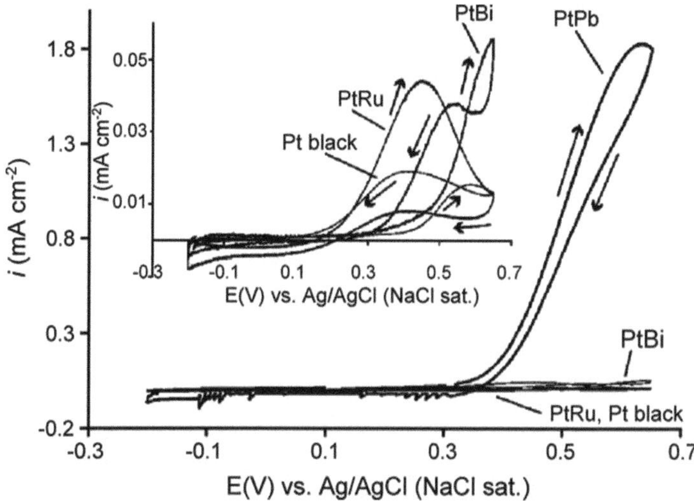

Fig. 7.17 Cyclic voltammogram for MOR using carbon-supported Pt and PtM (M=Bi, Pb, and Ru) alloy nanoparticles (reprinted with permission from [52], copyright 2006 American Chemical Society)

surface composition in affecting the FAOR by studying PtAg alloy electrocatalyst (Fig. 7.18) [98]. In the study $Pt_{18}Ag_{82}$ alloy nanoparticles were first synthesized by reducing Pt acetylacetonate and Ag stearate in a mixture of 1,2-hexadecanediol, oleylamine, oleic acid, and dipenyl ether at 200 °C, and then put on a C support for preparing $Pt_{18}Ag_{82}$/C. By adjusting the potential range used for electrochemical dissolution of Ag from the $Pt_{18}Ag_{82}$, the surface composition of the particles can be well controlled. The optimal surface composition was discovered to be Pt_3Ag, which leads to an overall particle composition of $Pt_{34}Ag_{66}$, for the FAOR, with the activity being more than one order higher than that using pure Pt. In comparison, the Pt–Ag structure with a pure Pt surface, which was produced using the $Pt_{18}Ag_{82}$ by altering the electrochemical treatment potential, exhibits much less activity comparing to $Pt_{34}Ag_{66}$.

7.5.3.3 Ethanol Oxidation Reaction

Ethanol oxidation reaction (EOR) is an even more complex electrochemical reaction comparing to MOR. A complete EOR process involves C–C bond breakage, transfer of 12 electrons, and generation of many intermediate species. Figure 7.19 shows some suggested reaction routes for EOR from mechanistic studies [84]. In situ FTIR and on-line DEMS studies show that ethanol cannot be effectively electro-oxidized into CO_2 on Pt. The CO_2 yield is only about a few percent, with the major products being acetaldehyde and acetic acid. The low yield of CO_2 is caused by inefficient breakage of the C–C bond, which creates a big energy barrier and requires high energy input.

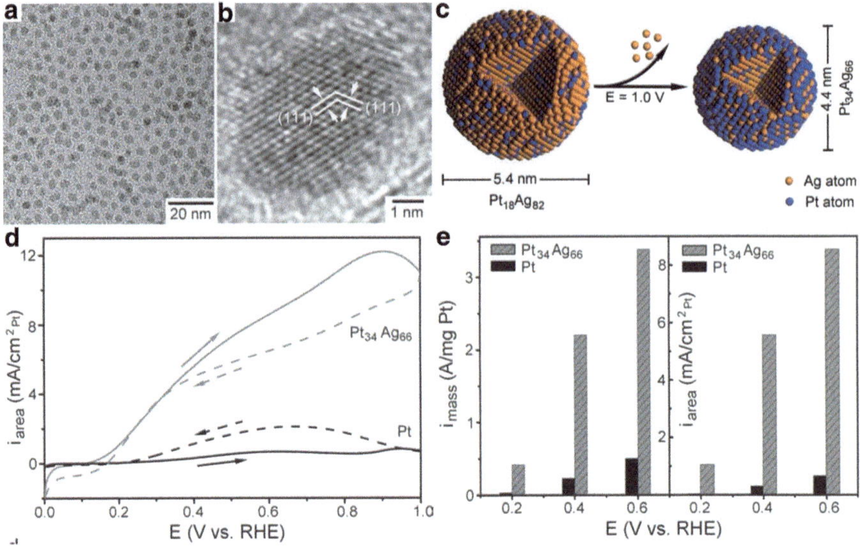

Fig. 7.18 **a** TEM and **b** HRTEM of as-synthesized $Pt_{18}Ag_{82}$ alloy nanoparticles, **c** preparation of Pt surface-rich $Pt_{34}Ag_{66}$ nanostructure by controlled electrochemical Ag dissolution from $Pt_{18}Ag_{82}$, and **d**, **e** cyclic voltammetry and FAOR activity using the Pt surface-rich $Pt_{34}Ag_{66}$ and pure Pt electrocatalysts (adapted with permission from [98], copyright 2010 Wiley-VCH)

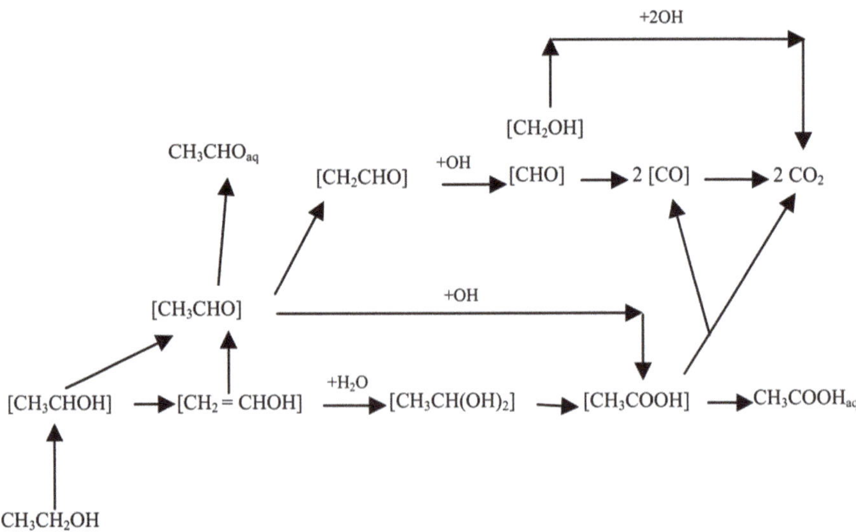

Fig. 7.19 Multipathway scheme for EOR on metal surface (reprinted with permission from [84], copyright 2008 Springer)

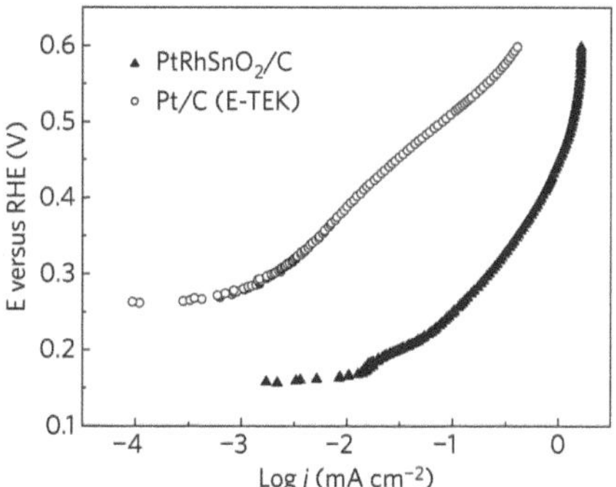

Fig. 7.20 EOR polarization curves on PtRhSnO$_2$/C and Pt/C (reprinted with permission from [99], copyright 2009 Nature Publishing Group)

Adzic and his coworkers have recently researched a ternary PtRhSnO$_2$/C electrocatalyst and claimed high effectiveness of the catalyst in splitting the C–C bond [99]. They prepared the PtRhSnO$_2$/C nanoparticles using a controllable deposition called the cation-adsorption-reduction-galvanic-displacement method. The EOR activity for the PtRhSnO$_2$/C is more than two orders of magnitude higher than that of the commercial Pt/C at 0.3 V vs. RHE. The dramatically improved activity was attributed to a synergistic effect, in which Pt facilitates ethanol dehydrogenation and modifies the electronic structure of Rh, Rh facilitates C–C bond breaking and ethanol oxidation, and SnO$_2$ provides adsorbed OH species to oxidize the dissociated CO at Rh sites (Fig. 7.20).

7.5.4 *Electrochemistry of Nitrogen-Containing Compounds*

7.5.4.1 Ammonia Oxidation Reaction

Ammonia oxidation reaction (AOR) electrocatalyzed by metals has attracted considerable interest for many potential applications, including direct ammonia fuel cells [100–102], electrochemical detection of ammonia [103], and wastewater treatment [104, 105]. Mechanistic study discovers that AOR on Pt occurs via stepwise dehydrogenation and generation of $NH_{x,ad}$ ($x=1$, 2) species (Fig. 7.21) [106, 107]. The $NH_{x,ad}$ undergoes dimerization to produce $N_2H_{y,ad}$ ($y=2$, 3, 4), which is the rate-limiting step. Further dehydrogenation of $N_2H_{y,ad}$ produces final product, N_2 [108–110]. Meanwhile, N_{ad} can be generated via complete dehydrogenation of $NH_{x,ad}$. The produced N_{ad} block the active sites due to strong chemisorption to Pt and largely suppress AOR activity [109]. Studies on Pt single crystals find that the reaction is geometric structure sensitive. Pt (100) is much more active than Pt (111) and (110) [111–113]. Both experimental and theoretical results also suggest less generation and easier removal of N_{ad} on Pt (100) than on other surfaces [112, 114–117]. The synergistic effect leads to a dramatically higher AOR activity.

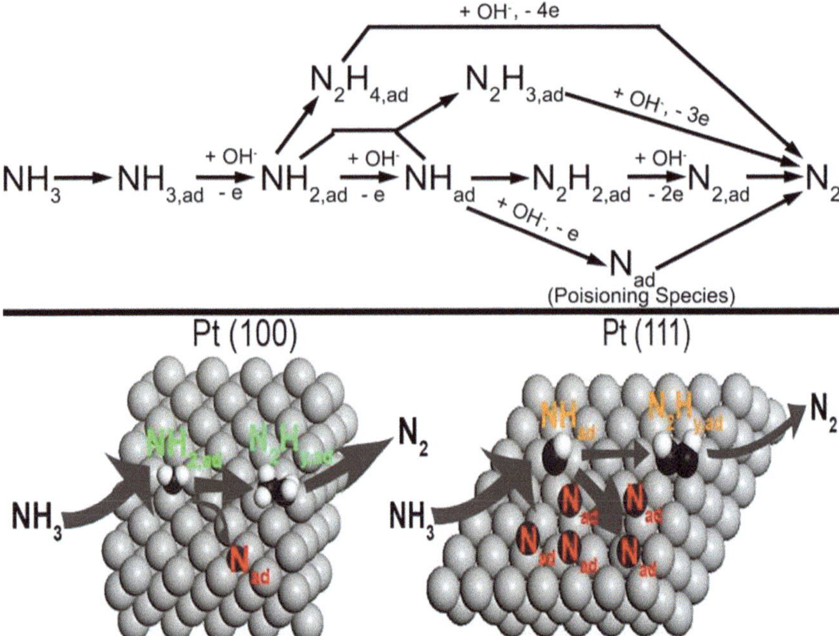

Fig. 7.21 Schematic illustration of AOR pathways on Pt (100) and (111) surfaces (reprinted with permission from [35], copyright 2013 Royal Society of Chemistry)

The finding suggests that cubic Pt nanoparticles, which are enclosed exclusively by the Pt (100) surfaces, are good AOR electrocatalyst. Peng and coworkers have recently produced cubic Pt nanoparticles on C support (PtNCs/C) using a similar solid-state chemistry method as for preparing octahedral Pt–Ni/C [35]. The PtNCs/C exhibits an AOR current density of 1.44 mA/cm² Pt at 0.6 V vs. RHE, which is about five times of that using the commercial Pt/C (0.30 mA/cm²) at the same potential. The significant enhancement in the current density represents much improved reaction kinetics and demonstrates the geometric effect of Pt nanoparticles on the AOR property (Fig. 7.22).

7.5.4.2 Hydrazine Oxidation Reaction

Hydrazine oxidation reaction (HZOR) has been applied in many industrial applications, including metal plating, corrosion protection, and direct hydrazine fuel cells. It has been proposed that HZOR on metal surfaces proceeds through direct hydrazine electrochemical dehydrogenation into N_2. The reaction involves no N–N bond breakage and less electron transfer, and is thus intrinsically more active comparing to AOR. The reaction pathway in an acidic electrolyte can be depicted as follows [118]:

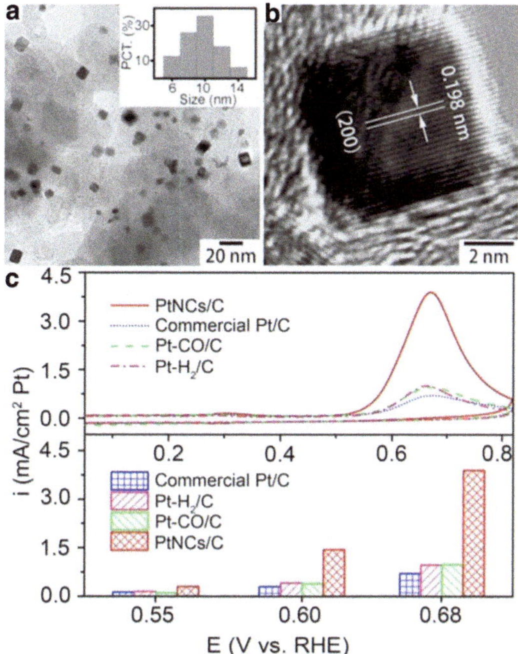

Fig. 7.22 **a** TEM and **b** HRTEM of cubic Pt/C nanoparticles (PtNCs/C) prepared using solid-state chemistry method, and **c** the AOR activity using the PtNCs/C and the commercial Pt/C (reprinted with permission from [35], copyright 2013 Royal Society of Chemistry)

$$N_2H_5^+ + S \Leftrightarrow N_2H_3 \cdot S + 2H^+ + e^- \tag{7.43}$$

$$N_2H_3 \cdot S + S \Leftrightarrow N_2H_2 \cdot 2S + H^+ + e^- \quad \text{(rate-determining step)} \tag{7.44}$$

$$N_2H_2 \cdot 2S \Leftrightarrow N_2 + 2H^+ + 2e^- \tag{7.45}$$

Step (7.44) seems to be the rate-determining step and is discovered to be both electronic and geometric structure sensitive. Dissimilar to AOR, for which Pt is the most active among all metals, Rh exhibits the highest activity for AZOR. It indicates Rh has the most optimal electronic structure for promoting the rate-limiting process. The AZOR activity on different Rh surfaces follows the trend of Rh (100) > Rh (111) > Rh (110) (Fig. 7.23). The geometric structure-dependent activity could be attributed to the active sites requirement for $N_2H_2 \cdot 2S$ species generation, in which two adjacent active sites with an optimal distance are needed. The studies on single Rh crystals suggest that cubic Rh nanoparticles could be a good AZOR electrocatalyst.

Fig. 7.23 Cyclic voltammetry of HZOR in 0.1 M $HClO_4$ on different Rh single crystal electrodes (reprinted with permission from [118], copyright 2002 Electrochemical Society)

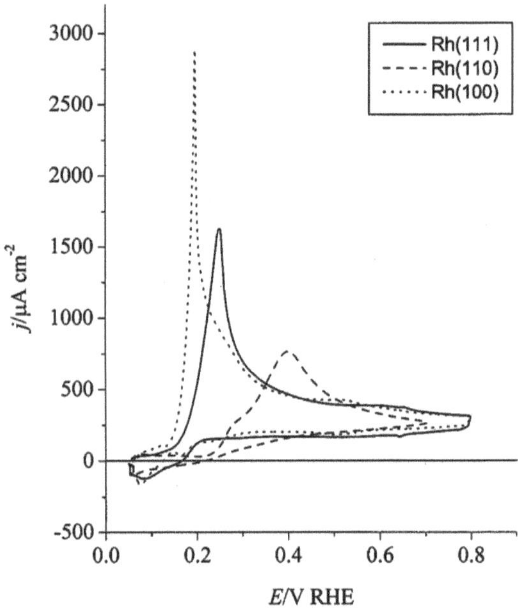

References

1. Peng Z, Yang H (2009) Designer platinum nanoparticles: Control of shape, composition in alloy, nanostructure and electrocatalytic property. Nano Today 4:143–164
2. Chan KY, Ding J, Ren JW et al (2004) Supported mixed metal nanoparticles as electrocatalysts in low temperature fuel cells. J. Mater. Chem. 14:505–516
3. Guo D-J, Ding Y (2012) Porous nanostructured metals for electrocatalysis. Electroanal. 24:2035–2043
4. Kelly TG, Chen JG (2012) Metal overlayer on metal carbide substrate: Unique bimetallic properties for catalysis and electrocatalysis. Chem. Soc. Rev. 41:8021–8034
5. Spendelow JS, Wieckowski A (2004) Noble metal decoration of single crystal platinum surfaces to create well-defined bimetallic electrocatalysts. Phys. Chem. Chem. Phys. 6:5094–5118
6. Chen A, Chatterjee S (2013) Nanomaterials based electrochemical sensors for biomedical applications. Chem. Soc. Rev. 42:5425–5438
7. Bing Y, Liu H, Zhang L et al (2010) Nanostructured Pt-alloy electrocatalysts for PEM fuel cell oxygen reduction reaction. Chem. Soc. Rev. 39:2184–2202
8. Lee K, Zhang JJ, Wang HJ et al (2006) Progress in the synthesis of carbon nanotube- and nanofiber-supported pt electrocatalysts for PEM fuel cell catalysis. J. Appl. Electrochem. 36:507–522
9. Shao Y, Liu J, Wang Y et al (2009) Novel catalyst support materials for PEM fuel cells: Current status and future prospects. J. Mater. Chem. 19:46–59
10. Shao Y, Yin G, Gao Y (2007) Understanding and approaches for the durability issues of Pt-based catalysts for PEM fuel cell. J. Power Sources 171:558–566
11. Zhang J, Xie Z, Zhang J et al (2006) High temperature PEM fuel cells. J. Power Sources 160:872–891
12. Cheng F, Chen J (2012) Metal-air batteries: From oxygen reduction electrochemistry to cathode catalysts. Chem. Soc. Rev. 41:2172–2192

13. Kraytsberg A, Ein-Eli Y (2013) The impact of nano-scaled materials on advanced metal-air battery systems. Nano Energy 2:468–480
14. Lee J-S, Kim ST, Cao R et al (2011) Metal-air batteries with high energy density: Li-air versus Zn-air. Adv. Energy Mater. 1:34–50
15. Rahman MA, Wang X, Wen C (2013) High energy density metal-air batteries: A review. J. Electrochem. Soc. 160:A1759–A1771
16. Diaz-Morales O, Calle-Vallejo F, de Munck C et al (2013) Electrochemical water splitting by gold: Evidence for an oxide decomposition mechanism. Chem. Sci. 4:2334–2343
17. Fujii K, Nakamura S, Sugiyama M et al (2013) Characteristics of hydrogen generation from water splitting by polymer electrolyte electrochemical cell directly connected with concentrated photovoltaic cell. Int. J. Hydrogen Energy 38:14424–14432
18. Lopes T, Andrade L, Ribeiro HA et al (2010) Characterization of photoelectrochemical cells for water splitting by electrochemical impedance spectroscopy. Int. J. Hydrogen Energy 35:11601–11608
19. Bagotsky VS (2006) Fundamentals of electrochemistry. John Wiley & Sons, Pennington, New Jersey
20. Gasteiger HA, Kocha SS, Sompalli B et al (2005) Activity benchmarks and requirements for Pt, Pt-alloy, and non-Pt oxygen reduction catalysts for PEMFCs. Appl. Catal. B-Environ. 56:9–35
21. Marin GB, Yablonsky GS (2011) Kinetics of chemical reactions. Wiley-VCH Verlag & Co., Weinheim
22. Bard AJ, Faulkner LR (2001) Electrochemical methods—fundamentals and applications. John Wiley & Sons, United States
23. Koper MTM, Santen RAv, Neurock M (2003) Theory and modeling of catalytic and electrocatalytic reactions. In: Wieckowski A, Savinova ERVayenas CG (ed) Catalysis and electrocatalysis at nanoparticle surfaces, Marcel Dekker, New York, p 1–34
24. Thomas JM, Thomas WJ (1996) Principles and practice of heterogeneous catalysis. John Wiley & Sons, Weinheim
25. Lima FHB, Zhang J, Shao MH et al (2007) Catalytic activity-d-band center correlation for the O_2 reduction reaction on platinum in alkaline solutions. J. Phys. Chem. C 111:404–410
26. Skoplyak O, Barteau MA, Chen JGG (2006) Reforming of oxygenates for H_2 production: Correlating reactivity of ethylene glycol and ethanol on Pt(111) and Ni/Pt(111) with surface d-band center. J. Phys. Chem. B 110:1686–1694
27. Alonso JA (2000) Electronic and atomic structure, and magnetism of transition-metal clusters. Chem. Rev. 100:637–677
28. Greeley J, Norskov JK, Mavrikakis M (2002) Electronic structure and catalysis on metal surfaces. Annual Rev. Phys. Chem. 53:319–348
29. Low GG (1969) Electronic structure of some transition metal alloys. Adv. Phys. 18:371–&
30. Ruban A, Hammer B, Stoltze P et al (1997) Surface electronic structure and reactivity of transition and noble metals. J. Mol. Catal. A-Chem. 115:421–429
31. Demirci UB (2007) Theoretical means for searching bimetallic alloy as anode electrocatalaysts for direct liquid-feed fuel cells. J. Power Sources 173:11–18
32. Garcia G, Rodriguez P, Rosca V et al (2009) Fourier transform infrared spectroscopy study of CO electro-oxidation on Pt(111) in alkaline media. Langmuir 25:13661–13666
33. Liu Y, Li D, Stamenkovic VR et al (2011) Synthesis of Pt_3Sn alloy nanoparticles and their catalysis for electro-oxidation of CO and methanol. ACS Catal. 1:1719–1723
34. McGrath P, Fojas AM, Reimer JA et al (2009) Electro-oxidation kinetics of adsorbed co on platinum electrocatalysts. Chem. Eng. Sci. 64:4765–4771
35. Zhang C, Hwang SY, Peng Z (2013) Shape-enhanced ammonia electro-oxidation property of a cubic platinum nanocrystal catalyst prepared by surfactant-free synthesis. J. Mater. Chem. A 1:14402–14408
36. Spendelow JS, Babu PK, Wieckowski A (2005) Electrocatalytic oxidation of carbon monoxide and methanol on platinum surfaces decorated with ruthenium. Curr. Opinion Solid State Mater. Sci. 9:37–48

37. Cohen JL, Volpe DJ, Abruna HD (2007) Electrochemical determination of activation energies for methanol oxidation on polycrystalline platinum in acidic and alkaline electrolytes. Phys. Chem. Chem. Phys. 9:49–77
38. Antolini E, Salgado JRC, Gonzalez ER (2006) The methanol oxidation reaction on platinum alloys with the first row transition metals—the case of Pt-Co and -Ni alloy electrocatalysts for DMFCs: A short review. Appl. Catal. B-Environ. 63:137–149
39. Wu J, Yang H (2013) Platinum-based oxygen reduction electrocatalysts. Acc. Chem. Res. 46:1848–1857
40. Wang C, Markovic NM, Stamenkovic VR (2012) Advanced platinum alloy electrocatalysts for the oxygen reduction reaction. ACS Catal. 2:891–898
41. Antolini E, Lopes T, Gonzalez ER (2008) An overview of platinum-based catalysts as methanol-resistant oxygen reduction materials for direct methanol fuel cells. J. Alloys and Compounds 461:253–262
42. Alonso-Vante N (2010) Platinum and non-platinum nanomaterials for the molecular oxygen reduction reaction. ChemPhysChem 11:2732–2744
43. Kim KT, Hwang JT, Kim YG et al (1993) Surface and catalytic properties of iron-platinum carbon electrocatalysts for cathodic oxygen reduction in PAFC. J. Electrochem. Soc. 140:31–36
44. Hwang JT, Chung JS (1993) The morphological and surface-properties and their relationship with oxygen reduction activity for platinum-iron electrocatalysts. Electrochimica Acta 38:2715–2723
45. Tegou A, Papadimitriou S, Armyanov S et al (2008) Oxygen reduction at platinum- and gold-coated iron, cobalt, nickel and lead deposits on glassy carbon substrates. J. Electroanal. Chem. 623:187–196
46. Tegou A, Papadimitriou S, Pavfidou E et al (2007) Oxygen reduction at platinum- and gold-coated copper deposits on glassy carbon substrates. J. Electroanal. Chem. 608:67–77
47. Alia SM, Jensen K, Contreras C et al (2013) Platinum coated copper nanowires and platinum nanotubes as oxygen reduction electrocatalysts. ACS Catal. 3:358–362
48. Spendelow JS, Lu GQ, Kenis PJA et al (2004) Electrooxidation of adsorbed co on Pt(111) and Pt(111)/Ru in alkaline media and comparison with results from acidic media. J. Electroanal. Chem. 568:215–224
49. Peng Z, Yang H (2009) PtAu bimetallic heteronanostructures made by post-synthesis modification of Pt-on-Au nanoparticles. Nano Res. 2:406–415
50. Morimoto Y, Yeager EB (1998) CO oxidation on smooth and high area Pt, Pt-Ru and Pt-Sn electrodes. J. Electroanal. Chem. 441:77–81
51. Calle-Vallejo F, Koper MTM, Bandarenka AS (2013) Tailoring the catalytic activity of electrodes with monolayer amounts of foreign metals. Chem. Soc. Rev. 42:5210–5230
52. Roychowdhury C, Matsumoto F, Zeldovich VB et al (2006) Synthesis, characterization, and electrocatalytic activity of ptbi and ptpb nanoparticles prepared by borohydride reduction in methanol. Chem. Mater. 18:3365–3372
53. Matsumoto F (2012) Ethanol and methanol oxidation activity of PtPb, PtBi, and $PtBi_2$ intermetallic compounds in alkaline media. Electrochem. 80:132–138
54. Tao F, Dag S, Wang L-W et al (2009) Restructuring of hex-Pt(100) under CO gas environments: Formation of 2-D nanoclusters. Nano Lett. 9:2167–2171
55. Yajima T, Uchida H, Watanabe M (2004) In-situ ATR-FTIR spectroscopic study of electrooxidation of methanol and adsorbed co at Pt-Ru alloy. J. Phys. Chem. B 108:2654–2659
56. Radmilovic V, Gasteiger HA, Ross PN (1995) Structure and chemical-composition of a supported Pt-Ru electrocatalyst for methanol oxidation. J. Catal. 154:98–106
57. Iwasita T, Hoster H, John-Anacker A et al (2000) Methanol oxidation on PtRu electrodes. Influence of surface structure and Pt-Ru atom distribution. Langmuir 16:522–529
58. Gojkovic SL, Vidakovic TR, Durovic DR (2003) Kinetic study of methanol oxidation on carbon-supported PtRu electrocatalyst. Electrochimica Acta 48:3607–3614
59. Chu D, Gilman S (1996) Methanol electro-oxidation on unsupported Pt-Ru alloys at different temperatures. J. Electrochem. Soc. 143:1685–1690

60. Gattrell M, MacDougall B (2003) The oxygen reduction/evolution reaction. In: Vielstich W, Lamm A Gasteiger HA (ed) Handbook of fuel cells— fundamentals technology and applications, vol. 2: Electrocatalysis, John Wiley & Sons, Chichester, p 443–464

61. Zhang C, Hwang SY, Trout A et al (2014) Solid-state chemistry-enabled scalable production of octahedral Pt–Ni alloy electrocatalyst for oxygen reduction reaction. J. Am. Chem. Soc. 136:7805–7808

62. Stamenkovic VR, Fowler B, Mun BS et al (2007) Improved oxygen reduction activity on $Pt_3Ni(111)$ via increased surface site availability. Science 315:493–497

63. Carpenter MK, Moylan TE, Kukreja RS et al (2012) Solvothermal synthesis of platinum alloy nanoparticles for oxygen reduction electrocatalysis. J. Am. Chem. Soc. 134:8535–8542

64. Lim B, Jiang M, Camargo PHC et al (2009) Pd-Pt bimetallic nanodendrites with high activity for oxygen reduction. Science 324:1302–1305

65. Wang D, Xin HL, Hovden R et al (2013) Structurally ordered intermetallic platinum–cobalt core–shell nanoparticles with enhanced activity and stability as oxygen reduction electrocatalysts. Nature Mater. 12:81–87

66. Wu J, Qi L, You H et al (2012) Icosahedral platinum alloy nanocrystals with enhanced electrocatalytic activities. J. Am. Chem. Soc. 134:11880–11883

67. Hong JW, Kang SW, Choi B-S et al (2012) Controlled synthesis of Pd–Pt alloy hollow nanostructures with enhanced catalytic activities for oxygen reduction. 6:2410–2419

68. Stephens IEL, Bondarenko AS, Perez-Alonso FJ et al (2011) Tuning the activity of Pt(111) for oxygen electroreduction by subsurface alloying. 133:5485–5491

69. Wang JX, Inada H, Wu L et al (2009) Oxygen reduction on well-defined core–shell nanocatalysts: Particle size, facet, and pt shell thickness effects. J. Am. Chem. Soc. 131:17298–17302

70. Zhang J, Lima FHB, Shao MH et al (2005) Platinum monolayer on nonnoble metal–noble metal core–shell nanoparticle electrocatalysts for O_2 reduction. J. Phys. Chem. B 109:22701–22704

71. Zhang Y, Hsieh Y-C, Volkov V et al (2014) High performance Pt monolayer catalysts produced via core-catalyzed coating in ethanol. 738–742

72. Shao M, Shoemaker K, Peles A et al (2010) Pt monolayer on porous Pd–Cu alloys as oxygen reduction electrocatalysts. J. Am. Chem. Soc. 132:9253–9255

73. Zhang L, Iyyamperumal R, Yancey DF et al (2013) Design of Pt-shell nanoparticles with alloy cores for the oxygen reduction reaction. ACS Nano 7:9168–9172

74. Kibsgaard J, Gorlin Y, Chen Z et al (2012) Meso-structured platinum thin films: Active and stable electrocatalysts for the oxygen reduction reaction. J. Am. Chem. Soc. 134:7758–7765

75. Strasser P, Koh S, Anniyev T et al (2010) Lattice-strain control of the activity in dealloyed core–shell fuel cell catalysts. Nature Chem. 2:454–460

76. Debe MK, Schmoeckel AK, Vernstrom GD et al (2006) High voltage stability of nanostructured thin film catalysts for PEM fuel cells. J. Power Sources 161:1002–1011

77. Ross PN (2003) The oxygen reduction/evolution reaction. In: Vielstich W, Lamm A Gasteiger HA (ed) Handbook of fuel cells—fundamentals technology and applications, vol. 2: Electrocatalysis, John Wiley & Sons, Chichester, p 465–480

78. Cui C, Gan L, Heggen M et al (2013) Compositional segregation in shaped Pt alloy nanoparticles and their structural behaviour during electrocatalysis. Nature Mater. 12:765–771

79. Choi S-I, Xie S, Shao M et al (2013) Synthesis and characterization of 9 nm Pt–Ni octahedra with a record high activity of 3.3 A/mgPt for the oxygen reduction reaction. Nano Lett. 13:3420–3425

80. Zhang J, Yang H, Fang J et al (2010) Synthesis and oxygen reduction activity of shape-controlled Pt_3Ni nanopolyhedra. Nano Lett. 10:638–644

81. Wu J, Zhang J, Peng Z et al (2010) Truncated octahedral Pt_3Ni oxygen reduction reaction electrocatalysts. J. Am. Chem. Soc. 132:4984–4985

82. Trasatti S (1984) Electrocatalysis in the anodic evolution of oxygen and chlorine. Electrochimica Acta 29:1503–1512

83. Li H, Lee K, Zhang J (2008) Electrocatalytic H_2 oxidation reaction. In: Zhang J (ed) PEM fuel cell electrocatalysts and catalyst layers, Springer, London, p 135–164

84. Gyenge E (2008) Electrocatalytic oxidation of methanol, ethanol and formic acid. In: Zhang J (ed) PEM fuel cell electrocatalysts and catalyst layers, Springer, London, p 165–287

85. Baldauf M, Kolb DM (1996) Formic acid oxidation on ultrathin Pd films on Au(hkl) and Pt(hkl) electrodes. J. Phys. Chem. 100:11375–11381

86. Hoshi N, Kida K, Nakamura M et al (2006) Structural effects of electrochemical oxidation of formic acid on single crystal electrodes of palladium. J. Phys. Chem. B 110:12480–12484

87. Lee H, Habas SE, Somorjai GA et al (2008) Localized Pd overgrowth on cubic Pt nanocrystals for enhanced electrocatalytic oxidation of formic acid. J. Am. Chem. Soc. 130:5406–+

88. Mazumder V, Sun S (2009) Oleylamine-mediated synthesis of Pd nanoparticles for catalytic formic acid oxidation. J. Am. Chem. Soc. 131:4588–+

89. Wang R, Liao S, Ji S (2008) High performance Pd-based catalysts for oxidation of formic acid. J. Power Sources 180:205–208

90. Choi J-H, Jeong K-J, Dong Y et al (2006) Electro-oxidation of methanol and formic acid on PtRu and PtAu for direct liquid fuel cells. J. Power Sources 163:71–75

91. Obradovic MD, Tripkovic AV, Gojkovic SL (2009) The origin of high activity of Pt-Au surfaces in the formic acid oxidation. Electrochimica Acta 55:204–209

92. Zhang S, Shao Y, Liao H-g et al (2011) Graphene decorated with PtAu alloy nanoparticles: Facile synthesis and promising application for formic acid oxidation. Chem. Mater. 23:1079–1081

93. Casado-Rivera E, Gal Z, Angelo ACD et al (2003) Electrocatalytic oxidation of formic acid at an ordered intermetallic PtBi surface. ChemPhysChem 4:193–199

94. Tripkovic AV, Popovic KD, Stevanovic RM et al (2006) Activity of a PtBi alloy in the electrochemical oxidation of formic acid. Electrochem. Comm. 8:1492–1498

95. Alden LR, Han DK, Matsumoto F et al (2006) Intermetallic PtPb nanoparticles prepared by sodium naphthalide reduction of metal-organic precursors: Electrocatalytic oxidation of formic acid. Chem. Mater. 18:5591–5596

96. Matsumoto F, Roychowdhury C, DiSalvo FJ et al (2008) Electrocatalytic activity of ordered intermetallic PtPb nanoparticles prepared by borohydride reduction toward formic acid oxidation. J. Electrochem. Soc. 155:B148–B154

97. Zhang LJ, Wang ZY, Xia DG (2006) Bimetallic PtPb for formic acid electro-oxidation. J. Alloys and Compounds 426:268–271

98. Peng Z, You H, Yang H (2010) An electrochemical approach to PtAg alloy nanostructures rich in Pt at the surface. Adv. Func. Mater. 20:3734–3741

99. Kowal A, Li M, Shao M et al (2009) Ternary Pt/Rh/SnO$_2$ electrocatalysts for oxidizing ethanol to CO$_2$. Nature Mater. 8:325–330

100. Cairns EJ, Simons EL, Tevebaug AD (1968) Ammonia-oxygen fuel cell. Nature 217:780–781

101. Rees NV, Compton RG (2011) Carbon-free energy: A review of ammonia- and hydrazine-based electrochemical fuel cells. Energy Environ. Sci. 4:1255–1260

102. Schuth F, Palkovits R, Schlogl R et al (2012) Ammonia as a possible element in an energy infrastructure: Catalysts for ammonia decomposition. Energy Environ. Sci. 5:6278–6289

103. de Mishima BAL, Lescano D, Holgado TM et al (1998) Electrochemical oxidation of ammonia in alkaline solutions: Its application to an amperometric sensor. Electrochim. Acta 43:395–404

104. Diaz LA, Botte GG (2012) Electrochemical deammonification of synthetic swine wastewater. Ind. Eng. Chem. Res. 51:12167–12172

105. Feng C, Sugiura N, Shimada S et al (2003) Development of a high performance electrochemical wastewater treatment system. J. Hazard. Mater. 103:65–78

106. Gerische H, Mauerer A (1970) Studies on anodic oxidation of ammonium on platinum electrodes. J. Electroanal. Chem. 25:421–433

107. Rosca V, Duca M, de Groot MT et al (2009) Nitrogen cycle electrocatalysis. Chem. Rev. 109:2209–2244

108. Gootzen JFE, Wonders AH, Visscher W et al (1998) A DEMS and cyclic voltammetry study of NH$_3$ oxidation on platinized platinum. Electrochim. Acta 43:1851–1861

109. de Vooys ACA, Mrozek MF, Koper MTM et al (2001) The nature of chemisorbates formed from ammonia on gold and palladium electrodes as discerned from surface-enhanced Raman spectroscopy. Electrochem. Commun. 3:293–298
110. Endo K, Katayama Y, Miura T (2005) A rotating disk electrode study on the ammonia oxidation. Electrochim. Acta 50:2181–2185
111. Vidal-Iglesias FJ, Solla-Gullon J, Montiel V et al (2005) Ammonia selective oxidation on pt(100) sites in an alkaline medium. J. Phys. Chem. B 109:12914–12919
112. Rosca V, Koper MTM (2006) Electrocatalytic oxidation of ammonia on pt(111) and pt(100) surfaces. Phys. Chem. Chem. Phys. 8:2513–2524
113. Vidal-Iglesias FJ, García-Aráez N, Montiel V et al (2003) Selective electrocatalysis of ammonia oxidation on pt(100) sites in alkaline medium. 5:22–26
114. Novell-Leruth G, Valcarcel A, Clotet A et al (2005) DFT characterization of adsorbed NHx species on Pt(100) and Pt(111) surfaces. J. Phys. Chem. B 109:18061–18069
115. Novell-Leruth G, Ricart JM, Perez-Ramirez J (2008) Pt(100)-catalyzed ammonia oxidation studied by DFT: Mechanism and microkinetics. J. Phys. Chem. C 112:13554–13562
116. Novell-Leruth G, Valcarcel A, Perez-Ramirez J et al (2007) Ammonia dehydrogenation over platinum-group metal surfaces. Structure, stability, and reactivity of adsorbed NHx species. J. Phys. Chem. C 111:860–868
117. Offermans WK, Jansen APJ, van Santen RA et al (2007) Ammonia dissociation on Pt{100}, Pt{111}, and Pt{211}: A comparative density functional theory study. J. Phys. Chem. C 111:17551–17557
118. Alvarez-Ruiz B, Gomez R, Orts JM et al (2002) Role of the metal and surface structure in the electro-oxidation of hydrazine in acidic media. J. Electrochem. Soc. 149:D35–D45

Chapter 8
Metallic Nanostructures for Catalytic Applications

W. David Wei, Brendan C. Sweeny, Jingjing Qiu and Joseph S. DuChene

Abstract The discovery of substantial catalytic activity over metallic nanostructures has established a modern industrialized society, largely sustained by heterogeneous catalysts for supplying a wide variety of consumer and industrial goods. Heterogeneous catalysts allow for highly efficient and selective chemical processes, while simultaneously reducing the energy costs associated with chemical manufacturing by lowering the activation energy of the desired product pathway. The remarkable catalytic activity of metallic nanostructures is intimately linked to their unique physical and electronic properties. Unfortunately, the catalytic efficiency of modern heterogeneous catalysts is often hindered by poor control over active catalytic sites and a substantial thermal energy requirement for reaction, which leads to a significant reduction in catalyst lifetime and imposes a growing strain on the environment. Recent breakthroughs involving the precise control over the size, shape, and electronic structure of metallic nanostructures as well as the optimization of metal-support interactions have paved the way for highly efficient catalytic materials poised to ease these environmental burdens. The continued optimization and development of photoresponsive metallic nanostructures using Earth-abundant materials is expected to further reduce both the energetic and financial requirements for obtaining highly efficient and selective heterogeneous catalysts. Deeper fundamental understanding of the physical and electronic properties governing the catalytic properties of these metallic nanostructures promises to provide a means of engineering highly efficient catalysts capable of meeting the pressing material needs of the growing industrial world while achieving a sustainable economic landscape.

8.1 Introduction

The rapid expansion of the industrialized world has placed tremendous pressure on chemical and manufacturing industries to keep pace with rising global demand for consumer goods. In order to meet the ever-increasing appetite of expanding

W. D. Wei (✉) · B. C. Sweeny · J. Qiu · J. S. DuChene
Department of Chemistry and Center for Nanostructured Electronic Materials,
University of Florida, Gainesville, FL 32611, USA
e-mail: wei@chem.ufl.edu

© Springer International Publishing Switzerland 2015
Y. Xiong, X. Lu (eds.), *Metallic Nanostructures,* DOI 10.1007/978-3-319-11304-3_8

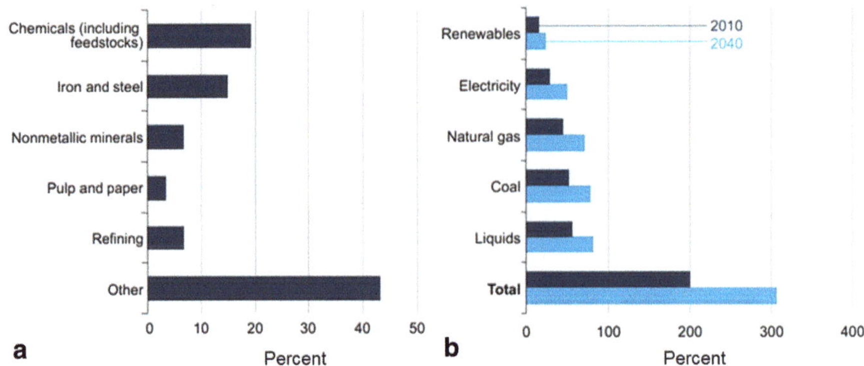

Fig. 8.1 Worldwide industrial sector energy consumption by **a** industry and **b** energy source (current and projected) [1]

markets for raw chemical feedstocks, catalysts are employed to improve the inherently slow rates of many chemical reactions. Catalysts play a critical role in lowering the activation energy of a chemical reaction, which would otherwise occur with lower efficiency or at a slower rate. Therefore, it is becoming increasingly rare that a modern industrial chemical process is performed in the absence of some form of catalyst. Despite the ubiquitous use of industrial catalysts, the fundamental chemical mechanisms governing catalytic efficiency remain a highly active area of both pure and applied chemical research. Although, advancements in catalytic performance continue to occur, to date, most industrial processes still require significant thermal energy inputs in order to achieve sufficient catalytic efficiency. As a result, the chemical industry remains the single largest energy consumer in the industrial sector at nearly 20 % as of 2013, and is expected to rise in the coming decades (Fig. 8.1; [1]). Further improvements in catalytic selectivity and efficiency, coupled with innovative methods to reduce thermal demands, are required in order to continually improve the performance of industrial catalysts.

Catalyst design involves consideration of numerous factors in order to achieve an appropriate balance between conversion efficiency and product selectivity. Metals offer particularly intriguing opportunities for catalysis due to their unique electronic structure, which results from partially filled d-bands. The occupancy (or vacancy) of these d-bands, coupled with the spatial orientation of these electronic states, is often responsible for the varied catalytic activity of different metal nanocrystals, which provide highly-localized electron density for reduction reactions, or conversely, spatially-localized vacancies for adsorbing reactants. Furthermore, size-dependent electronic properties arise as a consequence of the band structure of the metal itself, which depend sensitively on the diameter (d) of the metal catalyst. For instance, very small metal clusters, ranging in size from ~10–200 atoms (e.g. $d<2$ nm), exhibit quantized electronic levels similar to that of semiconductors with a true band gap [2], whereas classical electronic properties emerge at larger sizes ($d>2$ nm) as a continuous band structure is established within the metal [3]. The ability to transition between these regimes through recent advancements in controlled nanocrystal growth have provided new "synthetic knobs" to tune the physicochemical properties of the resulting metal catalyst for a particular application.

Catalytically active metal nanocrystals are often deposited on another material support in order to increase the active surface area available for reactions. In the context of heterogeneous catalysis, it is important to distinguish between so-called "passive" and "active" supports in evaluating the role of both constituents in the overall catalytic process. Certain reactions are dramatically enhanced in the presence of these active supports, such as carbon monoxide (CO) oxidation [4], as the presence of unique active sites at the interface between the metal nanocrystal and the support enhance adsorption and activation of the reactant molecules. These types of unique active sites exert significant influence over the catalytic activity observed in many industrially important reactions and must be considered when designing catalytic materials for a given purpose.

As mentioned previously, the majority of these catalytic processes require substantial thermal energy to drive the reaction in order to achieve sufficient conversion efficiencies at an industrial scale. Unfortunately, this thermal requirement represents an undesirable industrial practice, as additional thermal inputs add significant production costs to the overall catalytic cycle. Moreover, operating heterogeneous catalysts at elevated temperatures accelerates deactivation of the catalysts [5, 6]. Depending on the reaction of interest, several potential products may be expected, and these unwanted byproducts are often increased at elevated temperatures, further reducing the overall selectivity of the catalyst [7]. The development of high efficiency catalysts, that mitigate thermal inputs, constitute a vital requirement for more sustainable industrial practices. Two critical criteria that must be addressed to improve future catalytic materials involve mitigating these thermal requirements by improving catalytic efficiency and boosting reaction selectivity.

The industrial applications for heterogeneous metal catalysts are as varied as the reactions they catalyze. Archetypal catalytic reactions such as CO oxidation, ethylene epoxidation, and the water–gas shift reaction have shaped the modern paradigm of metal nanocrystal catalysis [4]. Although these reactions are historically and industrially important, perhaps the most interesting development in metal catalysis came with the realization that small Au nanoparticles (NPs) exhibit remarkably increased catalytic activity for CO oxidation at room temperature [8]. This discovery has had important implications for the future direction of metal nanocrystal catalytic research and has ushered a renewed interest in the unique catalytic properties of metal NPs for catalytic applications.

The use of metallic nanostructures for sustainable chemistry will become a greater emphasis in the future, with the realization that ever-increasing global demand can only be met by reducing energy inputs. The efficient use of renewable energy sources will be of vital importance to the chemical industry, as the search continues for new ways to improve catalytic efficiency and selectivity, while minimizing the thermal requirements for industrial synthesis. In particular, the development of both alloyed and core-shell nanostructures offer improved selectivity for challenging chemical reactions, such as the oxygen reduction reaction [9, 10], a demanding catalytic reaction pivotal to the eventual realization of a sustainable hydrogen economy. Heterogeneous metal catalysts for electrolytic reactions, such as the hydrogen evolution reaction from water [11, 12], or electroreduction of CO_2 [13] are also expected to receive substantial attention in the coming decades, as these reactions represent a sustainable means of procuring chemical energy and raw chemical feedstocks for further synthesis of fine chemicals.

Recently, in an effort to reduce energy inputs, researchers have turned to plasmonic-metal NPs as potential photocatalysts for a variety of photochemical reactions [6, 14–18]. The unique optical properties of plasmonic-metal nanostructures stem from their ability to support collective electronic oscillations known as surface plasmons. Optical excitation at the surface plasmon resonance (SPR) frequency of the metal nanoparticle induces a coherent oscillation of the electron gas, generating an enhanced electromagnetic (EM) field spatially localized at the surface of the plasmonic-metal nanostructure [19]. The SPR frequency is highly sensitive to the composition, shape, and size of the metal nanostructure, as well as the local dielectric environment, offering a means of tuning the optical response for a given application [20, 21]. The large optical cross sections of plasmonic-metal nanostructures make them attractive candidates for solar-powered applications, with tunable SPR properties throughout the entire solar spectrum. Efficiently harvesting solar energy for catalyzing chemical reactions offers a sustainable route to power industrial reactions, while reducing the thermal inputs required for achieving sufficient reaction efficiencies. The synergistic use of both light and thermal energy offers intriguing possibilities for driving industrially relevant catalytic processes, while minimizing extraneous thermal inputs. It is anticipated that these plasmonic nanocrystals may offer a new approach to reduce external energy inputs, while maintaining suitable catalytic efficiency.

8.2 Traditional Small Metallic Nanostructure Catalysis

Metals possess inherent material properties that make them unique from other forms of matter and position them as promising materials for catalysis. These properties are derived from the nature of the electronic interactions between atoms as they combine to form a discrete metal particle. As the number of atoms of a metal increase to form a periodic solid lattice, continuous energy bands form as a result of orbital overlap of the quantum mechanical wavefunctions (Fig. 8.2). For metals, the highest-energy band is only partially filled, allowing for transport of these electrons between the unoccupied electronic states throughout the nanocrystal lattice at room temperature, facilitating electronic conduction. For metals in group VIII–XI, the catalytic activity is partially due to the degree of vacancy of the d-bands, as this occupancy determines the adsorption strength of reactants and molecular intermediates for a particular reaction. For this reason, the metals found to be most active, and consequently most commonly studied for heterogeneous catalysis are: the 3d metals of Fe, Co, Ni, and Cu; the 4d metals of Rh, Pd, and Ag; and the 5d metals, Pt and Au [22]. A brief overview of commonly used metals is presented below.

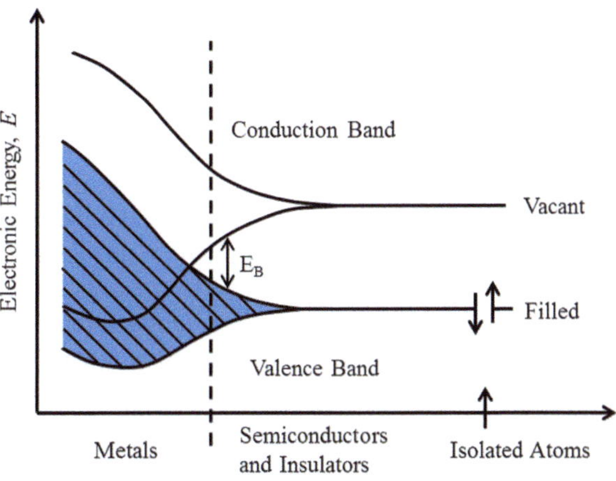

Fig. 8.2 Density of states as a function of interatomic distance. (Adapted from reference [80])

The choice of metal has a significant effect on the resulting selectivity and activity for certain reactions because of the intrinsic differences in material properties of the metals. Properties such as the lattice parameters, work function, and ionization energy affect overall catalytic activity. For example, the relatively low ionization energy of Ag engenders this metal with high selectivity for ethylene epoxidation, resulting in an important industrial feedstock, ethylene oxide (EO), rather than unwanted H_2O and CO_2 byproducts [23, 24]. It has been suggested that Ag can readily stabilize atomic oxygen on its surface, whereas Au cannot regenerate surface oxygen without oxidative treatment [24].

By comparison, Au nanocrystals are thoroughly studied for CO oxidation despite their inability to readily adsorb molecular oxygen (O_2) [25]. The relatively weak chemisorption strength of CO on the Au surface allows for easy desorption following oxidation to CO_2. This process is much more difficult on the Pt nanostructure surface because the relatively strong adsorption of CO_2 on the Pt surface prevents desorption after reaction, leading to catalyst poisoning [22]. Hence, Au has been the subject of many catalytic studies because of its unusual activity and selectivity for aerobic oxidation reactions [26–28].

Pt nanocrystals are known to be highly active for water reduction, and are often the benchmark by which new hydrogen-evolving electrocatalysts are measured [29]. The Pt surface is highly active because of the appropriate mix between active adsorption of the molecular reactant and ready desorption of the molecular product [30]. The prolific hydrogen evolution reaction (HER) activity of Pt NPs make them attractive candidates for augmenting photoelectrochemical cells to improve the conversion of solar energy into chemical energy in the form of H_2 gas. This approach has shown to yield substantial benefits in overall photoelectrochemical activity, while requiring a minimal amount of precious metal cocatalyst [29, 31]. Further research has centered around involving the more plentiful third-row transition metals into metal nanocrystal catalytic platforms in order to achieve an appropriate balance between overall catalytic activity and catalyst cost.

Pd nanostructures excel in a particular type of C–C coupling reaction between a halide and a boronic acid known as the Suzuki–Miyaura cross-coupling reaction. This coupling reaction has proven to be a very versatile reaction for the industrial-scale formation of C–C bonds in aryl compounds such as styrene [32], and has been so important that A. Suzuki, R. F. Heck, and E. Negishi shared the 2010 Nobel Prize in Chemistry for their pioneering work in this process. Pd atoms excel at this reaction because of their ability to add to the organohalide reactant, the rate-limiting step, relative to other metals [33]. Originally, this procedure involved homogenous Pd(0) organometallic catalysts dispersed in organic solvents. Unfortunately, the use of a homogeneous catalyst is inconvenient due to the inherent difficulty associated with separating the Pd catalyst from the final product [34]. When Pd metal nanostructures were used in place of the originally-developed organometallic catalyst, improved stability and separation from the metal catalyst was realized [33]. The key to develop highly selective and efficient catalysts is to exploit the inherent chemical properties of the metal nanostructure, and tailor the physical and electronic properties in order to minimize the activation energy at an active site, while maximizing the density of those sites.

8.2.1 Size and Facet-dependent Catalysis

Current industrial needs require the rational design of heterogeneous catalysts. The utilization of nanostructured metals offers multiple practical advantages over bulk materials: (1) physically, decreasing the size increases the surface to volume ratio and the density of the active sites [5]. Additionally, the exposed facets can be manipulated via shape-controlled synthesis methods to preferentially expose a greater proportion of catalytically active surfaces [30, 35]; (2) electronically, the metal band structure can be considerably different at the nanoscale and is modified to facilitate the formation of discrete bands through quantum confinement [36]. It is through quantum confinement or SPR that these optical properties can be tailored for photocatalytic reactions. Supported single-crystal metal nanostructures generally assume a semi-spherical shape, and for FCC metals in groups X and XI are often bound by low index {100}, {111}, and even {110} facets (Fig. 8.3). These periodic arrangements of atoms vary in surface energy and typically have a distinct effect on catalytic activity. Recently, it has been demonstrated that Pd nanocubes exclusively bound by {100} facets show the highest activity for the Suzuki coupling of 4–bromoanisole and phenylboronic acid [33]. The basis for this improved activity was found to be the leaching of Pd through reaction with adsorbed oxygen, which occurred more readily on the {100} facet surface than the others. Exposed crystal facets can also affect the surface energy and lead to a reduced dissociation barrier for some reactant species. In the case of nanoporous Au catalysts, it was observed that the selective exposure of high-energy facets and surface defects

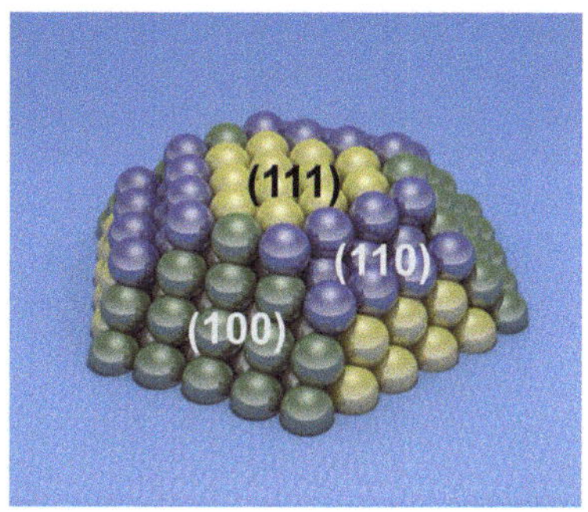

Fig. 8.3 Geometric representation of common low-energy facets found in FCC metals

including kinks and steps, while simultaneously preventing the formation of low energy {111} facets, leads to a higher activity for the aerobic oxidation of CO [37]. These results highlight a common problem in nanocrystal catalysis, in that these high-energy surface facets are highly active for catalysis, but often reorganize to less active, lower energy crystal facets during (or prior to) catalytic reaction, reducing catalytic activity and overall catalyst lifetime [38].

The exposed crystal facet also plays a major role in selectivity for EO formation. By comparing Ag nanorods, which are bound by both {100} and {111} facets, to nanocubes bound only by {100} facets (Fig. 8.4) and multifaceted nanospheres, Linic et al. determined the influence of exposed {100} surface area on selectivity for EO [35]. The selectivity for EO was observed to be directly proportional to the surface area of the exposed {100} facet (Fig. 8.4). This facet-dependent activity was further confirmed based on the geometry of the nanorods, which exhibited {111} surfaces on the ends of the rods and {100} surfaces on the sides, as lengthening the nanorod led to higher selectivity.

In many cases, the facet-dependent activity is inseparable from the size dependence, since nanocrystal size limits the exposed facet surface area and dictates the types of exposed surface sites. For instance, Goodman et al. successfully correlated the number of low-coordination number Au atoms (located at the perimeter or corner of the NP with the CO oxidation activity (Fig. 8.5; [39]). More recently, Murray et al. demonstrated a similar phenomenon in both Pd and Pt NPs adsorbed on CeO_2 for CO oxidation activity [40]. It was observed that when normalized for metal surface area, the sample with the smallest size had the highest activity because this catalyst had the largest proportion of perimeter and corner atoms.

Further manipulation of metal nanocrystal size leads to unique electronic properties. Nanoparticles in the $d < 2$ nm size range exhibit band structures that begin to reflect quantum mechanical properties. At this size, transition metals exhibit a valence band (VB) and conduction band (CB) with an intermediate band gap similar to semiconductors or quantum dots. This attribute was apparent early on in Goodman's study of size dependent aerobic oxidation over Au on TiO_2 supports. While it has been widely accepted that sub-5 nm Au nanostructures are inherently active for CO oxidation, Goodman et al. highlighted that the onset of catalytic activity in supported Au catalysts occurred at ~1 nm thickness (~300 atom cluster), at nearly the

Fig. 8.4 Ethylene epoxidation over **a** Ag nanorods and **b** Ag nanocubes exhibit remarkable differences in selectivity **c** due to differences in exposed surface facets. (Reprinted with permission from reference [35])

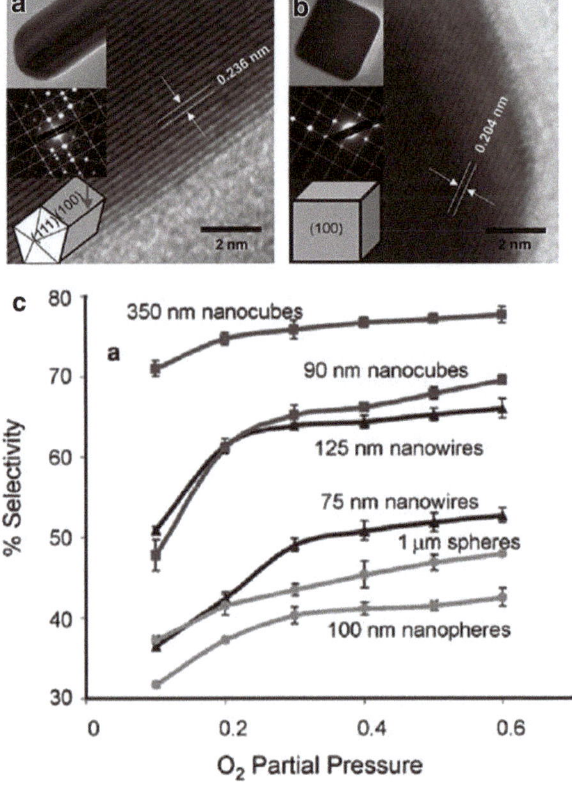

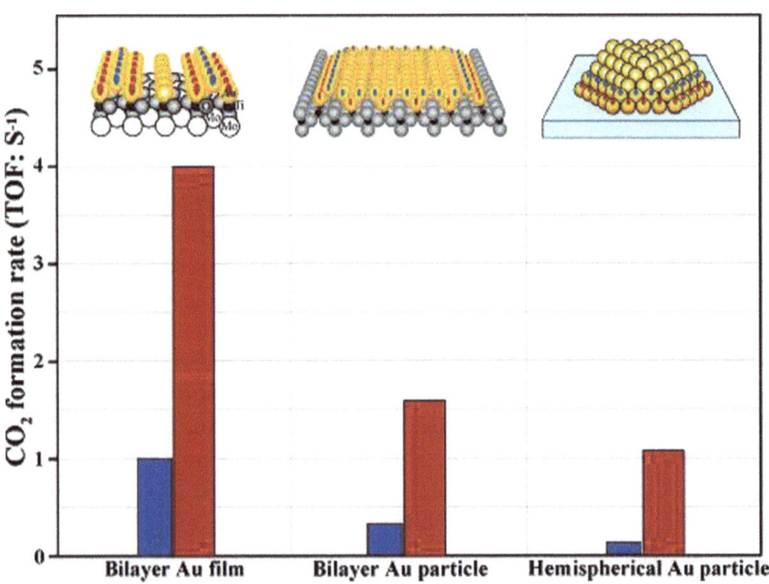

Fig. 8.5 Dependence of CO oxidation activity on specific active sites (*red* and *blue*) on Au bilayer films (*left*), bilayer particles (*center*), and hemispherical particles (*right*). (Reprinted with permission from reference [39])

same point at which quantum confinement is observed [36]. However, it should be noted that the surface catalytic activity of the metallic nanostructures persists into the large NP regime ($d > 20$ nm) in most cases.

8.2.2 Support Effect

The metal nanostructure itself can function as a catalyst for certain reactions. However, there are many physical and electronic advantages of using a support material in conjunction with metallic nanostructures, including: (1) improved stability of smaller nanostructures ($d < 5$ nm) by minimizing aggregation/coalescence, (2) no requirement for surface ligands to stabilize the nanocrystal, offering unhindered access to the catalyst surface, (3) can alter the electronic structure of the supported metal nanocrystal, and (4) so-called "active" supports can actually participate in the reaction mechanism. Another important result from the work of Murray et al. was that the density of under-coordinated Pd or Pt atoms only affected the activity when adsorbed on CeO_2 supports and not Al_2O_3 [40]. This was explained by the fact that CeO_2, like TiO_2, α-Fe_2O_3, ZnO, $SrTiO_3$, and MgO, is considered an "active" support, meaning that they are involved in the adsorption and activation of molecular O_2 to either atomic oxygen or a reactive oxygen species such as singlet oxygen (O_2^*) or superoxide (O^-) radicals. The properties of the catalyst support affect all aspects of a catalytic reaction, including adsorption, surface reaction, and desorption. When complimentary to the reactivity of the metallic nanostructure, the catalyst support can provide synergistic catalytic activity otherwise unattainable with unsupported nanostructures.

8.3 Catalytic Reactions Using Traditional Metallic Nanostructure Catalysts

Small metal nanostructures have been employed for a wide variety of catalytic reactions. The aim of this section is to highlight specific reactions that are actively studied, have significant industrial application, and have advanced the fundamental understanding of surface catalysis in metallic nanostructures.

8.3.1 Carbon Monoxide Oxidation

The scission of the C–O bond during CO oxidation is regarded as the rate limiting step in the Fischer–Tropsch reaction, which forms hydrocarbons from the hydrogenolysis of CO [41]. This reaction has also been the target of catalytic converters in automobiles since the 1970s in order to lower the toxicity of tailpipe emissions by converting CO to hydrocarbons and CO_2 [42]. Haruta et al. first observed catalytic

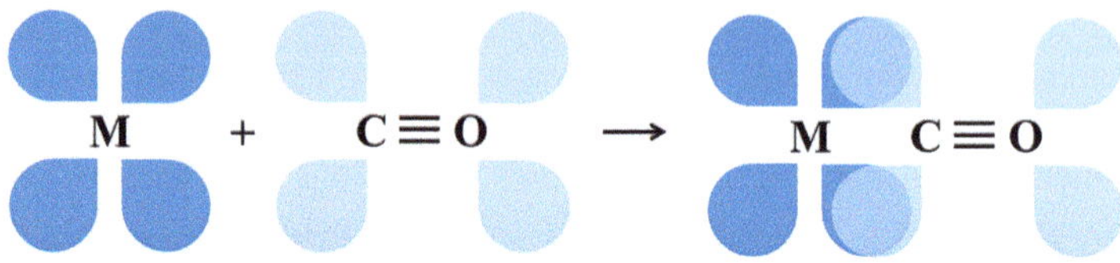

Fig. 8.6 Orbital interaction between metal surface and CO that leads to chemisorption (**a**) and overlap that leads to electron back-donation (**b**)

activity for CO oxidation over small ($d<5$ nm) supported Au NPs in the 1980s [8]. This observation was remarkable for several reasons. First, bulk Au catalysts are known to be inactive for CO oxidation. Second, the oxidation activity of Au nanocrystals was observed at surprisingly low temperatures ($T<0\,°C$), which is not reproduced using other metallic nanostructures.

As with many surface catalytic processes, low temperature CO oxidation is suggested to occur through a Langmuir–Hinshelwood mechanism involving four primary steps: (1) reactant adsorption, (2) surface diffusion, (3) surface reaction, and (4) product desorption [43]. The process is initiated by the chemisorption of CO onto the surface of a Au NP, during which a CO molecule chemisorbs to the Au surface through overlap of the molecular 2σ orbital of carbon with the hybrid d_σ and $d_z{}^2$ orbitals of Au (Fig. 8.6a; [44]). When chemisorbed, electron density is spontaneously transferred to the vacant hybrid Au orbitals. This transfer of electrons leads to an unsustainably high electron density, resulting in the back donation of electrons to the C $2\pi^*$ orbital (Fig. 8.6b), weakening the C–O bond, and strengthening the Au–C bond [44]. For low temperature CO oxidation, it is suggested that the active sites for Au NPs supported on a metal oxide are at the nanocrystal perimeter, consisting of under-coordinated Au atoms, and therefore, the adsorbed CO and O_2 species must diffuse to this active site for reaction before intermediate formation and desorption of CO_2 occurs (Fig. 8.7; [22, 28, 40, 45, 46]).

The activation of O_2 is widely considered the rate limiting step in the overall CO oxidation mechanism and unlike Ag or Pd, adsorption of molecular O_2 onto low energy facets of Au is unfavorable [22]. Therefore, this reaction is far more dependent on the catalyst support for at least the partial adsorption and activation of O_2 species. It has been suggested that reducible metal oxides (e.g., TiO_2, $SrTiO_3$) can activate O_2 because of their ability to both readily adsorb and transfer electrons to O_2 molecules [47]. Recent studies have suggested that the activation of O_2 on a reducible oxide can occur via a Mars–van Krevelen pathway in which lattice oxygen from the support can reduce to superoxide (O^-) and react with CO adsorbed to a perimeter Au atom, leaving behind an oxygen vacancy to which molecular O_2 can

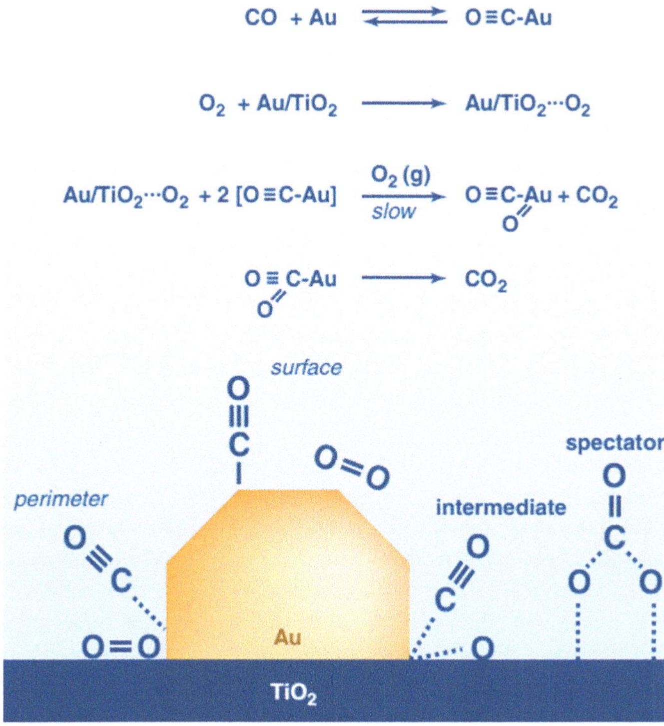

Fig. 8.7 Proposed CO oxidation mechanism involving the chemisorption of CO on the Au surface followed by diffusion to a perimeter Au atom where oxidation by activated oxygen species occurs. (Reprinted with permission from reference [22]).

then adsorb, thereby filling the oxygen vacancy in the support and completing the catalytic cycle [47].

While molecular O_2 does not readily adsorb to Au, it is possible for O_2 to bridge the interface of the Au and metal oxide support in addition to adsorption on the metal oxide. Yates et al. recently published results suggesting that molecular O_2 species adsorb to the interface between the Au nanocrystal and the TiO_2 support, "bridging" between a perimeter Au atom and a five-coordinated Ti site, as identified by in situ Fourier transform infrared (FTIR) spectroscopy and further supported by density functional theory (DFT) calculations [28]. In addition to providing evidence for the adsorption of CO to the Au surface, the authors suggest that the CO can adsorb to TiO_2 as well and that CO at both adsorption sites subsequently diffuse to the perimeter of the Au at $T > 120$ K, where oxidation occurs from a bridge-site oxygen species.

The oxidation state of the supported metal is vitally important to CO oxidation activity [45]. The inherent properties of the exposed facets of the metal-oxide can either donate or withdraw electron density from the supported metal. This observation is particularly apparent in small Au clusters, as Landman et al. demonstrated with Au_8 clusters supported on different MgO facets [48]. Monitoring the $\nu(CO)$

stretch via in situ FTIR, the authors determined that the activity of Au clusters adsorbed to the MgO {110} facet with a high density of F-center defects was attributable to the donation of electron density from the MgO defects to the Au NPs, resulting in a more electron-rich nanocluster as compared to the defect free surface. The surface charge on the Au NP affects back transfer of electrons to the π^* orbital, either increasing chemisorption strength or C–O bond strength.

Additionally, the oxidation state can be manipulated using oxidative and reductive treatment. As Tauster et al. first demonstrated, thermal treatment of Au on TiO_2 under H_2 atmosphere leads to the partial encapsulation of the adsorbed Au NPs but also electron transfer from the TiO_2 support to the metal NP [49]. In a recent study, an oxidative strong metal-support interaction (SMSI) phenomenon was demonstrated over Au on ZnO, showcasing a reversible encapsulation and electron transfer from Au to ZnO when treated in an O_2 atmosphere [50]. The resulting NPs possessed a partial positive charge (δ^+) as measured by FTIR that could be transformed to a partial negative charge (δ^-) by further treating the catalyst with H_2 and subsequently accompanied by the reversal of NP encapsulation (Fig. 8.8). These observations showcase a universal approach to the tailoring of metallic nanostructure catalysts for gas phase reactions and have been pursued as a means of tuning reaction activity. However, this approach is costly and energy intensive, as it requires expensive equipment and operates under high pressure and temperature, motivating researchers to seek other pathways to manipulate the surface charge of a metal nanocrystal. These recent approaches will be discussed in greater detail in the sections to follow.

8.3.2 Ethylene Epoxidation

The epoxidation of ethylene is an industrially relevant reaction for the production of EO, a precursor for detergents, plastics, and other chemical feedstocks. While EO can be produced with approximately 87.5 % selectivity over Ag-based industrial catalysts, the global market for EO is expected to surpass 27 million t by 2017 [24, 51]. Because of this projection, even seemingly small improvements in catalytic activity are expected to have major financial consequences. Challenges for the production of EO include process temperatures in the 500–600 K range and resulting catalyst instability [52].

The epoxidation of ethylene involves molecular O_2 adsorption over an Ag catalyst whose reaction is given by:

$$C_2H_4 + \tfrac{1}{2}\,O_2 \rightarrow C_2H_4O \tag{8.1}$$

Ag is the most studied and feasible catalyst for ethylene epoxidation, given that other metallic nanostructures result in the complete oxidation to CO_2 or are more selective for undesirable byproducts. The Ag surface is easily oxidized at room

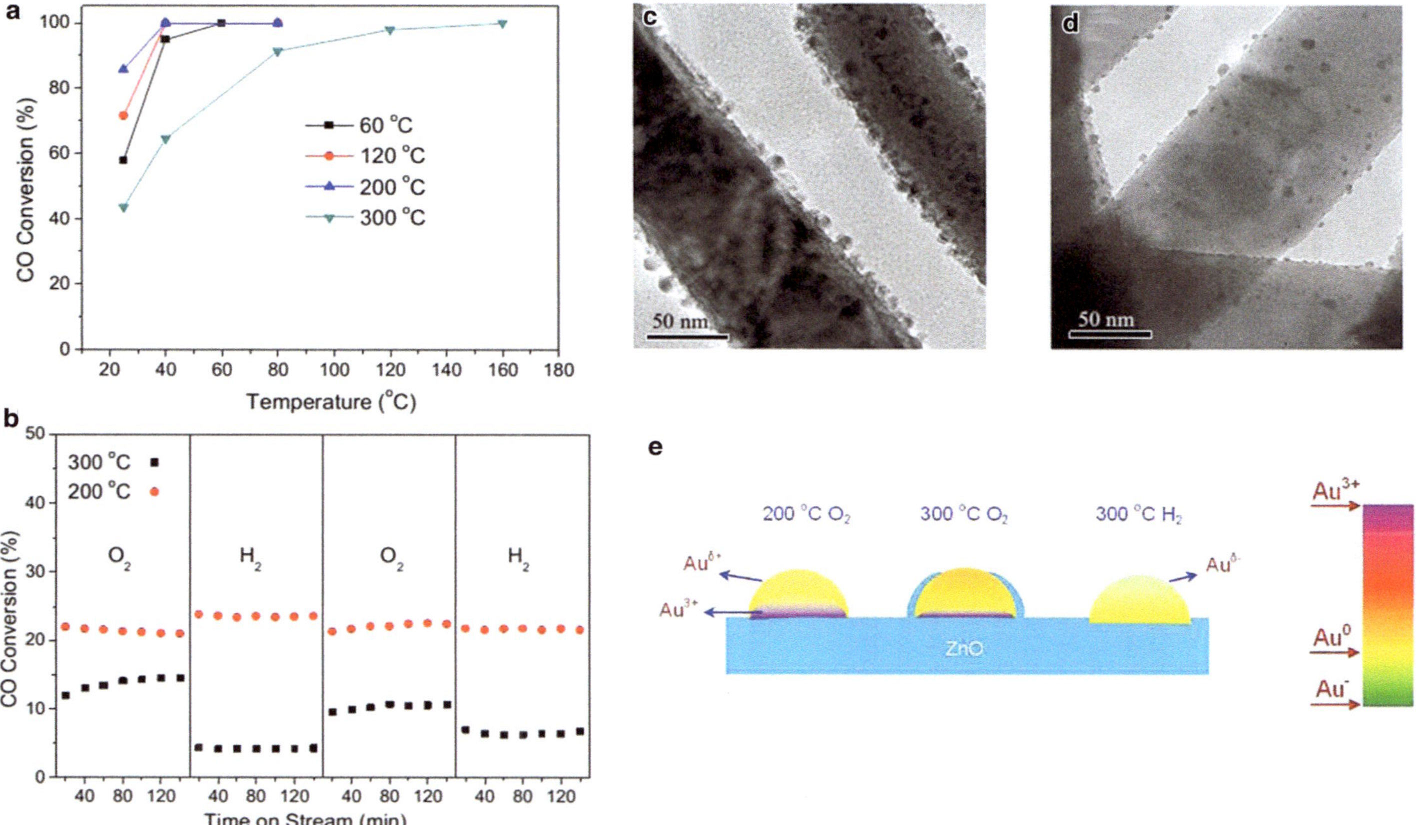

Fig. 8.8 CO conversion over Au–ZnO heterostructures following oxygen pretreatment at varying temperatures (**a**) and during alternation between oxidizing/reducing treatments (**b**). Encapsulation of Au following oxidative treatment (**c**) is reversible after treatment in reductive atmospheres (**d**). The corresponding change in oxidation states of supported Au nanoparticles following pretreatments (**e**). (Reprinted with permission from reference [50])

temperature under atmospheric pressure but the weak Ag–O bond is not stable at higher temperatures, leading to dissociation and oxidation of adsorbed ethylene [24]. It has been suggested that electrophilic oxygen ($O^{\delta+}$) selectively attacks the C=C double bond, whereas nucleophilic surface oxygen will preferentially attack the C–H bond, removing a bonded hydrogen atom [24]. DFT calculations suggest that out of oxygenated surfaces of Au, Ag, and Cu metals, Ag exhibits the lowest energy barrier, whereas the Au surface cannot be easily reoxidized, while the Cu surface very readily oxidizes, but retains oxygen on the surface at higher temperatures, leading to a high activation energy [23].

Optimization of the physical properties of metallic nanostructures for a variety of catalytic reactions is ongoing and heavily studied in the field of catalysis. However, manipulation of both the physical and electronic properties of the catalyst is critical for achieving fully optimized materials with high yield and selectivity. One promising approach that has been the focus of recent studies involves the application of visible light in the catalytic process to further manipulate the oxidation state of supported metal nanocrystals. In this regard, harnessing the SPR properties of plasmonic-metal nanostructures has shown significant promise as a means of reducing thermal energy inputs, while simultaneously boosting catalytic activity.

8.4 Large Metallic Nanostructure Catalysis

8.4.1 *Photocatalysis Model*

As progress is made to optimize the catalytic activity of metallic nanostructures through manipulation of their physical and electronic structure, significant effort has been put toward expanding upon the inherent catalytic activity of metallic nanoparticles through heterostructure formation with semiconductors. Since the pioneering work of Honda and Fujishima in the 1970s, which demonstrated the photolysis of water to generate H_2 and O_2 by ultraviolet (UV) irradiation of TiO_2 [53], semiconductor synthesis and design continues to constitute active areas of fundamental photocatalytic research. Unlike a metal's band structure, semiconductors exhibit discrete band gaps that allow for absorption of incident photons with energies larger than the band gap, which is given by the energy difference between the VB maximum and CB minimum. In traditional semiconductor photocatalysis, there are three sequential processes to consider: (1) absorption of light to generate electron-hole pairs, (2) diffusion of these excited-state charge carriers to the semiconductor surface, and finally, (3) transfer of these charges to reactant molecules adsorbed on the photocatalyst surface to catalyze a chemical reaction [54]. Specifically, incident photons can generate excitons in the semiconductor through the process of exciting VB electrons to the CB, (Fig. 8.9a) leaving behind a charge vacancy, or hole, in a process known as charge-separation. If charge-separation occurs sufficiently close to the semiconductor–electrolyte interface, electrons and holes may diffuse

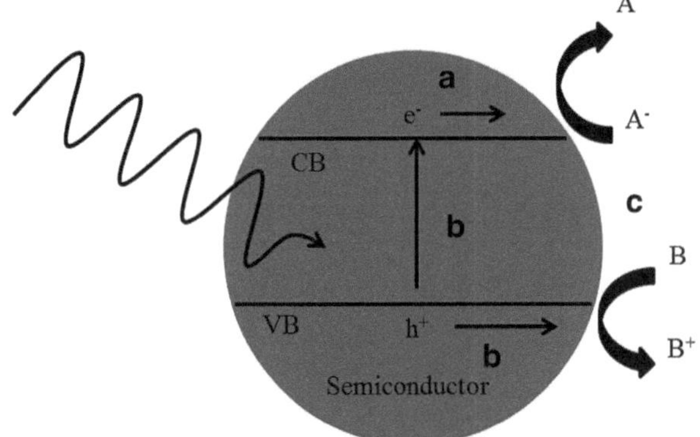

Fig. 8.9 Photocatalytic model involving **a** photon absorption, **b** charge–carrier diffusion, and **c** surface reaction

(Fig. 8.9b) to the surface of the semiconductor and react as part of a reduction or oxidation half-reaction, respectively (Fig. 8.9c). Efficient photocatalysis requires that the semiconductor balances several properties simultaneously, including: (1) appropriate band edge potentials, (2) suitable band gap range to harvest solar insolation, (3) photochemical stability, and (4) minimal charge carrier recombination. This last parameter is often a significant limitation in semiconductor photocatalysis because of the significant mismatch between the typical length scales required for effective photon absorption (~μm) and the typical diffusion distance of excited-state charge carriers (~nm) in most metal oxides (e.g., TiO_2, ZnO, Fe_2O_3, etc.) [54]. A commonly studied approach to overcome this limitation involves heterostructure formation with a metal cocatalyst.

8.4.2 Metallic Nanoparticles as an Electron Sink

Similar to the catalysts used in traditional surface catalysis, metallic nanostructures are physically anchored to the semiconductor surface, forming a metal–semiconductor heterostructure. The interface between these two materials comprises the critical region of the entire heterostructure for determining photocatalytic activity. The electronic properties of the substituent materials determine the type of interface that is formed, which can either promote or hinder efficient electron transfer [55]. These contacts can be either rectifying or Ohmic depending on the Fermi level of the metal relative to the semiconductor, and therefore, impede or facilitate electron transfer between these materials.

A Schottky barrier is a type of rectifying potential barrier formed at the interface between a metal and an n-type semiconductor. Assuming an ideal interface, without Fermi-level pinning due to semiconductor surface states, a metal with a high work function relative to the work function of the semiconductor can induce band bending and a potential energy barrier in the semiconductor. When the Fermi level of the metal is below the Fermi level of the semiconductor, there will be a spontaneous

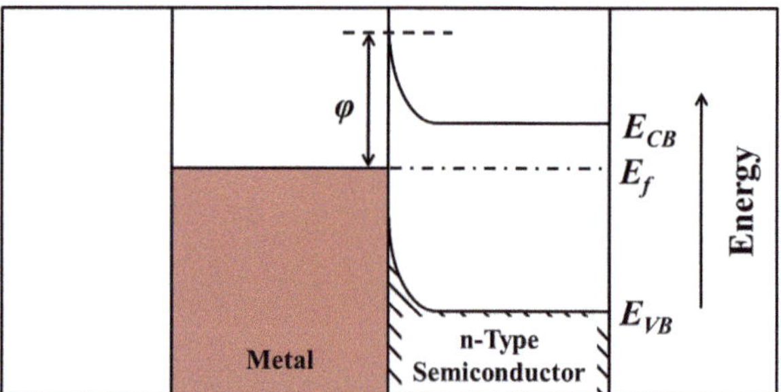

Fig. 8.10 Schottky barrier diagram exhibiting band bending upon interfacing of metal and n-type semiconductor where φ represents the barrier height, E_{CB} and E_{VB} the conduction band and valence band of the semiconductor, respectively, and E_f, the Fermi level of the heterostructure [56]

transfer of electrons from the semiconductor to the metal upon physical contact until a single Fermi level is established within the composite material. This transfer of charge leads to the formation of a positive space charge region within the semiconductor near the metal–semiconductor interface, which establishes a potential energy barrier within the semiconductor known as a Schottky barrier (Fig. 8.10; [55, 56]). The potential energy barrier, or Schottky barrier height, is given by the difference between the metal work function (ϕ) and the semiconductor electron affinity (χ). For example, at the interface of Au ($\phi=5.2$ eV) and anatase phase TiO_2 ($\chi=4.2$ eV), the potential energy required for an electron to completely surmount the Schottky barrier and transfer from metal to semiconductor would be ca. 1.0 eV. Using Au, and other metals with high work functions (e.g., Pt, 5.9 eV or Pd, 5.6 eV), can lead to a build-up of negative charge at particular reactive sites on the metal nanocrystal, while simultaneously minimizing recombination within the semiconductor and lowering reaction overpotential.

The creation of a metal–semiconductor heterostructure and subsequent Schottky barrier formation alleviates multiple limitations of traditional semiconductor-based photocatalysis. First, the overpotential for many catalytic processes, such as H_2 evolution, is much lower in metallic nanostructures such as Pt or Au as compared to a pristine semiconductor surface. Secondly, a rectifying barrier arises from the fact that excited-state electrons in the semiconductor CB can easily transfer to the metal, but the reverse transfer back to the semiconductor is energetically unfavorable. Therefore, the use of a rectifying barrier enables efficient charge separation, while simultaneously providing a means of shuttling excited-state electrons to reactive metal catalysts. For instance, this strategy has been employed since the late 1970s with Bard et al.'s initial observation of methane evolution from acetic acid over UV-irradiated $Pt–TiO_2$ heterostructures [57].

Since the Fermi level of the resultant heterostructure equilibrates upon conjoining two distinct materials, Kamat et al. studied the relative Fermi energy of Au–TiO_2 heterostructures by varying the Au NP size between 3, 5, and 8 nm [58]. Using flash photolysis and a (C_{60}/C_{60}^{-}) redox mediator, it was demonstrated that the effective Au–TiO_2 Fermi level shifts to more negative potentials with UV irradiation of the TiO_2 support, rendering the composite photocatalyst more reductive than bare TiO_2 (Fig. 8.11). Furthermore, the size of the Au NPs strongly affects the magnitude of the Fermi level shift, with 3 nm Au NPs inducing a Fermi level shift of -60 mV

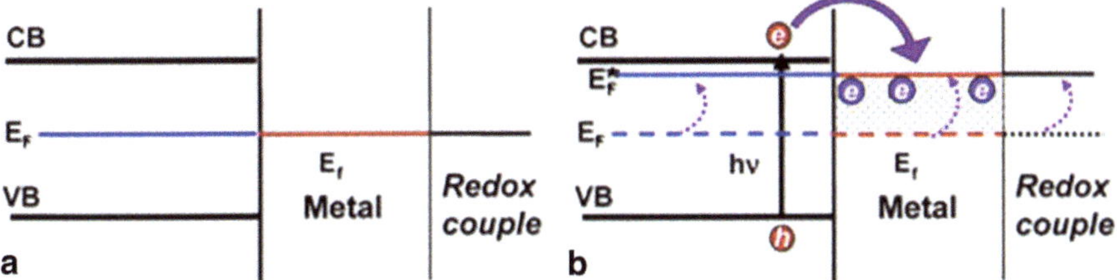

Fig. 8.11 **a** Fermi level equilibration in the dark and **b** Fermi level shift upon UV irradiation. (Used with permission from reference [58])

under UV irradiation (closer to the TiO_2 CB), whereas 8 nm Au NPs only shifted the Fermi level by -20 mV. The distinct shift in Fermi level was attributed to the differences in the density of states (DOS) of the different Au NP sizes. In an 8 nm Au NP, the potential energy increase per transferred electron is smaller than in 3 nm Au NPs due to the quantum size effect of Au, allowing electrons transferred to smaller Au NPs to reach higher unoccupied levels and more negative reduction potentials relative to larger Au NPs. This demonstrates one aspect of the synergistic effects between the physical and electronic properties of metallic nanostructures and effectively allows for the tailoring of their electronic properties for specific catalytic reactions. Another approach to minimize recombination through charge separation is by exploiting the inherent optical properties of the metallic nanostructures themselves.

8.4.3 Charge Transfer to the Support

Despite the ubiquitous use of TiO_2 and other wide band gap semiconductors over the past few decades, their application for solar photocatalysis is ultimately limited by their narrow optical response. Unmodified anatase TiO_2 primarily absorbs UV light ($\lambda < 415$ nm), which accounts for only ~4% of the solar energy reaching Earth's surface [59]. In order to maximize the potential for solar energy conversion in catalytic applications and reduce the remaining thermal requirements in common industrial processes, the absorption region must be expanded into the visible and infrared (IR) regimes, which account for more than 90% of the solar spectrum [59].

While many different approaches to expanding the absorption region have been explored (chemical doping, dye-sensitization, etc.), incorporating larger (10 nm $< d < 100$ nm) plasmonic-metal nanostructures has become a focus of photocatalysis research because of their unique optical properties [14]. Specifically, nanoscale noble metals of Au, Ag, Cu, Al, and Pd exhibit a strong optical response in the visible range (UV for Al and Pd) due to SPR [60]. The advantage of utilizing plasmonic nanostructures lies in the fact that the SPR response is highly tunable through manipulation of the size, shape, and local dielectric environment of the nanostructure, which means this phenomenon has the potential to be exploited

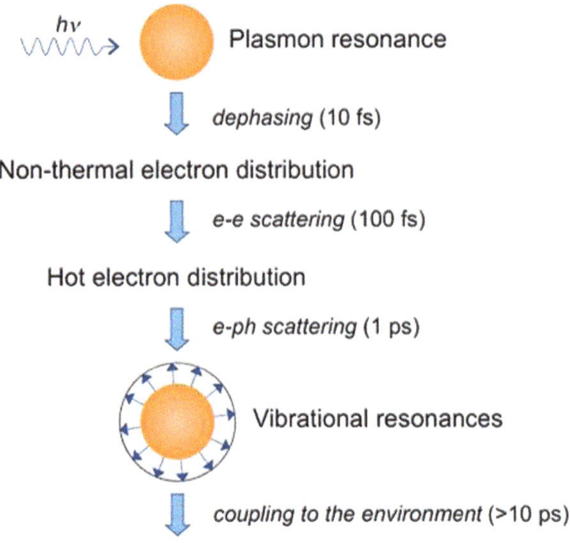

Fig. 8.12 Surface plasmon resonance decay dynamics. (Reprinted with permission from reference [62])

for panchromatic solar energy conversion throughout the entire solar spectrum [60, 61]. Following optical excitation, the SPR dephasing mechanism involves electron–electron scattering and eventually photothermal electron–phonon coupling (Fig. 8.12) [62]. The electron–electron scattering process generates a non-thermalized distribution of so-called "hot" electrons that are energetic enough to perform catalytic reactions on the surface of the metallic nanostructure, or overcome the Schottky barrier at the metal-semiconductor interface and occupy available CB states in an adjacent semiconductor [55, 56, 63, 64, 65].

The work by Tian and Tatsuma first demonstrated the feasibility of plasmon-mediated electron transfer (PMET) in metal–semiconductor heterostructures [66]. By using the photoelectrochemical characterization of Au on P25 (80 % anatase, 20 % rutile TiO_2), the authors were able to closely correlate the generated photocurrent to the SPR response of the supported Au NPs. To isolate the reduction half-reaction from the overall photocatalytic process, a sacrificial hole-scavenger (methanol) was used and the methanol oxidation products were found to be composed of a mixture of formic acid, 2-methoxymethanol, and ethyl formate. Not long after, the work by Silva et al. showcased a comparison of the UV and visible light model for photocatalysis using a colloidal heterostructure of Au NPs deposited on TiO_2 [64]. The study focused on the different charge transfer mechanisms between the two constituents depending on different irradiation conditions (UV or visible). The authors discovered that they could generate H_2 from water using only visible light incident upon the sample, corresponding to excitation of the Au NPs' SPR. The conclusion of these two early studies was that the distribution of electrons resulting from localized surface plasmon resonance (LSPR) excitation included a portion of electrons energetic enough to overcome or tunnel through the ~1.0 eV Schottky barrier at the $Au–TiO_2$ interface.

More recently, PMET has been demonstrated to be a viable approach to overall photoelectrochemical water splitting. Moskovits et al. showed that the charge vacancies left behind on Au following electron transfer to a semiconductor can be harnessed for oxygen evolution when coupled with an oxygen-evolving cocatalyst (cobalt oxide) [67]. Additionally, various reports have begun using anisotropic nanostructures combined with semiconductors in order to harness lower energy photons with the ultimate goal to harvest the entire solar spectrum through this PMET strategy [15, 68–70].

8.4.4 Direct Charge Transfer to Adsorbed Species

In addition to charge transfer to a semiconductor, it has recently been demonstrated that the excited electron distribution can directly transfer to unoccupied molecular orbitals of reactants adsorbed to the surface of the metallic nanostructure [18, 52, 71]. Conceivably, excited electrons in plasmonic materials can transfer to any materials with sufficient orbital mixing and sufficiently low potential energy barriers. This includes molecules chemisorbed or physisorbed to the metal surface [18, 52, 72]. While it is expected that direct electron transfer may be limited based on the relative density of unoccupied molecular orbitals, multiple studies have shown that the PMET approach can be applied to surface catalysis [18, 52, 71].

As discussed previously, Ag nanostructures are active for ethylene epoxidation. Linic et al. explored improvements in activity by utilizing the inherent plasmonic properties of Ag nanocubes [52]. Using {100} facet-bound Ag nanocubes supported on Al_2O_3, the authors showed that plasmonic excitation of the nanocubes with visible light could induce electron transfer to the adsorbed O_2 molecules on the surface, populating the $2\pi^*$ orbital of the O_2 molecules and effectively lowering the barrier for dissociation into atomic oxygen. The transfer of an electron to the anti-bonding $2\pi^*$ orbital increases the bond-length along the reaction coordinate, leading to a lower barrier to bond scission and eventual dissociation (Fig. 8.13). This observation was confirmed using the kinetic isotope effect with labelled $^{18}O_2$. While this did not completely alleviate the thermal requirement for reaction, it was observed that at 450 K, the steady-state activation of oxygen increased approximately 23 %, coupling the thermal and electromagnetic energy for surface catalytic transformation of ethylene [52]. Further study revealed that the dissociation of O_2 follows a superlinear dependence on light intensity at incident powers greater than 300 mW/cm^2 [73]. By partially reducing the thermal requirement, the incorporation of plasmonic excitation into surface catalysis has the potential to improve catalyst stability and therefore increase the longevity of the catalyst. Recently, Halas et al. demonstrated a similar mechanism as described above, in which the excited electrons from the supported Au NPs can populate the H_2 $1\sigma_u^*$ anti-bonding orbital leading to H_2 dissociation [18, 71].

Modification of the electronic state of the catalytically-active metal is another developing area of plasmonic research. The oxidation state is critical to the activity

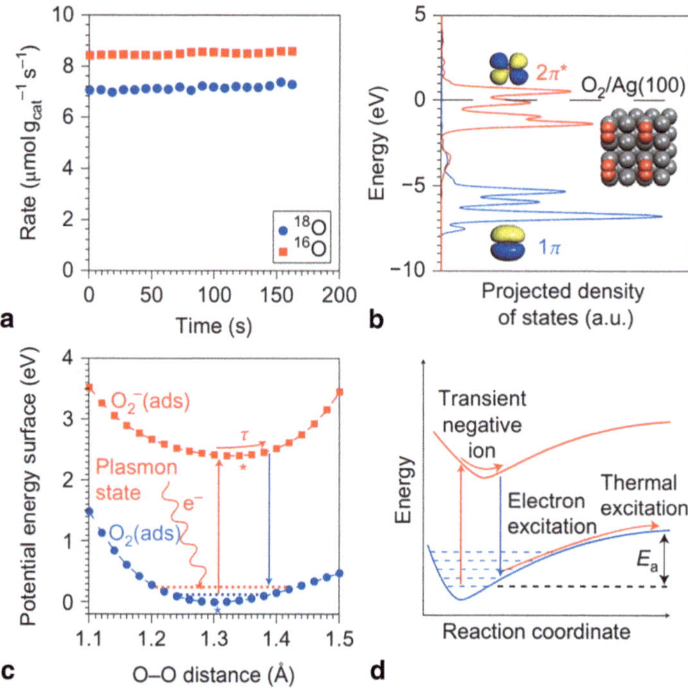

Fig. 8.13 **a** Kinetic isotope effect confirming electron-mediated O$_2$-dissociation process, **b** density of states for O$_2$ adsorbed to Ag{100}, **c** calculated potential energy surface for an electron transferred to O$_2$ 2π* molecular orbital from Ag, and **d** transient negative ion formation and resulting vibrational energy increase leading to O$_2$ bond dissociation. (Used with permission from reference [52])

of many catalytic processes such as propylene epoxidation over Cu NPs. Metallic Cu is selective for the epoxidation of multiple olefin species, yet is very susceptible to complete oxidation to CuO under reaction conditions, rendering it inactive with time [16]. Charge transfer to alloyed metal species has been shown to manipulate the surface electronic properties and improve catalyst stability by preventing oxidation [16, 74]. Shiraishi et al. fundamentally examined the oxidation of surface Cu atoms in a Au–Cu alloy using electron paramagnetic resonance (EPR) [74]. Using 2–propanol oxidation as a model reaction, the authors observed the continual increase in acetone formation under visible light irradiation, whereas the activity plateaued under dark conditions as the surface Cu atoms became fully oxidized over time. The EPR spectrum exhibited a signal for Cu^{2+} under dark conditions and Cu^0 under visible light irradiation (Fig. 8.14), supporting this proposed mechanism [74]. Linic et al. demonstrated the same plasmon-mediated oxidation state manipulation mechanism in supported Cu NPs for propylene epoxidation. The authors were able to observe drastic selectivity changes between dark and light conditions. The oxidation state changes were monitored through in situ UV-Vis spectroscopy and it was observed that with SPR excitation, the extinction features of metallic Cu^0 were

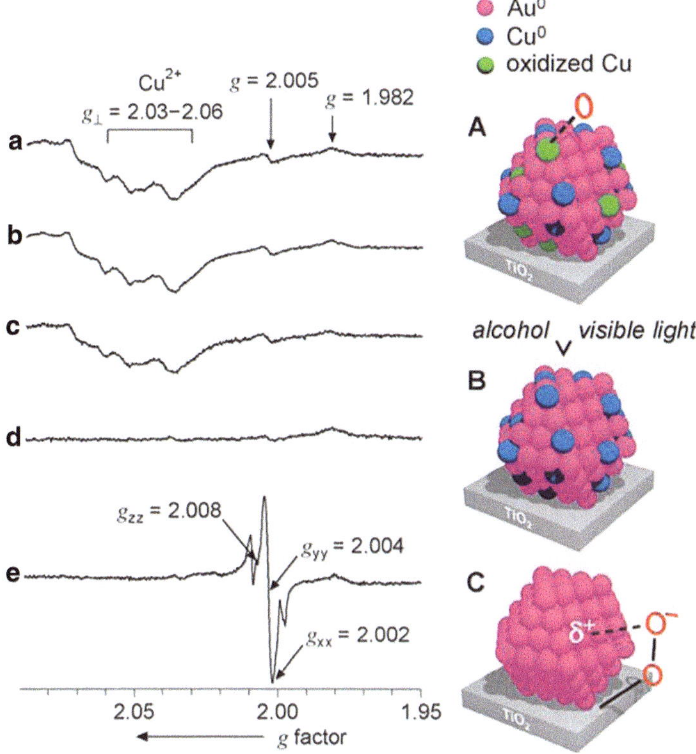

Fig. 8.14 Electron paramagnetic resonance analysis of Cu oxidation state. **a** Au/Cu–TiO$_2$ in the dark with g=2.005 and g=1.982 representing oxygen-vacancy-bound electrons, **b** irradiation of Au/Cu–TiO$_2$ for 3 h, **c** 2-propanol added to Au/Cu–TiO$_2$ in the *dark*, **d** Au/Cu–TiO$_2$ irradiated by visible *light* showing reduction of Cu^{2+} and **e** Au–TiO$_2$ confirming TiO$_2$ g-factor assignments. (Reprinted with permission from reference [74])

sustained during reaction, but were characteristic of Cu$_2$O under dark conditions [16]. These discoveries suggest a promising means of manipulating and stabilizing active oxidation states of metal nanocrystal catalysts while reducing energy inputs.

While the majority of recent research has focused on Au, Ag, and Cu, this technique is applicable to other metals such as Pd as well. Recent work by Xiong et al. has suggested that Pd nanocubes can also populate the O$_2$ 2π* anti-bonding orbital, leading to reactive singlet oxygen [75]. In this work, the resulting singlet oxygen was monitored by EPR in the presence of a spin trap, 2,2,6,6-tetramethyl-4-piperidone (4-oxo-TMP). A clear correlation between incident light power and catalytic activity was observed, however the relatively low absorption cross-section for Pd nanostructures ultimately limited the utility of such a photocatalyst for solar energy conversion. An additional strategy for leveraging the surface properties of materials with a small absorption cross-section such as Pd is via alloying with plasmonic nanostructures such as Ag or Au, which represents a promising approach to the controllable engineering of the physical and electronic properties of the plasmonic photocatalyst. Additionally, Yan et al. also successfully demonstrated a combination of the strong plasmonic response of Au with the surface propensity of Pd nanoparticles to improve the yield of certain Suzuki coupling reactions [72].

8.4.5 Photothermal Effect on Catalysis

In addition to the aforementioned PMET process, photothermal heating induced by the SPR effect of plasmonic NPs has also been harnessed to facilitate catalytic reactions [76]. It is known that most chemical reactions are adequately described by the Arrhenius equation and can thus be accelerated by increasing temperature [72, 77]. Additionally, the localized heating on the catalysts makes the catalytic reactions more energy efficient compared to those powered by bulk heating approaches [78].

Plasmonic nanoparticles, such as Au and Ag, have been used as nanoscale sources of heat in heterogeneous catalytic reactions [17, 79, 80–82]. Since radical species are often generated by thermal energy, localized photothermal heating has been investigated as a means of inducing radical reactions [78, 83]. The thermal decomposition of a common radical initiator, dicumyl peroxide, was investigated by irradiating Au NPs in a water droplet with a 532 nm laser [78]. This radical reaction was determined to be initiated by the photothermal effect of Au NPs. Surprisingly, the conversion rate of this chemical reaction reached 100 % within 1 min and the local temperature was estimated to be higher than 500 °C. In addition to radical polymerizations, non-radical polymerizations can also occur as a consequence of the plasmonic photothermal effect. For instance, caprolactam was polymerized to Nylon-6 on the surface of Ag NPs under resonant optical excitation [84] and polydimethylsiloxane (PDMS) has been photothermally cured into different morphologies by photothermal heating from irradiated Au NPs [85]. These recent examples highlight the potential opportunities afforded by local photothermal heating for catalyzing a variety of chemical reactions.

As an alternative to bulk system heating, localized photothermal heating is also effective for material fabrication *via* SPR-assisted NP deposition. To date, both metal and semiconductor architectures can be fabricated with sub-diffraction limit resolution by harnessing SPR-induced photothermal heating. Sub-10 nm Au NPs, which are well-known CO oxidation catalysts, have been deposited on a nanostructured Ag surface by using the intrinsic plasmonic properties of the Ag film [86]. Furthermore, common metal oxide supports, such as TiO_2, CeO_2, PdO, and carbon nanotubes (CNTs) have also been successfully deposited on Au NPs with the assistance of SPR-driven photothermal heating [81]. These nanomaterials can be further used as a support in SPR-catalyzed chemical reactions.

Au NPs are also effective in catalyzing important thermally-activated organic reactions. For example, an industrially important chemical reaction, the reforming of alcohol, was facilitated *via* the plasmon-driven photothermal effect [80]. Similar to the nano-reactor (Fig. 8.15), [87] this reaction was conducted within a

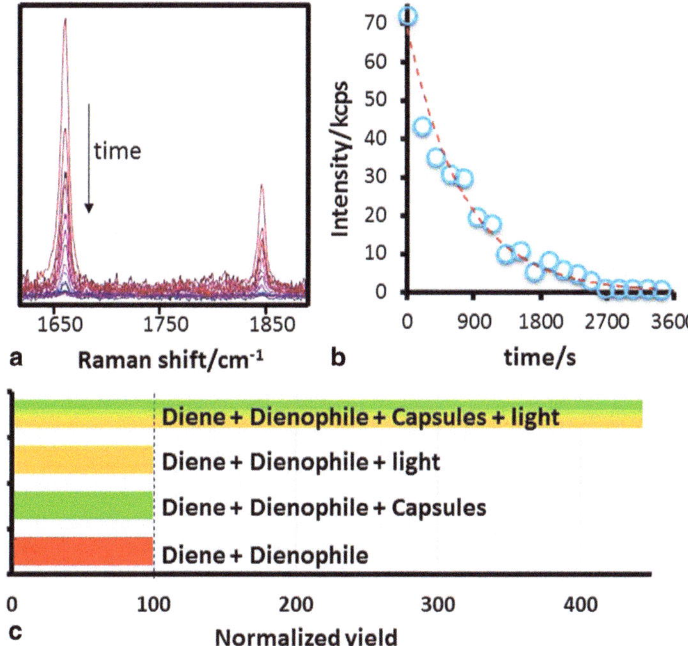

Fig. 8.15 SERS signal of maleic anhydride (dienophile) during reaction with 2,4-hexadienol (diene) to yield 4-isobenzofurancarboxylic acid (**a**, **b**). Comparison of product yields for different reaction conditions (**c**). (Reprinted with permission from reference [87])

microfluidic device to take better advantage of the localized photothermal effect. This also shows that plasmon-driven photothermal heating can catalyze chemical reactions in a nanometer volume when appropriately confined within small systems that are engineered to maximize the highly localized photothermal effect. Furthermore, such nanoheaters also can be combined with catalytically-active materials in other chemical reactions [17, 72, 79, 88, 89]. For instance, the photo-degradation of formaldehyde was studied over Au NPs dispersed on different metal oxides such as CeO_2, ZrO_2, SiO_2, and Fe_2O_3. All of these catalysts showed significant catalytic activities toward oxidation of formaldehyde in air at room temperature when irradiated with visible light [89]. A common model reaction, the oxidation of CO, has also been found to be enhanced with the photothermal effect of Au NPs when anchored on a Fe_2O_3 support (Fig. 8.16) [79]. These recent studies showcase the variety of catalytic applications driven by the photothermal effect.

Although encouraging, separation of numerous possible plasmonic enhancement mechanisms is challenging, and has impeded the conclusive demonstration of the photothermal effect as the sole reason for enhanced catalytic performance under visible-light irradiation of plasmonic-metal nanocrystals. Indeed, more complicated mechanisms invoking simultaneous photothermal heating and hot electron transfer have been proposed, as detailed in the previous section. It is anticipated that if these

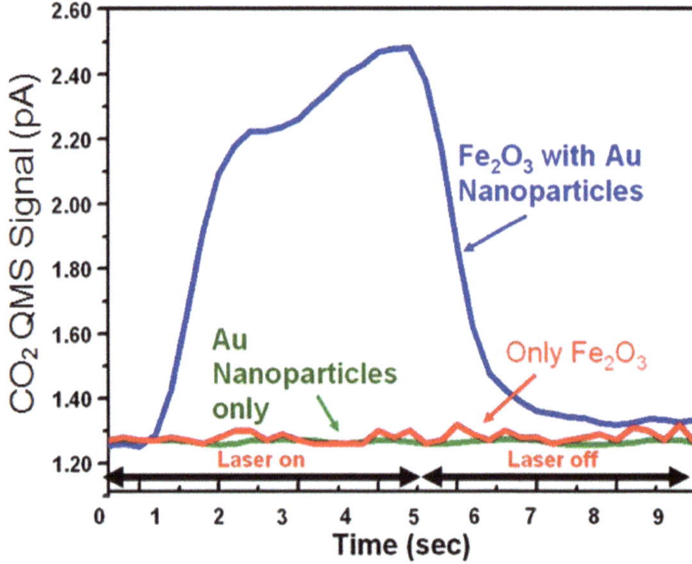

Fig. 8.16 Quadrupole mass spectrometry signals for CO_2 production over distinct catalysts under different reaction conditions. (Reprinted with permission from reference [79])

dual SPR effects of plasmonic-metal NPs can be appropriately harnessed to achieve synergistic enhancements in photocatalytic yield, new avenues of applied photochemical research may be realized.

8.5   Conclusion

Although significant progress has been made in heterogeneous surface catalysis, this field is still developing new catalytic materials that are optimized for industrial applications requiring high activity and selectivity. Most industrial catalytic reactions are still heavily reliant on expensive metals and significant thermal energy inputs to achieve suitable activities, and are therefore plagued by low catalyst lifetimes and poor control over reaction selectivity. In that regard, it is expected that continued development of nanocrystal metallic alloys will enable the synthesis of new materials with tailored reaction selectivity and improved longevity. Further improvements will come through a deeper mechanistic understanding of both surface catalysis and photocatalysis reaction pathways. The knowledge gained from these fundamental studies will aid the development of more Earth-abundant materials in order to leverage new insights for carefully engineering optimized heterostructures with improved reactivity. Particularly intriguing are the recent developments involving the use of Al and Cu plasmonic nanostructures as cost-effective alternatives to the more prominently used Ag or Au-based nanostructures. Furthermore, the controlled integration of plasmonic-metal nanostructures into semiconductor architectures is expected to facilitate the transfer of charge carriers to the site of reaction, improving the quantum yield of these processes while harnessing freely available sunlight. The motivations behind these developments are deeply rooted in the rapid expansion of global industrialization, as the global energy landscape and environmental health hang in the balance. Despite the pressing need for highly active industrial catalysts and their current limitations, application of metallic nanostructures to an ever-increasing range of catalytic systems remains promising.

References

1. M. Leahy, J.L. Barden, B.T. Murphy, N. Slater-Thompson, D. Peterson, International Energy Outlook 2013 (2013), http://www.eia.gov/forecasts/ieo/. Accessed Apr 2014
2. M. Zhu, C.M. Aikens, F.J. Hollander, G.C. Schatz, R. Jin, J. Am. Chem. Soc. **130**, 5883 (2008)
3. H. Qian, Y. Zhu, R. Jin, Proc. Natl. Acad. Sci. U. S. A. **109**, 696 (2012)
4. I.X. Green, W. Tang, M. Neurock, J.T. Yates, Acc. Chem. Res. **47**, 805 (2013)
5. Y. Lei, F. Mehmood, S. Lee, J. Greeley, B. Lee, S. Seifert et al., Science **328**, 224 (2010)
6. S. Linic, P. Christopher, H. Xin, A. Marimuthu, Acc. Chem. Res. **46**, 1890 (2013)
7. D.I. Enache, J.K. Edwards, P. Landon, B. Solsona-Espriu, A.F. Carley, A.A. Herzing et al., Science **311**, 362 (2006)
8. M. Haruta, N. Yamada, T. Kobayashi, S. Iijima, J. Catal. **115**, 301 (1989)
9. D. Wang, H.L. Xin, Y. Yu, H. Wang, E. Rus, D.A. Muller et al., J. Am. Chem. Soc. **132**, 17664 (2010)
10. D. Wang, H.L. Xin, R. Hovden, H. Wang, Y. Yu, D.A. Muller et al., Nat. Mater. **12**, 81 (2013)
11. N.P. Dasgupta, C. Liu, S. Andrews, F.B. Prinz, P. Yang, J. Am. Chem. Soc. **135**, 12932 (2013)
12. E.J. Popczun, C.G. Read, C.W. Roske, N.S. Lewis, R.E. Schaak, Angew. Chem. Int. Ed. **126**, 5531 (2014)
13. F. Studt, I. Sharafutdinov, F. Abild-Pedersen, C.F. Elkjær, J.S. Hummelshøj, S. Dahl et al., Nat. Chem. **6**, 320 (2014)
14. S. Linic, P. Christopher, D.B. Ingram, Nat. Mater. **10**, 911 (2011)
15. Y. Nishijima, K. Ueno, Y. Kotake, K. Murakoshi, H. Inoue, H. Misawa, J. Phys. Chem. Lett. **3**, 1248 (2012)
16. A. Marimuthu, J. Zhang, S. Linic, Science **339**, 1590 (2013)
17. F. Wang, C. Li, H. Chen, R. Jiang, L.-D. Sun, Q. Li et al., J. Am. Chem. Soc. **135**, 5588 (2013)
18. S. Mukherjee, L. Zhou, A.M. Goodman, N. Large, C. Ayala-Orozco, Y. Zhang et al., J. Am. Chem. Soc. **136**, 64 (2013)
19. K.A. Willets, R.P. Van Duyne, Annu. Rev. Phys. Chem. **58**, 267 (2007)
20. K.M. Mayer, J.H. Hafner, Chem. Rev. **111**, 3828 (2011)
21. X. Lu, M. Rycenga, S.E. Skrabalak, B. Wiley, Y. Xia, Annu. Rev. Phys. Chem. **60**, 167 (2009)
22. M. Haruta, Cattech **6**, 102 (2002)
23. M.O. Özbek, I. Onal, R.A. Van Santen, J. Catal. **284**, 230 (2011)
24. M.O. Özbek, R.A. Santen, Catal. Lett. **143**, 131 (2013)
25. R. Meyer, C. Lemire, S.K. Shaikhutdinov, H.J. Freund, Gold Bull. **37**, 72 (2004)
26. I.X. Green, W. Tang, M. Neurock, J.T. Yates, Angew. Chem. Int. Ed. **50**, 10186 (2011)
27. Y. Liu, H. Tsunoyama, T. Akita, S. Xie, T. Tsukuda, ACS Catal. **1**, 2 (2010)
28. I.X. Green, W. Tang, M. Neurock, J.T. Yates, Science **333**, 736 (2011)
29. J. Kiwi, M. Grätzel, J. Am. Chem. Soc. **101**, 7214 (1979)
30. R. Narayanan, M.A. El-Sayed, Nano Lett. **4**, 1343 (2004)

31. J. Kiwi, M. Grätzel, Nature **281**, 657 (1979)
32. N. Miyaura, A. Suzuki, Chem. Rev. **95**, 2457 (1995)
33. G. Collins, M. Schmidt, C. O'Dwyer, J.D. Holmes, G.P. McGlacken, Angew. Chem. Int. Ed. **53**, 4142 (2014)
34. Y. Li, X.M. Hong, D.M. Collard, M.A. El-Sayed, Org. Lett. **2**, 2385 (2000)
35. P. Christopher, S. Linic, ChemCatChem **2**, 78 (2010)
36. M. Valden, X. Lai, D.W. Goodman, Science **281**, 1647 (1998)
37. T. Fujita, P. Guan, K. McKenna, X. Lang, A. Hirata, L. Zhang et al., Nat. Mater. **11**, 775 (2012)
38. C.T. Campbell, S.C. Parker, D.E. Starr, Science **298**, 811 (2002)
39. M. Chen, D.W. Goodman, Chem. Soc. Rev. **37**, 1860 (2008)
40. M. Cargnello, V. Doan-Nguyen, T.R. Gordon, R.E. Diaz, E.A. Stach, R.J. Gorte et al., Science **341**, 771 (2013)
41. F. Gao, D.W. Goodman, Annu. Rev. Phys. Chem. **63**, 265 (2012)
42. A.T. Bell, Science **299**, 1688 (2003)
43. S.D. Lin, M. Bollinger, M.A. Vannice, Catal. Lett. **17**, 245 (1993)
44. F.A. Cotton, G. Wilkinson, *Advanced Inorganic Chemistry*, 5th edn. (Wiley, New York, 1988)
45. G.C. Bond, D.T. Thompson, Gold Bull. **33**, 41 (2000)
46. M.C. Kung, R.J. Davis, H.H. Kung, J. Phys. Chem. C **111**, 11767 (2007)
47. D. Widmann, R.J. Behm, Acc. Chem. Res. **47**, 740 (2014)
48. B. Yoon, H. Häkkinen, U. Landman, A.S. Wörz, J.-M. Antonietti, S. Abbet et al., Science **307**, 403 (2005)
49. S.J. Tauster, S.C. Fung, R.L. Garten, J. Am. Chem. Soc. **100**, 170 (1978)
50. X. Liu, M.-H. Liu, Y.-C. Luo, C.-Y. Mou, S.D. Lin, H. Cheng et al., J. Am. Chem. Soc. **134**, 10251 (2012)
51. Ethylene Oxide (EO) (Complete Report), A Global Strategic Business Report. Global Industry Analysts. (2011)
52. P. Christopher, H.L. Xin, S. Linic, Nat. Chem. **3**, 467 (2011)
53. A. Fujishima, K. Honda, Nature **238**, 37 (1972)
54. M.R. Hoffmann, S.T. Martin, W.Y. Choi, D.W. Bahnemann, Chem. Rev. **95**, 69 (1995)
55. Z. Zhang, J.T. Yates, Chem. Rev. **112**, 5520 (2012)
56. M.W. Knight, H. Sobhani, P. Nordlander, N.J. Halas, Science **332**, 702 (2011)
57. B. Kraeutler, A.J. Bard, J. Am. Chem. Soc. **100**, 2239 (1978)
58. V. Subramanian, E.E. Wolf, P.V. Kamat, J. Am. Chem. Soc. **126**, 4943 (2004)
59. Z. Zou, J. Ye, K. Sayama, H. Arakawa, Nature **414**, 625 (2001)
60. M. Rycenga, C.M. Cobley, J. Zeng, W.Y. Li, C.H. Moran, Q. Zhang et al., Chem. Rev. **111**, 3669 (2011)
61. J.S. Duchene, W. Niu, J.M. Abendroth, Q. Sun, W. Zhao, F. Huo et al., Chem. Mater. **25**, 1392 (2013)
62. G.V. Hartland, Chem. Rev. **111**, 3858 (2011)
63. L. Liu, P. Li, B. Adisak, S. Ouyang, N. Umezawa, J. Ye et al., J. Mater. Chem. A **2**, 9875 (2014)
64. C. Gomes Silva, R. Juárez, T. Marino, R. Molinari, H. García, J. Am. Chem. Soc. **133**, 595 (2010)
65. Y. Tian, T. Tatsuma, J. Am. Chem. Soc. **127**, 7632 (2005)
66. N. Tian, Z.-Y. Zhou, S.-G. Sun, Y. Ding, Z.L. Wang, Science **316**, 732 (2007)
67. S. Mubeen, J. Lee, N. Singh, S. Krämer, G.D. Stucky, M. Moskovits, Nat. Nanotechnol. **8**, 247 (2013)
68. Y.-C. Pu, G. Wang, K.-D. Chang, Y. Ling, Y.-K. Lin, R.C. Fitzmorris et al., Nano Lett. **13**, 3817 (2013)
69. N. Zhou, L. Polavarapu, N. Gao, Y. Pan, P. Yuan, Q. Wang et al., Nanoscale **5**, 4236 (2013)
70. K. Ueno, S. Takabatake, Y. Nishijima, V. Mizeikis, Y. Yokota, H. Misawa, J. Phys. Chem. Lett. **1**, 657 (2010)

71. S. Mukherjee, F. Libisch, N. Large, O. Neumann, L.V. Brown, J. Cheng et al., Nano Lett. **13**, 240 (2013)
72. S. Sarina, S. Bai, Y. Huang, C. Chen, J. Jia, E. Jaatinen et al., Green Chem. **16**, 331 (2014)
73. P. Christopher, H. Xin, A. Marimuthu, S. Linic, Nat. Mater. **11**, 1044 (2012)
74. Y. Sugano, Y. Shiraishi, D. Tsukamoto, S. Ichikawa, S. Tanaka, T. Hirai, Angew. Chem. Int. Ed. **52**, 5295 (2013)
75. R. Long, K. Mao, M. Gong, S. Zhou, J. Hu, M. Zhi et al., Angew. Chem. Int. Ed. **53**, 3205 (2014)
76. M. Gonzalez-Bejar, K. Peters, G.L. Hallett-Tapley, M. Grenier, J.C. Scaiano, Chem. Commun. **49**, 1732 (2013)
77. P. Hervés, M. Pérez-Lorenzo, L.M. Liz-Marzán, J. Dzubiella, Y. Lu, M. Ballauff, Chem. Soc. Rev. **41**, 5577 (2012)
78. C. Fasciani, C.J. Bueno Alejo, M. Grenier, J.C. Netto-Ferreira, J.C. Scaiano, Org. Lett. **13**, 204 (2011)
79. W.H. Hung, M. Aykol, D. Valley, W. Hou, S.B. Cronin, Nano Lett. **10**, 1314 (2010)
80. J.R. Adleman, D.A. Boyd, D.G. Goodwin, D. Psaltis, Nano Lett. **9**, 4417 (2009)
81. D.A. Boyd, L. Greengard, M. Brongersma, M.Y. El-Naggar, D.G. Goodwin, Nano Lett. **6**, 2592 (2006)
82. A. Biesso, W. Qian, X. Huang, M.A. El-Sayed, J. Am. Chem. Soc. **131**, 2442 (2009)
83. J.M. Walker, L. Gou, S. Bhattacharyya, S.E. Lindahl, J.M. Zaleski, Chem. Mater. **23**, 5275 (2011)
84. K.G. Stamplecoskie, C. Fasciani, J.C. Scaiano, Langmuir **28**, 10957 (2012)
85. M. Fedoruk, M. Meixner, S. Carretero-Palacios, T. Lohmüller, J. Feldmann, ACS Nano **7**, 7648 (2013)
86. J. Qiu, Y.-C. Wu, Y.-C. Wang, M.H. Engelhard, L. McElwee-White, W.D. Wei, J. Am. Chem. Soc. **135**, 38 (2013)
87. C. Vázquez-Vázquez, B. Vaz, V. Giannini, M. Pérez-Lorenzo, R.A. Alvarez-Puebla, M.A. Correa-Duarte, J. Am. Chem. Soc. **135**, 13616 (2013)
88. Z. Zhang, W. Wang, E. Gao, S. Sun, L. Zhang, J. Phys. Chem. C **116**, 25898 (2012)
89. X. Chen, H.-Y. Zhu, J.-C. Zhao, Z.-F. Zheng, X.-P. Gao, Angew. Chem. Int. Ed. **47**, 5353 (2008)

Chapter 9
Metallic Nanostructures for Electronics and Optoelectronics

Shan Zhou and Yujie Xiong

Abstract Metallic nanostructures have played an important role in electronic and optoelectronic devices. This chapter summarizes the related progress with a focus on metallic nanowires. We first outline the fabrication of transparent conductive electrodes with silver nanowires and copper nanowires, followed by various methods to improve their performance in electronic devices. We then discuss another perspective of metallic nanowires—plasmonic waveguiding that can be potentially implemented in optoelectronic devices and can overcome limitation of traditional electronic circuits in the era of "big data". In this section, light coupling as well as plasmon propagation in silver nanowire waveguides and functional parts of nanophotonic circuits are overviewed. At the end, a few well-established protocols for synthesizing metallic nanowires are presented, allowing the readers to further explore the related research.

9.1 Introduction

Electronic and optoelectronic devices are the central components in modern instrumentation and equipment for display, computation, communication, and sensing [1–3]. Over the past decades, significant reduction in critical dimensions and substantial growth in component integration of devices have led to enormous increases in speed, functionality, and computing capacity. This trend is likely to continue for some years, leading to devices and systems with exceptional operating performance. Notably, progress in such trend is dominated by a development path that involves increasing the number of device components, following the Moore's Law [4]. Certainly there exist many challenges to the development in materials science and engineering, among which the electrodes need to be fabricated in high-density, transparent, and flexible forms [5–7]. During the past years, metallic nanowires have been demonstrated as excellent candidate to meet this strong demand, enabling various applications especially in optoelectronics [8–10]. However, the emerging

Y. Xiong (✉) · S. Zhou
School of Chemistry and Materials Science, University of Science and Technology of China,
Hefei, 230026 Anhui, P. R. China
e-mail: yjxiong@ustc.edu.cn

Y. Xiong, X. Lu (eds.), *Metallic Nanostructures,* DOI 10.1007/978-3-319-11304-3_9

era of "big data" requires extremely high density of channels to achieve higher data transmission rates and capacity, which apparently cannot be afforded by traditional electronic circuits [11,12]. As a promising solution, metallic nanowires have shown their capabilities in plasmonic waveguiding which may potentially transmit data optically instead of electrically, within the circuits.

9.2 Transparent and Flexible Conductive Electrodes

Transparent conductive electrodes (TCEs) are critical components widely used in organic light emitting diodes (OLEDs), solar cells, and display technologies [8]. These optoelectronic devices require the electrodes to provide a low-resistance electrical contact while maintaining high transparency [13]. In the past decade, a wide variety of materials including metal oxide, carbon materials, metal grids, and metal nanowires, have been implemented to form transparent and electrically conductive electrodes [14–16].

9.2.1 Conventional Conductors

Transparent conducting oxides (TCOs) are a class of materials having perfect combination of high conductivity with excellent transparency in visible spectral region [15]. Indium oxide and indium–tin oxide (ITO) are two important TCOs which have been commonly utilized in optoelectronic devices. The ITO occupies 95 % of the market with $ 5 billion per year. With the ITO material, it is feasible to achieve sheet resistance of ~ 10 Ω/sq and light transmittance of > 90 % [17]. However, the scarcity of indium resources urgently calls for new material to replace ITO (~$ 800 per kg), and more importantly, the brittleness of metal oxides makes them easily crack when the substrate is bent [13]. From the perspective of device fabrication, the application of ITO is hindered by the high-temperature treatment involved in its production as well as slow vapour-based deposition (~0.01 m/s) [17].

For this reason, efforts have been made to develop new flexible and transparent conductors such as carbon nanotubes (CNTs), graphene, metal grids, and random networks of metallic nanowires [18]. Although CNTs and graphene have abundant material source, their sheet resistances are too high to replace the use of ITO in current-driven devices. The typical sheet resistance of CNT meshes on plastic is between 200 and 1000 Ω/sq with an optical transmittance of 80–90 %. This resistance value, caused by high contact resistance between CNTs, is significantly higher than that of ITO on plastic and Ag nanowires meshes [19]. Similarly, graphene has a relatively high sheet resistance (~200 Ω/sq) [20]. Such a high sheet resistance impedes the applications of carbon materials as the substitutes for ITO in current-based devices such as organic light-emitting diodes and solar cells [13,19].

Metal grid is another candidate for replacing ITO. It has been experimentally demonstrated that the Ag gratings that are patterned at the nanoscale can be employed as a transparent electrode for OLEDs achieving performance comparable to that of ITO [21]. However, the fabrication of nanopatterned silver gratings is too costly for wide applications in solar cells and OLEDs. This cost consideration has motivated people to develop more cost-effective methods for fabricating metal electrodes.

Notably, mesh electrodes can be made by simply casting suspensions of solution-synthesized metal nanowires. The metal nanowires, which have high conductivity and optical transmittance, have become the most appealing substitutes of ITO in TCEs. In this section, we mainly focus on discussing the use and performance of silver and copper nanowires as TCEs.

9.2.2 Silver Nanowires (Ag NWs)

9.2.2.1 Typical Synthesis and Electrode Fabrication

In 2008, Lee et al. demonstrated the potential application of Ag NWs in transparent mesh electrodes. In their work, the Ag NWs were synthesized by reducing Ag nitrate in ethylene glycol (EG) with the addition of poly(vinyl pyrrolidone) (PVP). The yielded Ag nanowires were 8.7 ± 3.7 μm in length and had a diameter of 103 ± 17 nm (see Fig. 9.1).

To fabricate the transparent electrodes, suspensions of Ag NWs were dropcast on glass substrate that had been prepatterned with 100-nm thick Ag contact pads, and were dried in air for 10 min. The electrodes were then thermally annealed prior

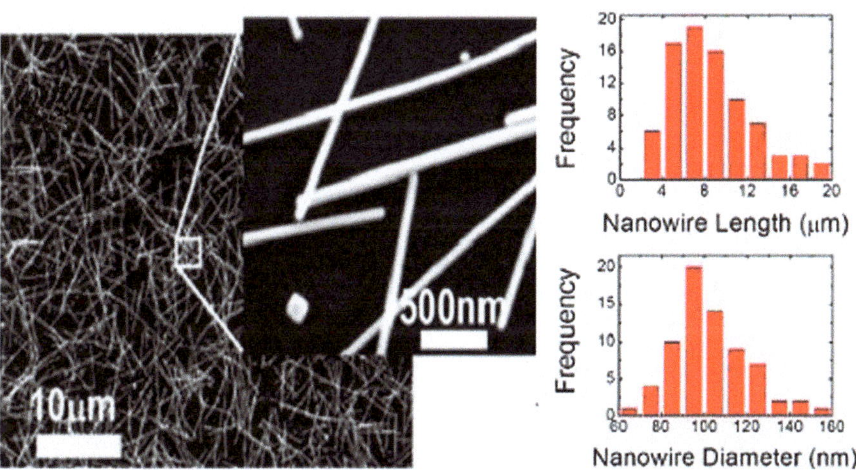

Fig. 9.1 SEM images of an as-synthesized Ag NW mesh on silicon substrate and the length and diameter histograms of Ag NWs [13]

to resistivity and transmittance measurements. The fabricated electrodes appeared to be random meshes consisting of Ag NWs. As shown in the scanning electron microscope (SEM) image in Fig. 9.1, the Ag NWs are uniformly dispersed on the surface. To facilitate discussions, we name the electrode fabricated via this method *typical Ag NW mesh electrode* in following sections.

9.2.2.2 Performance of Typical Ag NW Mesh Electrodes

There are four major parameters critical to assessing the performance of TCEs: sheet resistance, transmittance, stability, and surface roughness. The discussions in the following sections will focus on these four parameters, together with some technical approaches to improve these properties.

Sheet Resistance The sheet resistance (R_{sh}) of an Ag NW mesh originates from two factors—wire resistance (R_w) and wire–wire contact resistance (R_c), as well as depends on wire length (L) and wire aerial density (D). The R_{sh} of Ag NW meshes has been experimentally measured with varied wire density, as shown in Fig. 9.2. The average R_w can be determined to 18 Ω based on the bulk resistivity of silver. To analyze the roles of R_c, the measured data of R_{sh} are compared with model calculations assuming $R_c = 1$, 40, and 100 Ω. At $D < 0.3$ μm^{-2}, the data comply well with the model for $R_c = 40$ Ω. As the D is over 0.3 μm^{-2}, the circumstance of $R_c = 1$ Ω better matches the data.

Transmittance TCE requires high optical transmittance over a broad wavelength range [19]. Figure 9.3 shows spectrally resolved diffuse transmittance for Ag NW meshes. For the two Ag NWs meshes with R_{sh} of 10.3 and 22.1 Ω, their solar photon flux-weighted transmissivity (T_{solar}) can reach 84.7 and 88.3 %, respectively,

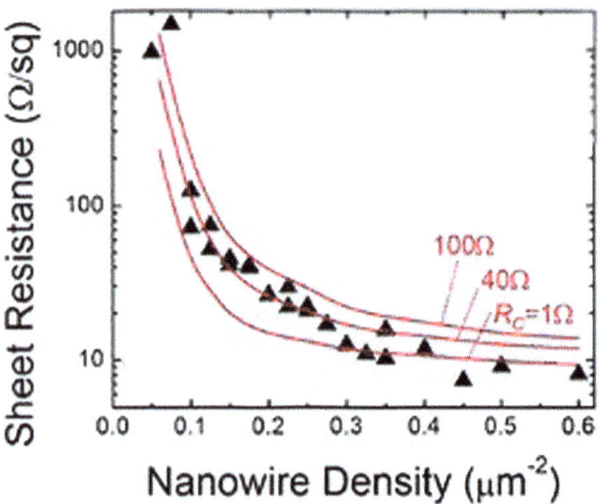

Fig. 9.2 Experimentally-determined sheet resistance vs. nanowire density (D) of the typical Ag NW mesh electrodes. *Red lines* represent the theoretical results for different contact resistance (R_c)[13]

Fig. 9.3 Transmittance vs. wavelength of two Ag nanowire meshes with different R_{sh}. The inset shows solar transmissivity vs. nanowire mesh aerial density (D) [13]

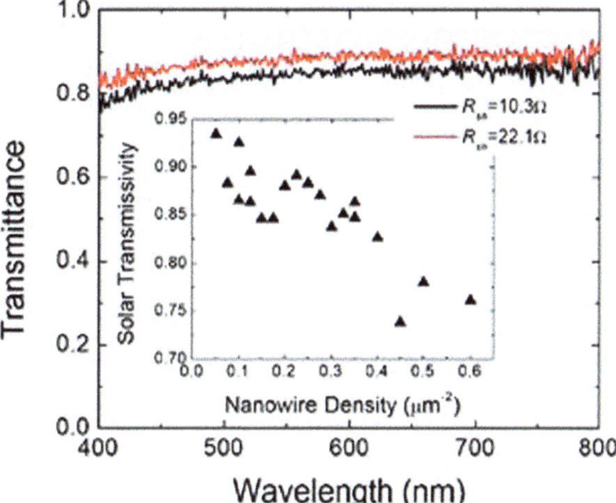

Fig. 9.4 Solar photon flux-weighted transmissivity vs. sheet resistance for Ag gratings (*blue*), ITO (*red dot*), CNT meshes ($\varDelta$), and Ag NW meshes (■) deposited on a glass substrate [13]

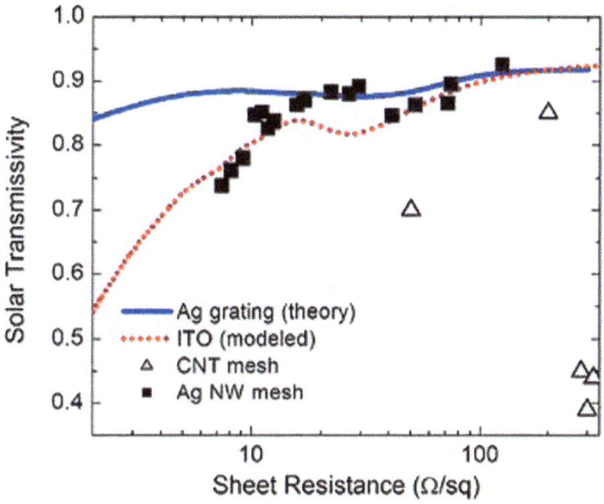

certainly depending on the wire density (the inset of Fig. 9.3). As shown in Fig. 9.4, Ag NW meshes on glass exhibit a T_{solar} comparable or even higher than that of ITO on glass for a given R_{sh}.

Stability Many factors may impact on the stability of electrodes: (1) the adhesion strength of Ag NWs with the substrate; (2) the robustness upon being bent; (3) the resistivity to chemical damage and heat. The last one is required to guarantee the long-term stability of devices integrated with Ag NW mesh electrodes.

The stability of Ag NW mesh electrodes has been tested by Hu et al. in 2010. First of all, the film adhesion to the substrate was assessed by a mechanical tape test. It turns out that Ag NW mesh electrode without encapsulation has a large increase in sheet resistance after the tape test. In comparison, ITO electrode well maintains its sheet resistance during the test. Second, bending tests show that the ITO on polyethylene terephthalate (PET) substrate is quite fragile while the Ag NW mesh electrode shows great flexibility. Third, the sheet resistance of Ag NWs that have been stored in air for a month does not change beyond 5 %. In parallel, it has also been demonstrated that Ag NW electrode can well maintain its sheet resistance when exposed to water, acetone, and isopropanol (IPA). Strikingly, the sheet resistance even decreases when the electrode is heated, because the contact between Ag NW and the substrate is improved through annealing.

Surface Roughness Surface roughness is another important factor determining the applications of electrodes. In certain applications such as OLED, 1–2 nm surface roughness is required.

9.2.2.3 Methods for Reducing Sheet Resistance

Annealing The initial resistance of the typical Ag NW mesh electrodes demonstrated above is > 1 kΩ/sq due to the presence of thick PVP capping layer. The PVP can be decomposed and flow during an annealing process, and meanwhile, this process allows the nanowires to contact or fuse together. As a result, the sheet resistance of mesh electrodes has a sharp drop to ~ 100 Ω/sq after annealing at 200 °C for 20 min. However, excessive annealing for over 40 min leads to an increase in R_{sh} due to the coalescence of Ag NWs (see Fig. 9.5).

Increasing Aerial Density The sheet resistance has been proven negatively correlated with the aerial density of Ag NWs [19]. Figure 9.6 shows the performance of mesh films consisting of Ag NWs with diameters of 40–100 nm but different densities on PET substrates. The Ag NW networks with less density turn out to have larger sheet resistance. The sheet resistance can reach ~ 10 Ω/sq by tuning the aerial

Fig. 9.5 Sheet resistance *vs.* annealing time of Ag NW mesh electrodes [13]

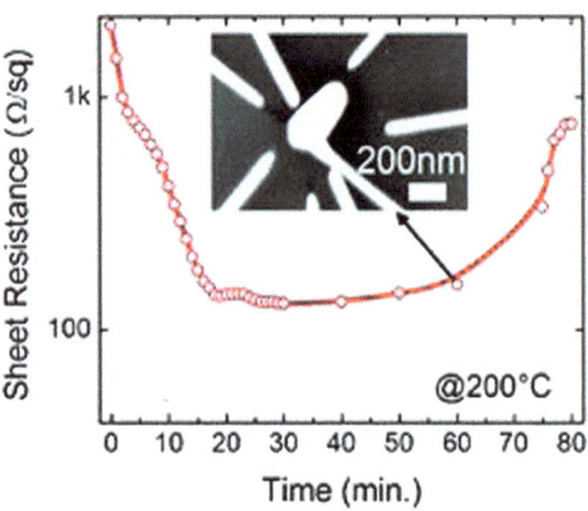

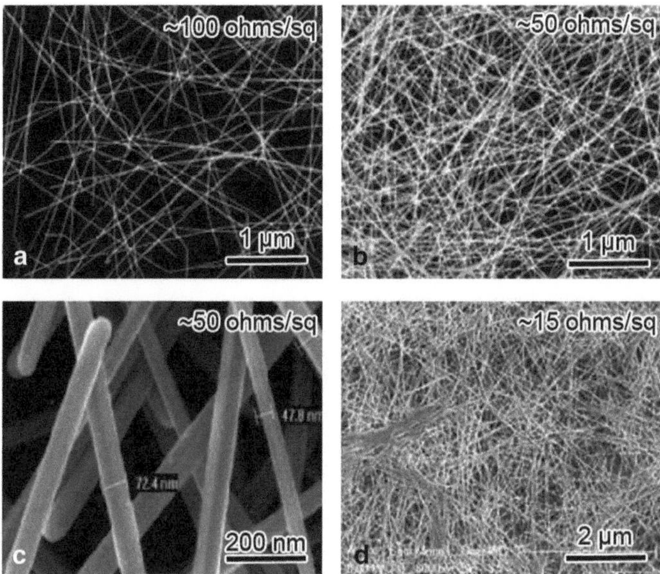

Fig. 9.6 SEM images of networks with different densities whose Ag NWs were synthesized by following the standard method. Different densities lead to varied sheet resistances shown in the *top right* corners [19]

density, which becomes comparable to that of ITO on plastic or glass substitutes for solar cell.

Modifying Parameters of Nanowires Hu et al. successfully synthesized longer and thinner Ag NWs by introducing KBr solution into the traditional synthesis. Such modifications on the parameters of nanowires enable a significant decrease in the resistance. Figure 9.7a illustrates the mechanism responsible for this modified synthesis. The addition of bromide results in the formation of AgBr, and thus constitutes a competition with the reduction of Ag^+ ions by EG. From the SEM images in Fig. 9.7b, one can see that the Ag NWs obtained via the modified method are about 50 % thinner than those by the standard process. The thinner Ag NWs typically have 20-Ω/sq sheet resistance and 80 % spectral transmittance at 500 nm, while the regular Ag NWs with the similar transmittance have much higher sheet resistance.

Replacing Surface With Au Within the NW network, charges can be transported along the wires or across the junctions. The measurements have demonstrated that the NW–NW junction resistance is higher than 1 GΩ while the resistance of single Ag NW is only about 300 Ω. Thus, the junction resistance holds the key to improving the electric performance of Ag NW mesh electrodes.

In principle, the sheet resistance of Ag NW electrode can be reduced by improving the conductance of the NW–NW junction. Au replacement has been proven as an effective method to achieve this goal. Experimentally the Au addition can be achieved by immersing the Ag NW film in boiling $HAuCl_4$ aqueous solution for

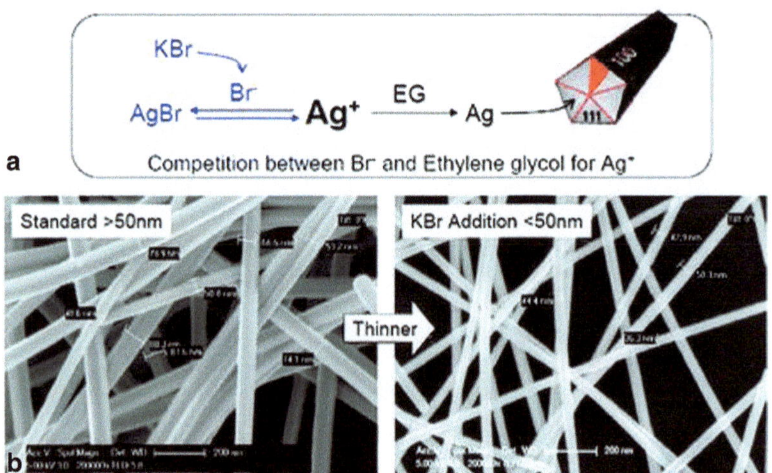

Fig. 9.7 **a** Schematic for synthesizing longer and thinner nanowires by utilizing the competition between Br⁻ and ethylene glycol for Ag⁺. **b** SEM images of Ag NWs synthesized via the standard protocol and Ag NWs with the addition of KBr. The diameters of Ag NWs shrunk from 50–100 nm to 30–50 nm with this modification to the synthesis [19]

several minutes and then drying with an air gun. As a result, the pristine Ag NW network has been coated by Au through galvanic displacement (Fig. 9.8a–c).

After the Au coating, the resistances have been probed on two typical nanowires with a cross junction to prove the efficiency of Au coating. It turns out that more significant resistance reduction occurs on the nanowire junction. The on-wire resistances for single nanowire are reduced from 260 to 220 Ω and from 320 to 300 Ω, respectively, and the junction resistance decreases from >1 GΩ to 450 Ω (Fig. 9.8d).

Pressing Electrodes Mechanically The sheet resistance of Ag NW electrode film can be reduced by holding a high mechanical pressure on the top surface. Figure 9.9 shows the surface of an Ag NW electrode after being pressed by an 81-GPa pressure for 50 s. The sheet resistance drops from hundreds of Ω/sq to tens of Ω/sq. The surface roughness also decreases from 110 to 47 nm after the pressure treatment.

Reducing Film Thickness Thickness has been proven as a critical parameter to the sheet resistance of Ag NW film, following a negative correlation [17]. As can be seen from Fig. 9.10, enabled by the increase of film thickness, the sheet resistance of Ag NW film can be reduced down to 10 Ω/sq.

9.2.2.4 Methods for Enhancing Optical Transmittance

Reducing Aerial Density The transmittances for Ag NW films with different aerial densities (thus exhibiting varied sheet resistance) are shown in Fig. 9.11. In general, the Ag NW films possess high transmittance in a very broad spectral

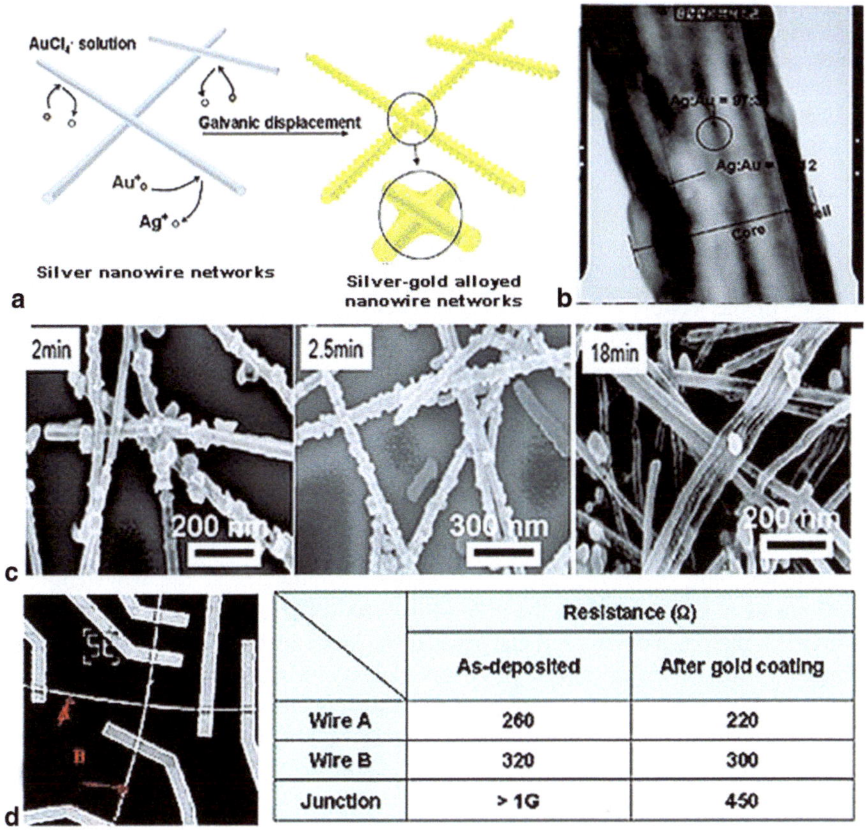

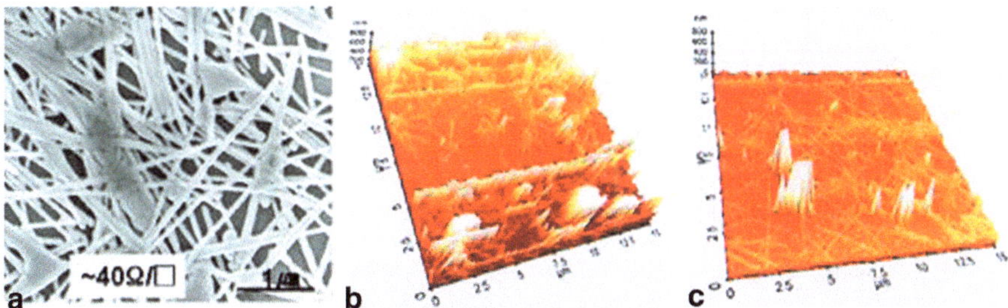

	Resistance (Ω)	
	As-deposited	After gold coating
Wire A	260	220
Wire B	320	300
Junction	> 1G	450

Fig. 9.8 **a** Schematic for galvanic displacement involved in Au replacement. **b** TEM image showing the NW treated with the Au replacement. **c** Surface evolution of NWs along with the Au addition. **d** Wire resistance and junction resistance of two 10-µm-long Ag NWs [19]

Fig. 9.9 **a** SEM image of Ag NW network after treated under pressure; **b** and **c** AFM images of the Ag NW network before and after pressing [19]

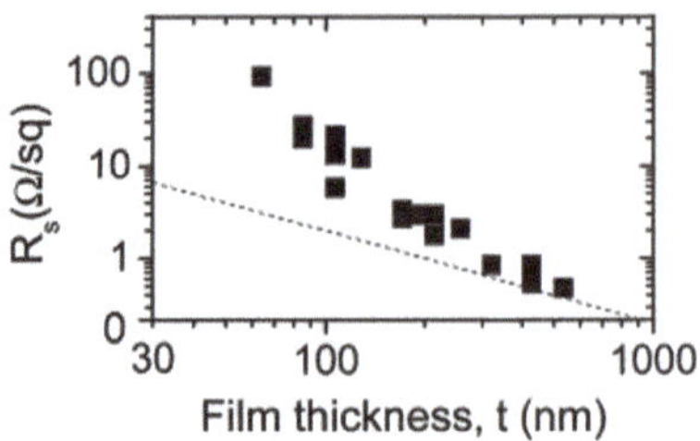

Fig. 9.10 Plots of sheet resistance varying with film thickness [17]

range. In the visible range, the transparency of Ag NW electrode with sheet resistance of 110 Ω/sq is very comparable to that of ITO electrode. However, the Ag NW electrode also exhibits high transparency in the near-infrared (NIR) range (close to that in the visible range), while the transmittance of ITO electrode decreases sharply as the light wavelengths switch towards the NIR. The unique broad-range high transmittance enables the Ag NW electrode to be used in infrared solar cells.

Reducing Film Thickness The optical transparency of Ag NW films has a strong correlation with their thickness. The light transmittances at the wavelength of 550 nm have been tested on the NW films with different thickness. As shown in Fig. 9.12, there is a clear trend that smaller thickness leads to a much higher transmittance. However, unfortunately as the thickness is reduced, the sheet resistance of Ag NW electrodes simultaneously increases. As a matter of fact, this feature can be also observed in Fig. 9.10. For this reason, a trade-off between optical transmittance and sheet resistance should be considered when the Ag NW meshes are employed as electrodes in various applications.

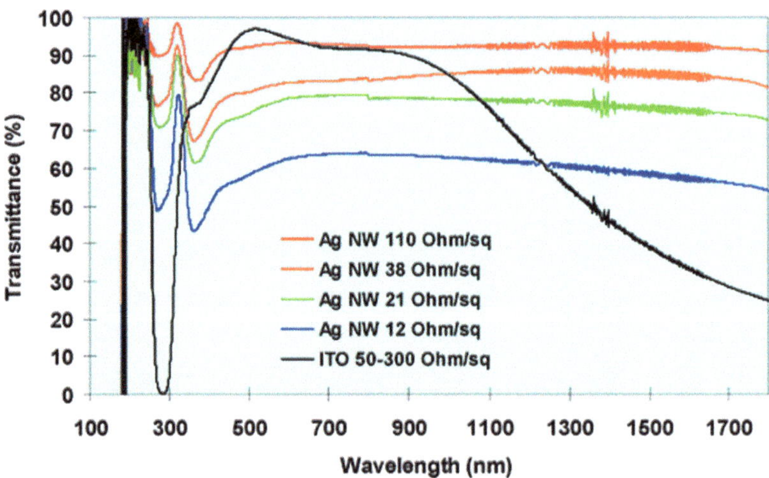

Fig. 9.11 Optical transmittance of transparent Ag NW electrodes compared with ITO on PET (excluding the influence from substrates) [19]

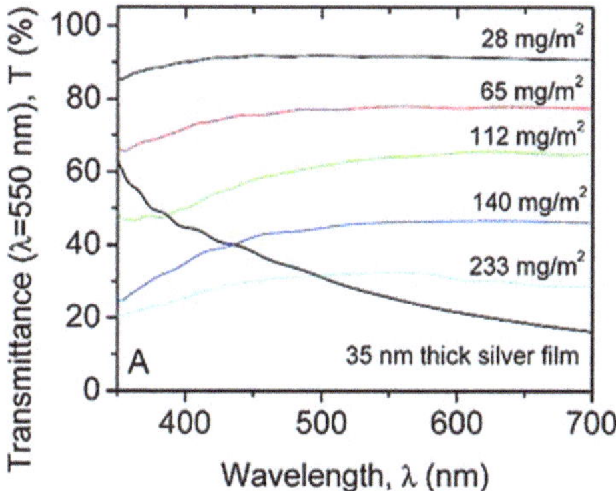

Fig. 9.12 Transmission spectra for Ag NW films with different thickness. A 35-nm-thick Ag film is presented for comparison [17]

9.2.2.5 Methods for Enhancing Stability

Encapsulating NW Film The sheet resistance of Ag NW electrode is doubled after a tape test, indicating failure to maintain adhesion of Ag NWs to plastic substrates. However, the electrode shows high stableness in sheet resistance after encapsulation with Teflon (Fig. 9.13a). It proves that the adhesion to the plastic substrates can be improved by performing encapsulation. In addition to the adhesion, resistance

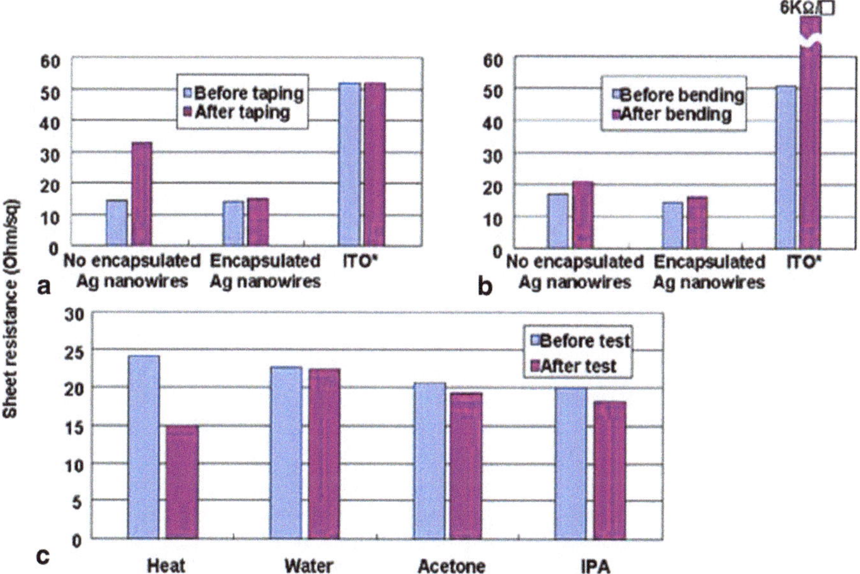

Fig. 9.13 **a** Tape tests on three electrodes: pristine Ag NW films, Ag NW films with Teflon encapsulation, and ITO on PET; **b** Bending tests on these three electrodes; **c** Chemical stability tests on Ag NWs: the film is exposed to heat (120 °C for 123 h), water for 10 min, acetone for 10 min, and IPA for 10 min, respectively [19]

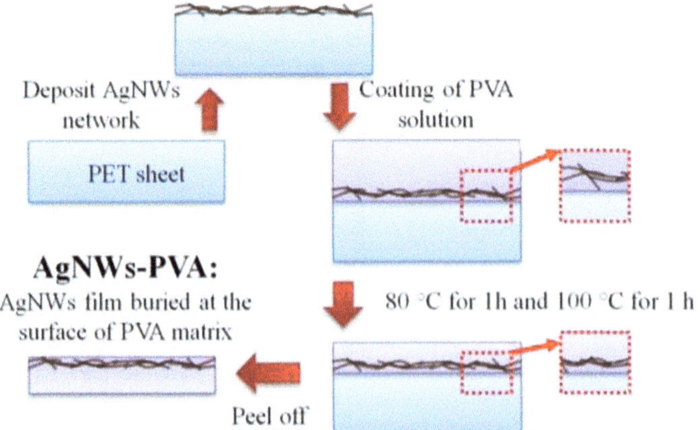

Fig. 9.14 Schematic for burying Ag NWs in PVA to fabricate the Ag NWs–PVA composite film [18]

to mechanical deformation and environmental changes is also critical to the stability of electrodes. The flexibility test indicates that the Ag NW electrodes are stable during bending regardless of encapsulation, while the ITO film cannot endure the bending (Fig. 9.13b). In terms of environmental reliability, the Ag NW electrodes can well survive under heating or in solvents without the need of encapsulation, and in the case of heating, they even show decrease of sheet resistance most likely owing to NW–NW junction and NW–substrate contact improvements (Fig. 9.13c).

There are some other encapsulation techniques to effectively improve the adhesion of Ag NW film to the substrate [18]. For example, polyvinyl alcohol (PVA) can be used as a transparent polymer matrix to fabricate Ag NWs–PVA composite film (see Fig. 9.14 for the process). This approach successfully enhances the electrical performance reliability of film, while its optical transmittance is not much reduced (Fig. 9.15). The sheet resistance of the Ag NWs–PVA film turns out to be even $10\sim30\%$ smaller than that of pristine Ag NW film on bare PET. As demonstrated in the work, a deposition density of 32 mg/m^2 can offer an optical transmittance of 88.0% (at 550 nm) and a sheet resistance of about 182 Ω. When the deposition density increases to 47.7 mg/m^2, the Ag NWs–PVA film can achieve a sheet resistance of 63 Ω; although, the optical transmittance is slightly reduced to 87.5% at 550 nm. This impressive performance record is very close to meet the criteria for touch-screen panels.

The resistance to mechanical deformation can be greatly improved by burying the Ag NWs film in PVA. The Ag NW film, that is directly deposited on PET intrinsically, has good robustness against bending; however, the Ag NWs cannot adhere to the substrate tightly. After being buried in PVA, Ag NW film has no obvious increase in sheet resistance after a standard tape test or being wiped (Fig. 9.16). Such a simple modification also enables excellent reliability in tensile and compressive tests, superior to the performance of ITO on PET substrate. As compared with the Ag NW film on PET, the modified electrode shows greater resistance to heat and chemical attack.

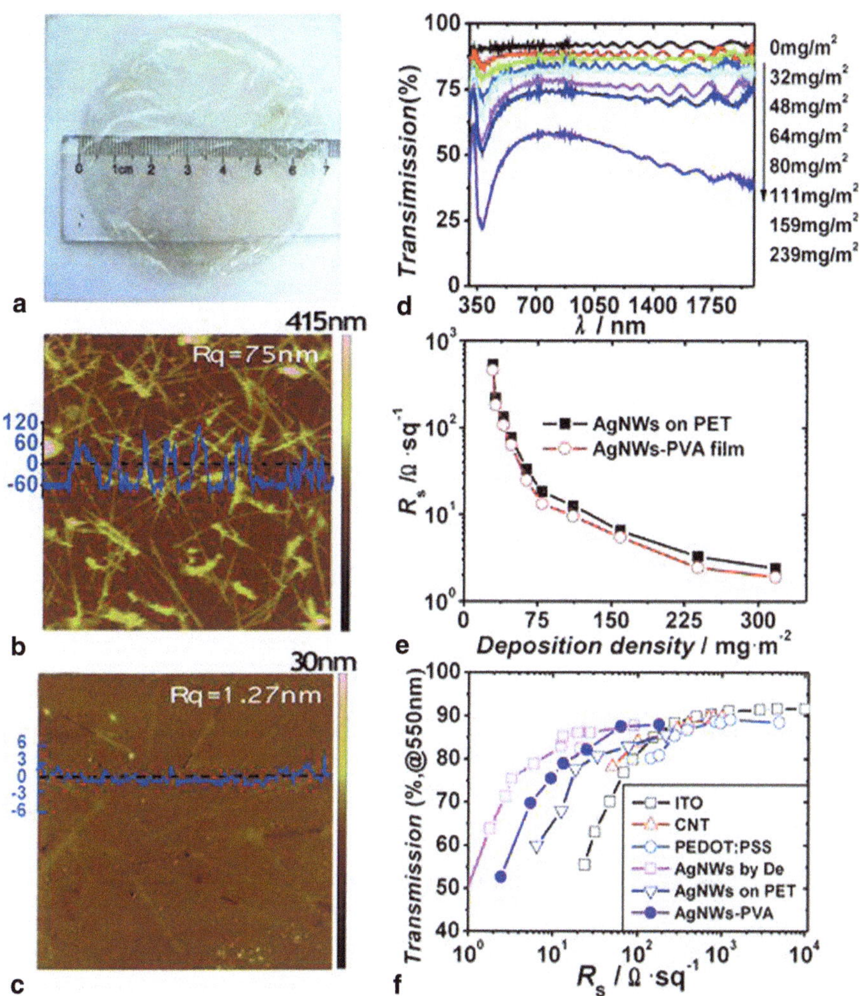

Fig. 9.15 **a** Photograph of the as-fabricated Ag NWs–PVA film. **b** and **c** AFM images of Ag NWs–PET and Ag NWs–PVA film. **d** UV-Vis-NIR transmission spectra and **e** sheet resistance curves for the Ag NWs–PVA film at different deposition densities. **f** Transmission (at 550 nm) vs. sheet resistance of the Ag NWs–PVA film compared with other typical transparent conductors [18]

Increasing Film Thickness De et al. prepared Ag NW films on PET at a number of thicknesses in 2009. Deposited mass per unit area (*M/A*) is used as a parameter to assess the thickness. The films are repeatedly bent from initial radius of 7.5 to 2.5 mm for many cycles. As shown in Fig. 9.17, the thinnest film (*M/A*=39 mg/m²) shows stable sheet resistance over the first 200 cycles, but the resistance is increased by 2 orders of magnitude in the next 2000 cycles. In stark contrast, the sheet resistance of thicker films remains almost unchanged over 2000 cycles. As the *M/A* reaches 79 mg/m², the sheet resistance has little change in both compressive and tensile bending. The thick Ag NW film shows reliability in terms of sheet resistance as high as bulk silver film, dramatically better than the ITO films.

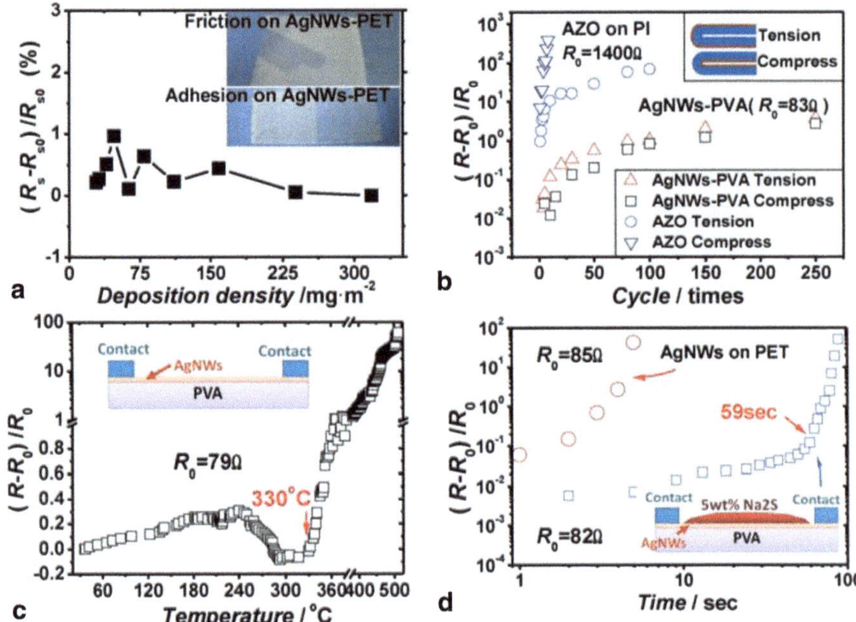

Fig. 9.16 **a** Sheet resistance change of Ag NWs–PVA after a tape test. The insets show that bare Ag NWs on PET are destroyed after friction and tape adhesion. **b** The increase in resistance after different cycles of folding up. 140-nm-thick Al-doped ZnO (AZO) film on 48-μm thick polyimide film (Kapton) is used as a comparison. **c** Sheet resistance change with temperature increase. **d** Sheet resistance change of Ag NWs–PVA electrode exposed to Na_2S solution [18]

Fig. 9.17 The change of sheet resistance with the increase of cycle number for Ag NW films at different thickness in compression or tension tests. A bulk silver film is used for comparison [17]

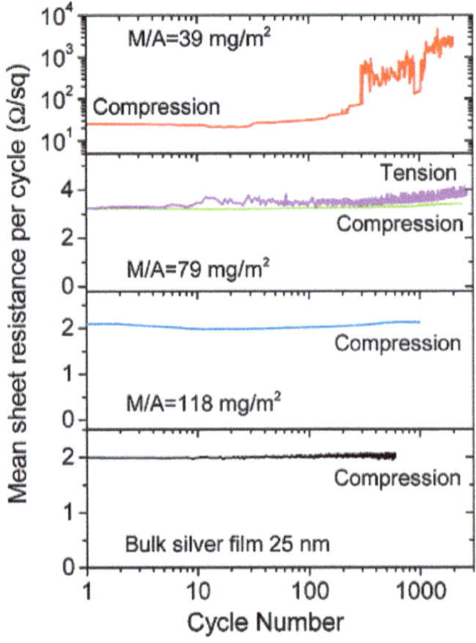

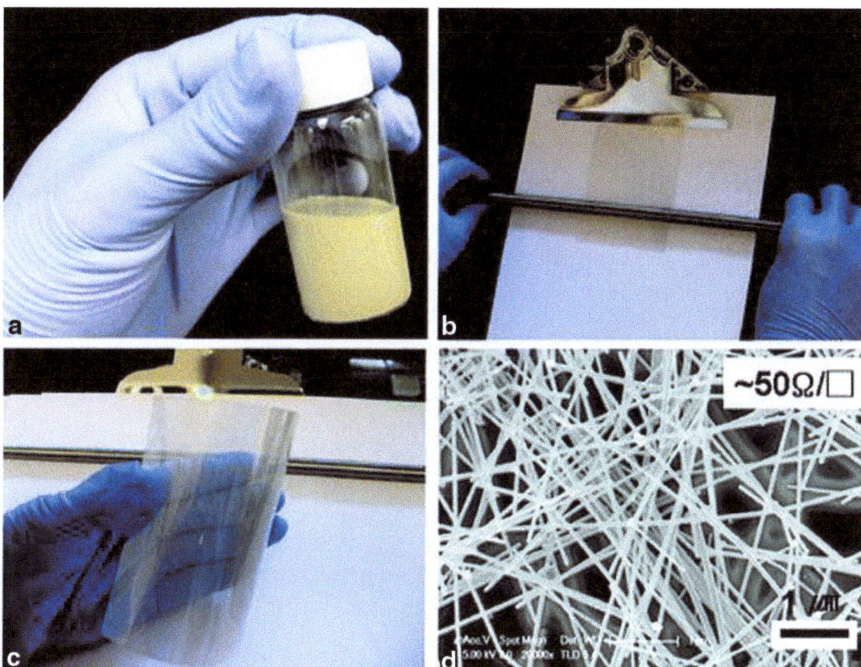

Fig. 9.18 **a** 10-mL Ag NW ink in methanol solution with a concentration of 2.7 mg/mL. **b** Meyer rod coating setup for scalable Ag NW coating on plastic substrate. **c** Ag NW film on PET substrate. **d** SEM image of as-synthesized Ag NWs [19]

9.2.2.6   Methods for Improving Electrode Smoothness

Improving Fabrication Process A Meyer rod coating setup has been demonstrated for lab-scale coating of Ag NWs in 2010 (Fig. 9.18) [19]. The Ag NWs are synthesized by following the protocol in literature with slight modifications [22]. The ink made from the suspension of Ag NWs is dropcast onto PET substrate, during which the solution is either pulled or rolled over by a Meyer rod. After the inking process, the wet coating is dried using an infrared lamp and then treated at 120 °C in an oven. In this technique, both the ink concentration and Meyer rod size influence the final thickness of Ag NW film. Methanol has been identified as the best solvent for uniformly coating the Ag NWs on PET substrate.

Increasing Film Thickness As shown in Fig. 9.19, the network of nanowires becomes less sparse with the increase of thickness. As a result, the substrate appears less transparent and macroscopically smoother. For the film with $M/A = 780$ mg/m^2, the NW film is so dense that the substrate below can be hardly seen.

Pressing Electrode Mechanically A mechanical pressure can not only reduce the sheet resistance, but also alter the morphology of Ag NW film. As shown in Fig. 9.9b, c, the surface roughness would be greatly decreased from 110 to 47 nm after pressing.

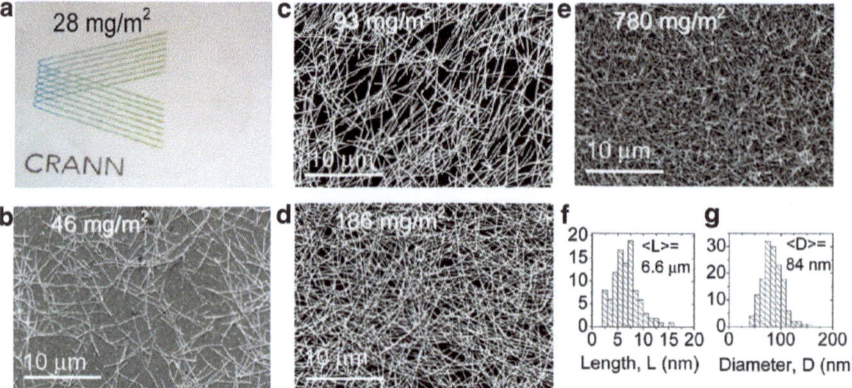

Fig. 9.19 **a** Photograph of an Ag NW film of PET. **b–e** SEM images of Ag NW films with different thickness: 28, 46, 83, 186, 780 mg/m². **f** Length and **g** diameter distribution diagrams of Ag NWs [17]

Encapsulating NW Film As discussed in the last section, when the Ag NW film is buried in PVA, the surface roughness can be significantly reduced from 30–100 nm to 1–5 nm (Fig. 9.15b, c). This surface roughness can basically meet the requirement for specific applications such as OLED.

9.2.2.7 Summary for Ag NW Mesh Electrodes

As demonstrated above, the performance of Ag NW mesh electrodes can be comparable to or even higher than that of ITO. The excellent flexibility and stability of Ag NW meshes in certain forms, together with their small surface roughness, make the mesh electrodes a promising candidate to replace conventional TCEs. It is not hard to figure out that many factors are interlinked together when people develop approaches to improve the performance of mesh electrodes in terms of four parameters—sheet resistance, optical transmittance, stability/reliability, and surface roughness. In order to fabricate optimal electrodes for specific applications, some factors have to be compromised to find a perfect combination.

9.2.3 Cu nanowires (Cu NWs)

Ag NW mesh electrodes can offer performance at the level up to or superior to conventional ITO electrodes, which have made them possible for commercializations; however, their further applications are still limited by some obvious drawbacks such as the scarcity and high price of silver [23]. Copper is 1000 times more abundant than silver and 100 times less expensive, so fabricating Cu NW electrodes would be an appealing alternative to Ag NWs or ITO.

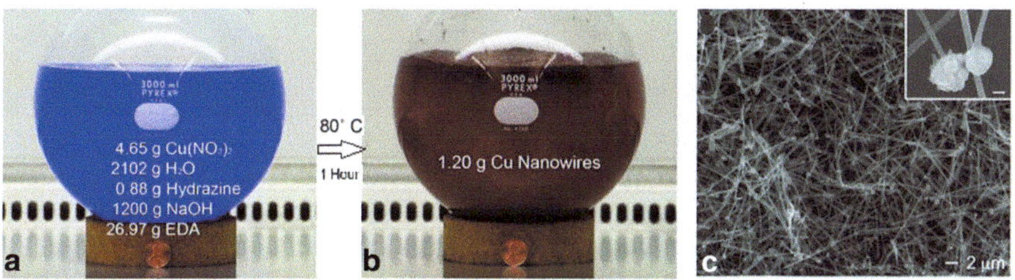

Fig. 9.20 Photographs for solution **a** before the synthesis and **b** after growth of Cu NWs at 80 °C for 1 h. **c** SEM image of as-synthesized Cu NWs. The inset shows that Cu NWs have a spherical Cu particle at one end (scale bar: 200 nm) [23]

9.2.3.1 Typical Synthesis of Cu NWs

Cu NWs can be synthesized on the gram scale to fabricate TCEs. The typical synthetic protocol is presented in Sect. 9.5.2, in which $Cu(NO_3)_2$ is reduced by hydrazine in NaOH and ethylenediamine (EDA) solution (Fig. 9.20). The Cu NWs have a diameter of 90 ± 10 nm and a length of 10 ± 3 µm.

9.2.3.2 Fabrication of Cu NW Electrodes

The product is first sonicated to form an aqueous suspension containing 3 wt% of hydrazine and 1 wt% PVP. The suspension is then diluted in a 10 wt% PVP solution, which allows the aggregates to sink to the bottom and leaves the well-dispersed NWs in the upper solution.

The Cu NWs are filtered onto polycarbonate membranes, and then printed onto glass microscope membranes (covered by Aleene's Clear Gel Glue). The membrane is then put in contact with a sticky gel film by hand, and finally peeled off to obtain the Cu NW film (Fig. 9.21a).

The film with 390 mg/m² of Cu NWs shown in Fig. 9.21a has a transmittance of 67 % (at the wavelength of 500 nm) and a sheet resistance of 61 Ω/sq. Both the compression and tensile bending tests (Fig. 9.21b, c) show that the Cu NW film has good reliability in sheet resistance after 1000 bending cycles. The stability of Cu NW film in air has been also demonstrated by 28-day exposure to air at room temperature. The test shows that the Cu NW film is still highly conductive with no degradation in optical transmittance (Fig. 9.21d). Unfortunately, the transparency of Cu NW film is lower than that of Ag NW and ITO films.

9.2.3.3 Methods for Improving Cu NW Electrode Performance

Synthesizing Longer and Thinner Cu NWs It has been identified that a few parameters of Cu NWs—short lengths, large diameters and aggregation—would cause the posterior performance of Cu NW film. Thus, it is imperative to develop synthetic

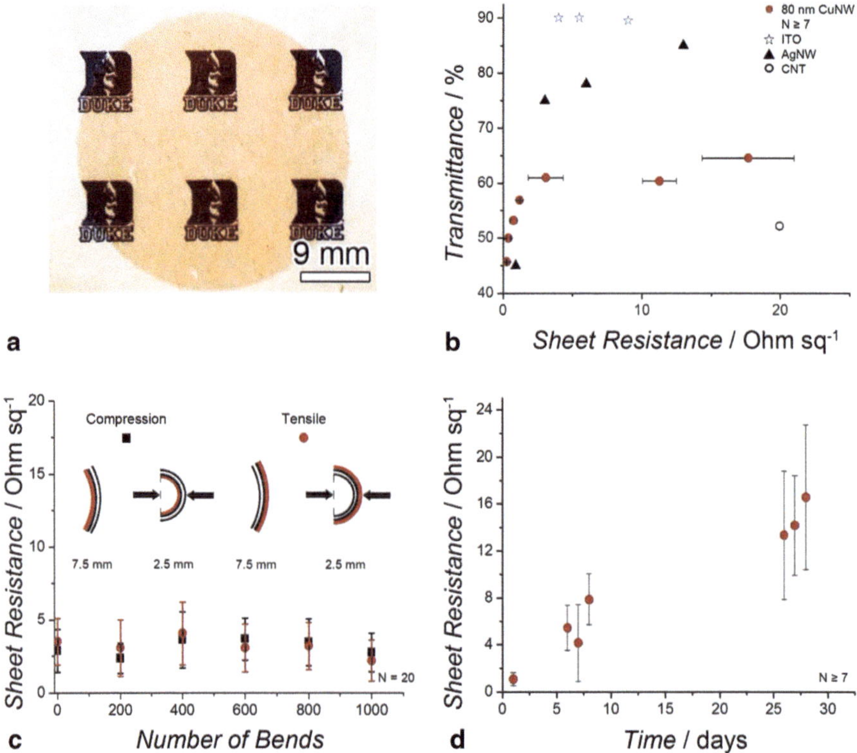

Fig. 9.21 a Photograph of a Cu NW film. **b** Transmittance vs. sheet resistance for Cu NW films, silver NWs, ITO, and CNTs. **c** Sheet resistance change of Cu NWs in compression and tensile bending tests. **d** Sheet resistance vs. time demonstrating the stability of the Cu NWs films in air [23]

methods for longer and thinner Cu NWs. In order to achieve this goal, a few modifications have been made: the heat time is shortened to 3 min, the PVP is used to replace the capping agent EDA to prevent the aggregation of Cu NWs, and the reaction solution is quickly cooled in ice bath [24]. In terms of electrode fabrication, a modified Meyer rod method is used where nitrocellulose is used to uniformly coat Cu NWs on PET substrate. After plasma clean and annealing at 175 °C, the final Cu NWs–PET network is ready for use. The as-synthesized Cu NWs have a length larger than 20 μm and a diameter smaller than 60 nm.

This Cu NW film turns out to exhibit greatly improved performance (Fig. 9.22). At a sheet resistance of 50 Ω/sq, the transmittance is higher than the Ag NW and CNT films made by Meyer rod coating and spray coating. Although ITO on glass with sheet resistance of 45 Ω/sq has a transmittance 7 % greater than the Cu NWs–PET film at a wavelength of 550 nm, the transmittance of Cu NW film appears to be much higher at the wavelength above 1200 nm. When the wavelength is 1550 nm, the transmittance of Cu NW film is 87 % while the ITO film only offers 54 %. The high transmittance of Cu NW films in the infrared region allows their applications in infrared photovoltaics.

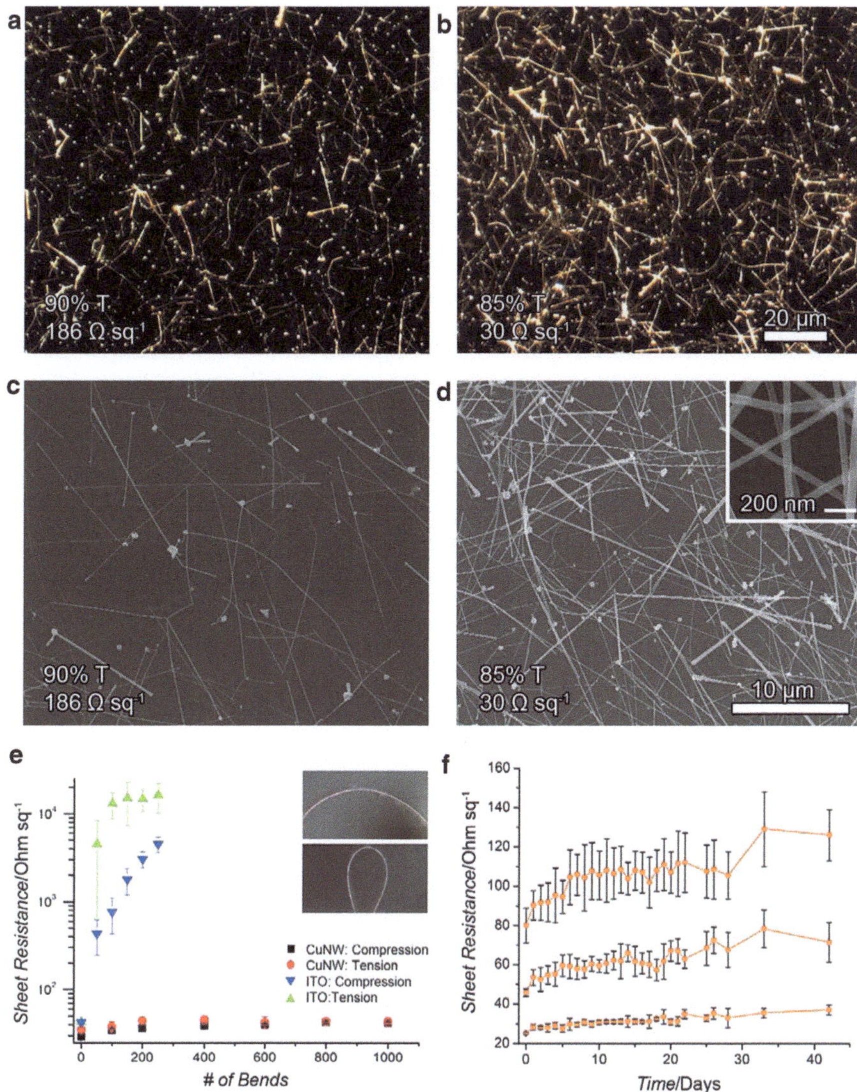

Fig. 9.22 Dark-field optical microscope images of Cu NW films: **a** transmittance of 90 % and sheet resistance of 186 Ω/sq, and **b** transmittance of 85 % and sheet resistance of 30 Ω/sq. **c, d** SEM images of Cu NW films shown in (**a**) and (**b**), respectively. The average length is 20 ± 5 µm, and the diameter is 52 ± 17 nm. **e** Sheet resistance vs. number of bends for Cu NW film (85 % transparent) and ITO on PET. The inset shows the Cu NW film in different curvature radii of 10 and 2.5 mm. **f** Sheet resistance vs. time of exposure to air for Cu NW films with transparency of 89, 86 and 82 % from top to bottom [24]

The modified Cu NW films retain their excellent reliability during bending. The sheet resistance stays unchanged during 1000 bends, while that of ITO is increased by 400 times after only 250 bends. The stability of Cu NW film in oxygen environment has also been demonstrated for a period of 42 days.

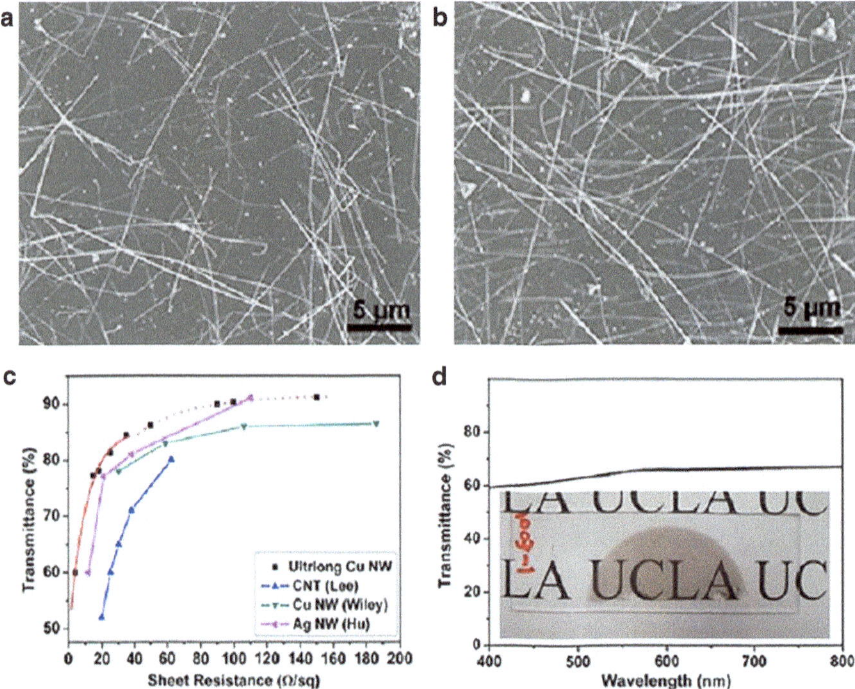

Fig. 9.23 SEM images of Cu NW electrodes with **a** 91 % and **b** 84.5 % transmittance. **c** Performance comparison of the electrodes made from the ultralong Cu NWs, normal Cu NWs [24], CNTs [26], and Ag NWs [19]. **d** Transmittance vs. wavelength of ultralong Cu NW electrode (sheet resistance ~5 Ω/sq) [25]

The electrical properties of electrodes can be further tuned by growing ultralong Cu NWs. For instance, Zhang et al reported that ultralong single-crystalline Cu NWs with good dispersity can be synthesized in the presence of hexadecylamine (HAD) and cetyltriamoninum bromide (CTAB), catalyzed by a thin layer of Pt [25]. The ultralong Cu NW film with transmittance of 90 % has a sheet resistance as low as 90 Ω/sq, exceeding the performance of some reported Ag NW films (Fig. 9.23) [19].

Covering Cu NWs with Ni Cu NW films are prone to oxidation as well as possess reddish-orange color, limiting their applications in display. In order to overcome the limitations, Cu–Ni alloy NWs have been developed by reducing $Ni(NO_3)_2 \cdot 6H_2O$ on Cu NWs [27]. The as-synthesized Cu–Ni NWs are then fabricated into mesh films by a Meyer rod method (Fig. 9. 24).

As compared with Cu NW film having the same sheet resistance, the transmittance of Cu–Ni NW film is somewhat reduced by the presence of Ni. For example, at a sheet resistance of 60 Ω/sq, the transmittance is reduced from 94.4 to 84.3 % at the wavelength of 550 nm as the Ni content rises from 0 to 54 %. Nevertheless, the addition of Ni enables much higher resistance to oxidation. This feature can be resolved by the tests on two films with similar transmittance in an oven at 85 °C. As

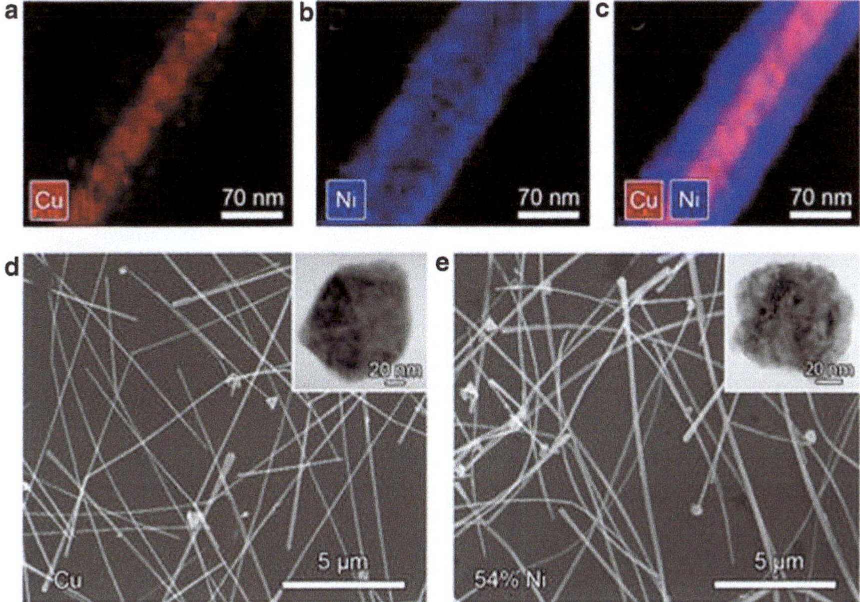

Fig. 9.24 **a–c** Energy dispersive X-ray spectroscopy mapping profiles of a Cu NW coated with 54 mol % Ni. SEM images of **d** Cu NWs before coating and **e** Cu NWs after coated with 54 mol % Ni. The diameters of the wires increase from 75 ± 19 to 116 ± 28 nm. Cross sections of pristine Cu NW and Ni-coated Cu NW are shown in the insets [27]

shown in Fig. 9.25, the sheet resistances of Cu NW film and Ag NW film increase by an order of magnitude after 5 days and 13 days, respectively. In sharp contrast, the sheet resistance of Cu–Ni NW film almost stays constant. An estimation based on the Arrhenius equation indicates that the Cu–Ni NW film is 1000 times and 100

Fig. 9.25 Sheet resistance vs. time for Ag NW film, Cu NW film and Cu–Ni NW film with transparency of 85–87 % stored at 85 °C [27]

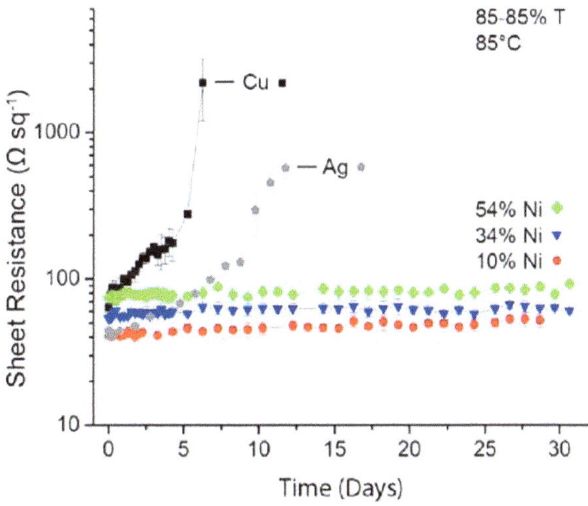

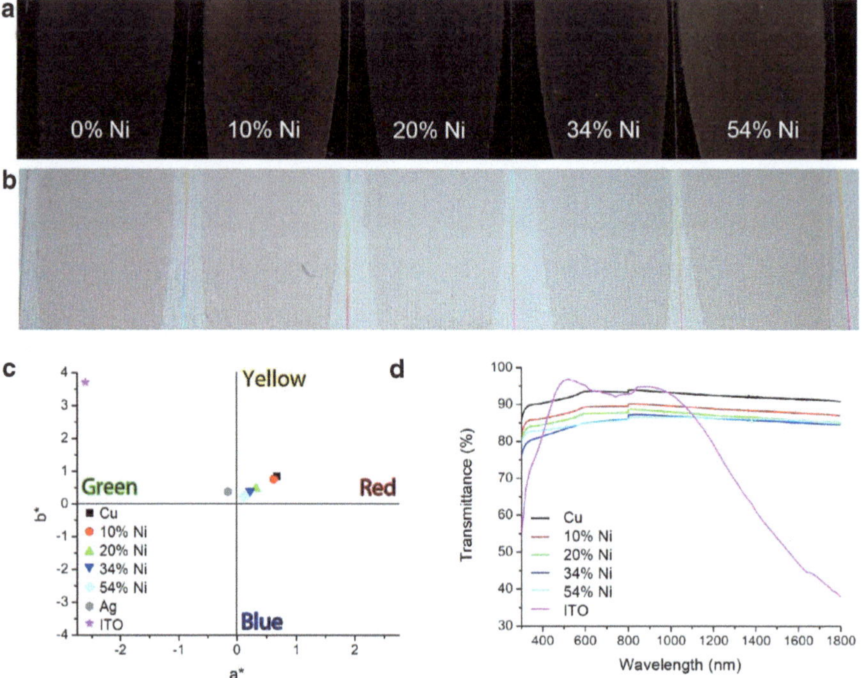

Fig. 9.26 Images of NW films with transparency of 87 % on a (a) black and (b) white backlit background. (c) Quantitative comparison of color on a Hunter color scale. (d) Transmittance vs. wavelength for ITO (11 Ω/sq), Cu NWs (60 Ω/sq), and CuNi NWs (60 Ω/sq), with different Ni contents [27]

times more resistant to oxidation than the Cu NW film and the Ag NW film at room temperature, respectively.

In addition to oxidation, the problem about the intrinsic color of Cu NW films has also been solved with Ni coating. As displayed in Fig. 9.26, the color of Cu–Ni NW films alters from reddish to gray with the amount of Ni increases. Given the yellow-green color of ITO, the Cu-Ni NW film displays more neutral color than the widely used ITO material.

9.2.3.4 Summary for Cu NW Electrodes

Although typical Cu NW electrodes do not have the performance comparable to ITO or Ag NW mesh electrodes, several methods have been developed to improve the transmittance at low resistance of Cu NWs electrodes. Such improvements result in enhanced performance even superior to some of the reported Ag NW mesh electrodes. Since copper is much cheaper than indium and silver, more efforts should be made to further improve the performance of Cu NW electrodes down the road.

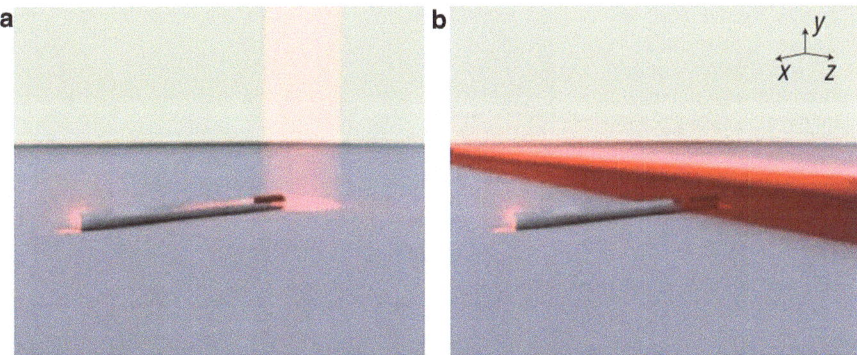

Fig. 9.27 Schematics for **a** coupling light to a NW by focusing a laser beam on the end of a NW and **b** guiding light with a polymer waveguide [12]

9.3   Plasmonic Waveguiding

The era of "big data" will require numerous input and output channels to increase data transmission rates and capacity. Traditional electronic circuits are limited when digital information needs to be sent from one point to the other end of a microprocessor. Although optical interconnects may play this role, the minimum size of photonic structures cannot meet the requirement for the integration with electronic chips due to the diffraction limit [28]. Intuitively, plasmonics can solve this problem by localizing and guiding light in subwavelength structures based on surface plasmons [28].

Nanowires have been found to be a suitable candidate for nanoscale plasmonic optical waveguides [29]. They can localize the electromagnetic energy at the nanoscale thus making it possible to miniaturize optical signal processing at subwavelength. Light coupling and plasmon propagation are two key components for such processing, and together with key functional parts of nanophotonic circuits, will make important contributions to plasmonic waveguides.

9.3.1   Light Coupling

There are several different approaches to coupling light to plasmon modes in Ag NWs. The Ag NWs are synthesized using the similar method outlined in Sect. 9.5.1. These NWs have smooth surface and tapered ends.

Approach 1: Laser Focusing on One End of a Nanowire By using one end of the Ag NW as the scattering center (Fig. 9.27a), the incident light can be coupled to propagating plasmon mode along the Ag NW [30]. As displayed in Fig. 9.28, no matter which end of an Ag NW is illuminated by a diffraction-limited laser spot, the plasmon propagation can occur along the axial direction of NW. However, when the laser is focused on the middle of the Ag NW, no plasmon modes can be launched

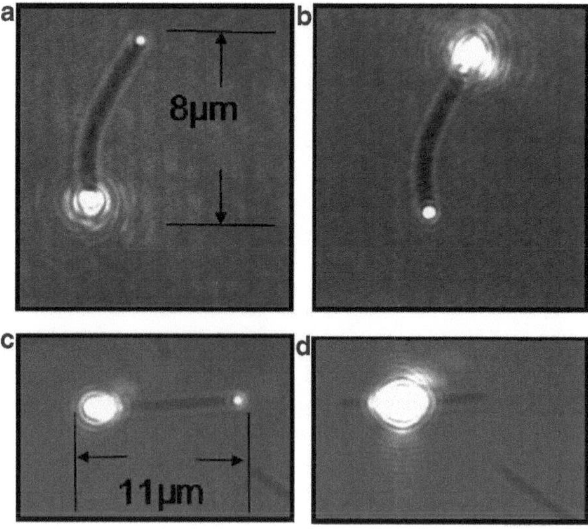

Fig. 9.28 Spatial sensitivity of launching plasmons in an Ag NW. **a** Excitation at the bottom end. **b** Excitation from the top end. **c** Excitation from the left end. **d** Laser focusing at the middle of the NW [30]

due to the cylindrically symmetric of the midsection that cannot scatter in the axial direction. Only on the tapered end or any discontinuity of nanowire where the symmetry is broken can light be coupled into plasmon modes [30].

Approach 2: Guiding Light with a Polymer Waveguide Polymer waveguiding is another method that has been developed to more conveniently excite plasmons in Ag NWs as shown in Fig. 9.27b [12]. In this method, the interconnect between photonic and plasmonic waveguides can be used to couple light into nanophotonic devices. To achieve the goal, Ag NWs are aligned perpendicularly to a polymer waveguide with one end exposed to the maximum light intensity in the photonic waveguide. In the design, the polymer waveguide has a SU-8 core with refractive index of 1.58 and 5-μm bottom cladding made of sol–gel with refractive index of 1.5.

When one end of the Ag NW is embedded in the polymer waveguide, light can be coupled into plasmon modes. In a well-designed device, experiment results clearly show that laser light propagates along the nanowire to the outside of polymer waveguide (Fig. 9.29). With the polarization of the laser light rotated by 90°, the light scattered from the nanowire is significantly decreased. It demonstrates that one can tune the light emitted from the nanowire simply by adjusting the polarization of the incident light. This method is also applicable to couple light from one polymer waveguide to multiple nanowires simultaneously.

Other Approaches Although the plasmon on the midsection of Ag NWs cannot be directly excited by coupling light to plasmon modes from free space, the coupling of plasmonic and photonic nanowires provides a possibility to solve this problem [31]. In the approach, light is first lens-coupled into a silica fiber and gradually tapered into a nanofiber. The nanofiber is then contacted with a ZnO or Ag NW to excite surface plasmons. In this way, plasmons can be excited in the midsection of Ag NWs.

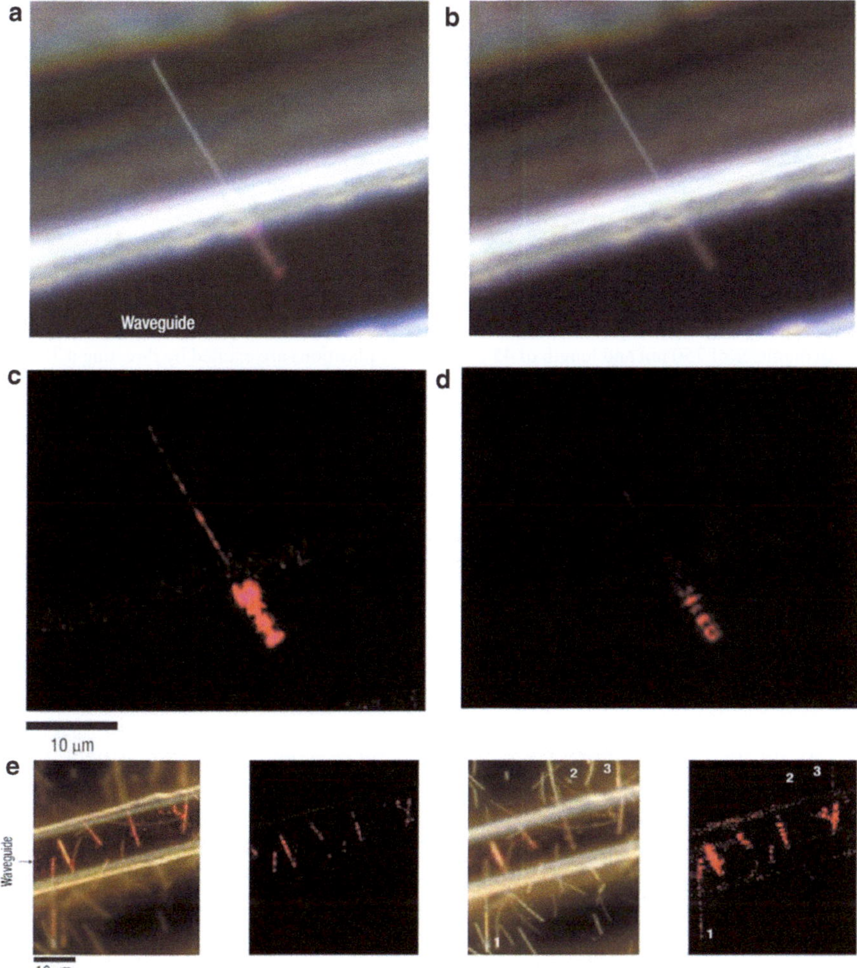

Fig. 9.29 Dark-field images of plasmon propagation in single Ag NW **a** with and **c** without white light illumination. The laser polarization is parallel to the nanowire. Dark-field images of plasmon propagation in the Ag NW **b** with and **d** without white light illumination after turning the polarization of incident laser light 90°. **e** Microscope images for coupling light from a polymer waveguide to multiple NWs [12]

9.3.2   *Plasmonic Propagation*

When light is coupled to plasmon modes in an Ag NW, the plasmon propagates along the NW. In general, two possible energy attenuation processes should be considered in the propagation—intrinsic ohmic loss and radiative energy loss. The energy loss due to the NW bending has been extensively investigated [29].

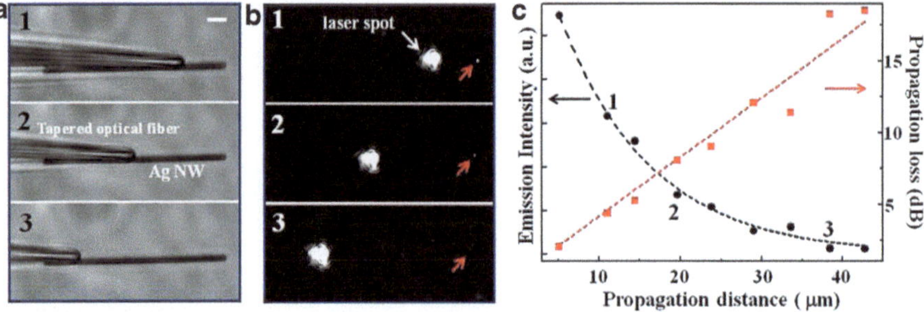

Fig. 9.30 **a** Bright-field optical images and **b** corresponding dark-field optical images of Ag NWs with diameter of 750 nm and length of 45 μm. Surface plasmons are excited by directing a 785-nm laser from the tapered optical fiber. (a1–a3) Different positions of the tapered optical fiber. (b1–b3) Emitted light at the output end of Ag NW (indicated by *red arrows*). The propagation distances are 11, 19 and 38 μm, respectively. The scale bar is 5 μm. **c** Emitted light intensity vs. propagation loss for different propagation distances [29]

Intuitively, intrinsic ohmic loss in the propagation can be facilely assessed in the system of straight Ag NWs. By changing the position of tapered optical fiber, light is coupled into an Ag NW at different locations which in turn change the propagation length (Fig. 9.30). As such, the propagation loss can be determined by measuring the emission light intensity. It turns out that different Ag NWs with varied diameters and lengths have a propagation loss from 0.35to 0.48 dB/μm due to the intrinsic ohmic loss.

As for the radiative energy loss, it occurs more when the Ag NW is bent due to the energy conversion from propagation to radiation. To probe this energy loss, an Ag NW with a diameter of 750 nm and length of 45 μm is bent to radii varied from 5 to 32 μm by tapered optical fiber manipulation. As shown in Fig. 9.31, the bending section keeps smooth, and no defects can be observed. In this bending system, the intensities of emitted light from the end of Ag NW are recorded (Fig. 9.30b). Given that it exists whenever Ag NW is bent or not, the propagation loss in this Ag NW is considered as constant value 18.5 dB. As such, the bending loss can be calculated by extracting propagation loss from the total energy loss. As summarized in Fig. 9.31c, the pure bending loss decreases exponentially with the increase of bending radii. This finding suggests that the emitted light intensity can be adjusted by changing the shape of Ag NW. A low bending loss can be readily achieved as long as the bending radius is large enough.

9.3.3 Functional Components in Nanophotonic Circuits

Branched Ag NWs have been exploited as the router and multiplexer for plasmon propagation [32]. In fact, the branched Ag NWs are occasionally recognized by spin-coating suspension of Ag NWs on ITO glass slides. The gaps separating the branches are tens of nanometers as displayed in Fig. 9.32.

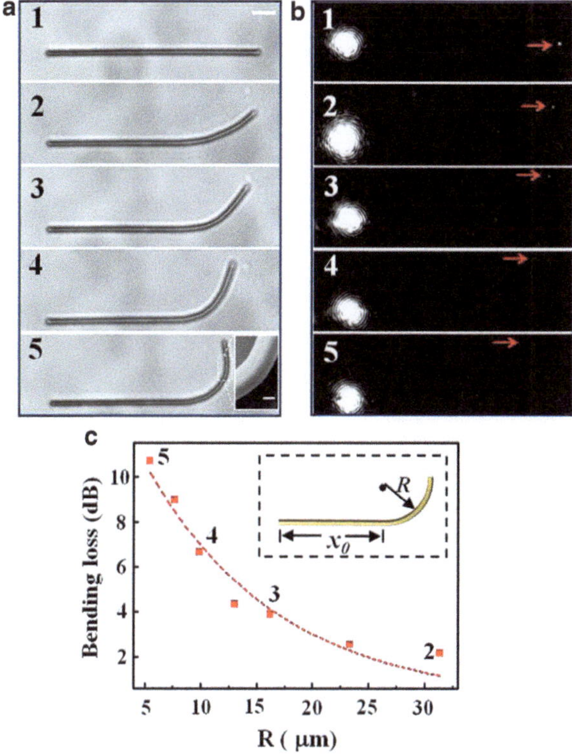

Fig. 9.31 **a** Bright-field optical images showing an Ag NW with different bending radii of infinite, 32, 16, 9 and 5 μm, respectively. The inset of (a5) is an SEM image of curved section with scale bar of 350 nm. **b** Dark-field optical images of Ag NW with different bending radii corresponding to (**a**). 785-nm laser is used for the excitation, and emitted light dots are indicated by *red arrows*. The scale bar is 5 μm in (**a**, **b**). **c** Change of pure bending loss with bending radius, well fitted with the red exponential curve. The inset shows the geometry of a bent wire [29]

When the end of one NW is excited, surface plasmons propagate along both the main wire and the branch wire. Interestingly, the relative intensities of the surface plasmons along the main and branch wires can be controlled by tailoring the polarization of the excited light. For instance, when the incident polarization is 140°, the light emission is strong at the wire end 2; however, as the polarization turns to 40°, the light emission at wire end 2 becomes very dim, leaving that at wire end 3 strong. It has been demonstrated that as the emission at the end of the main wire reaches the maximum, that at the end of the branched wire end gets to the minimum and vice versa. Thus surface plasmons can be routed to different destinations by maneuvering the incident polarization.

The branched Ag NW structure can also be used as a multiplexer. When this structure is excited by two different wavelengths, they can be routed differently. For example, 633- and 785-nm light are simultaneously focused on wire end 1 with polarization of 40°. The 633-nm light is almost completely routed to the branched

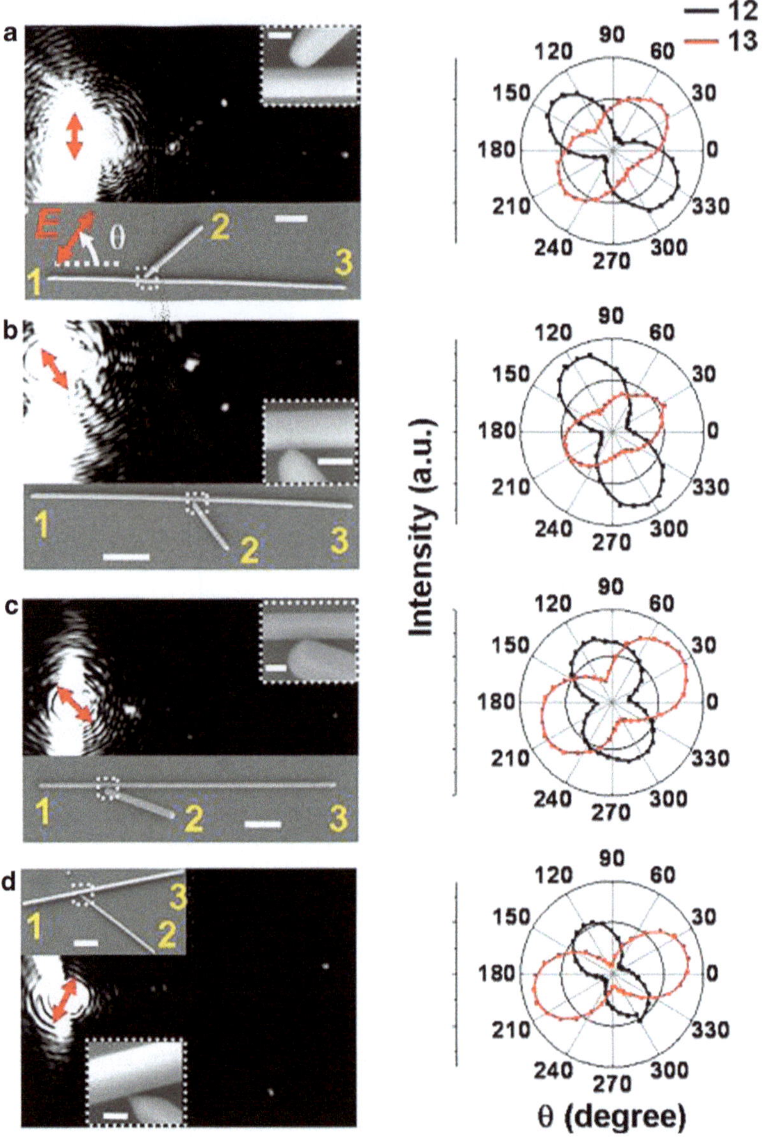

Fig. 9.32 (*Left column*) Optical images of the light emitted at the ends of NWs. Plasmons are excited at wire end 1 with 633-nm laser light. *Red arrows* represent the polarization orientations of the excitation laser. (*Insets*) SEM images of the router structures and the junctions. The scale bars are 2 μm and 200 nm, respectively. (*Right column*) Emission intensity as a function of polarization angle θ from wire ends 2 (*black*) and 3 (*red*), corresponding to the left column [32]

end, while the 785-nm one stays in the main wire. The switching and routing of surface plasmons can be separated for different wavelengths, and do not interfere with each other at the same time.

Other nanophotonic components such as Mach–Zehnder interferometers and microring cavities have also been achieved by using the coupling of photonic and plasmonic nanowires [31]. The development for assembling hybrid nanophotonic devices, based on different photonic and plasmonic components, has become an appealing strategy for enhancing the device performance and deserves more attention in the future [33].

9.3.4 Summary for Ag NW Plasmonic Waveguides

As plasmonic waveguides can carry optical and electric signals along the same circuitry beyond optical diffraction limit, they have the ability to combine the advantages of photonics and electronics on the same chip [28]. The promising properties of Ag NW plasmonic waveguides have been extensively studied. Nevertheless, more efforts will be needed to reduce energy loss in plasmonic waveguides and fulfill the waveguides in nanophotonic circuits.

9.4 Conclusion and Outlook

Electrical and photonic transports represent two different typical modes used in electronic and optoelectronic devices. Transparent conductors can be used in a wide variety of applications, such as touch-sensitive control panels and thin-film solar cells. Metallic nanowires have been demonstrated as promising materials to form flexible, transparent conducting films to replace ITO films, whose performance can be boosted by developing new materials and processing techniques. In parallel, the metallic nanowires can be employed in a new mode of subwavelength data transmission—plasmonic waveguiding for miniaturization to meet the demand in the era of "big data". The density and throughput of conventional diffraction-limited photonic integrated circuits are limited by the fairly large wavelength of light, which certainly are unable to provide sufficient input and output channels to increase data transmission rates and capacity. The metallic nanowires can not only carry optical and electrical signals along and across the nanowires, but also be used in hybrid nanophotonic components to form photo-plasmon coupling. It is envisioned that these properties and techniques for metallic nanowires will in turn drive their use in broader applications, which require more controllable and scalable synthesis to produce desired building blocks.

Certainly the applications of metallic nanostructures in electronics and optoelectronics are not limited to the developments outlined in this chapter. For instance, a simple approach has been developed to pattern silver microelectrodes in flexible, spanning, and stretchable form via omnidirectional printing of concentrated inks—suspensions of silver nanoparticles [34]. The microelectrodes have demonstrated their use for integration with semiconductor, plastic, and glass substrates, providing a brand new way to wire-bond three-dimensional interconnects for solar cells and LED arrays. Along this line, the bottom-up synthesis and assembly of metallic nanostructures can be perfectly integrated with top-down fabrication of materials, producing low-cost and high-performance devices.

9.5 Protocols

9.5.1 Typical Synthesis of Ag NWs

This protocol is adapted from [19] without further modifications.

A mixture of 0.334-g poly vinylpyrrolidone (PVP) and 20-mL ethylene glycol (EG) is heated and thermally stabilized at 170 °C in a flask. Once the temperature has been stabilized, 0.025 g of silver chloride (AgCl) is ground finely and added to the flask for initial nucleation of the silver seeds. After 3 min, 0.110 g of silver nitrate ($AgNO_3$) is titrated for 10 min. After sufficient silver sources have been supplied, the flask is heated for additional 30 min to ensure that the growth is complete. The cooled-down solution is then centrifuged three times at 6000 rpm for 30 min to remove solvent (EG), PVP, and other impurities in the supernatant. After the final centrifuge, the precipitate of Ag NWs is redispersed in 30 mL of methanol.

9.5.2 Typical Synthesis of Cu NWs

This protocol is adapted from [23] without further modifications.

Cu NWs are synthesized by adding NaOH (2000 mL, 15 M), $Cu(NO_3)_2$ (100 mL, 0.2 M), EDA (30 mL), and hydrazine (2.5 mL, 35 wt%) to a 3000-mL round bottom flask. This mixture is swirled by hand for 20 s after each addition to mix the reactants. The solution is then heated at 80 °C and stirred at 200 rpm for 60 min. After the reaction, the suspension is centrifuged at 4500 rpm for 5 min, and the supernate is decanted from the nanowires. The wires are then dispersed in a 3-wt% aqueous solution of hydrazine by vortexing for 30 s, and then centrifuged and decanted for three more cycles. The Cu NWs are stored in a 3-wt% hydrazine solution at room temperature under an argon atmosphere to minimize oxidation.

References

1. Lal, S., Link, S., Halas, N. J., Nat Photonics. **1**, 641 (2007)
2. Bae, S., Kim, H., Lee, Y., et al., Nat Nanotechnol. **5**, 574 (2010)
3. Lipomi, D. J., Vosgueritchian, M., Tee, B. C., et al., Nat Nanotechnol. **6**, 788 (2011)
4. Moore, G., Electronics. **38**, 114 (1965)
5. Stingelin-Stutzmann, N., Smits, E., Wondergem, H., et al., Nat Mater. **4**, 601 (2005)

6. Baca, A. J., Ahn, J. H., Sun, Y., et al., Angew Chem Int Edit. **47,** 5524 (2008)
7. Kim, D. H., Xiao, J., Song, J., et al., Adv Mater. **22,** 2108 (2010)
8. Kumar, A., Zhou, C., ACS Nano. **4,** 11 (2010)
9. Hu, L., Wu, H., Cui, Y., MRS Bull. **36,** 760 (2011)
10. Yu, Z., Li, L., Zhang, Q., et al., Adv Mater. **23,** 4453 (2011)
11. Ozbay, E., Science. **311,** 189 (2006)
12. Pyayt, A. L., Wiley, B., Xia, Y., et al., Nat Nanotechnol. **3,** 660 (2008)
13. Lee, J.-Y., Connor, S. T., Cui, Y., et al., Nano Lett. **8,** 689 (2008)
14. Gordon, R. G., MRS Bull. **25,** 52 (2000)
15. Murali, A., Barve, A., Leppert, V. J., et al., Nano Lett. **1,** 287 (2001)
16. Ellmer, K., Nat Photonics. **6,** 809 (2012)
17. De, S., Higgins, T. M., Lyons, P. E., et al., ACS Nano. **3,** 1767 (2009)
18. Zeng, X. Y., Zhang, Q. K., Yu, R. M., et al., Adv Mater. **22,** 4484 (2010)
19. Hu, L., Kim, H. S., Lee, J.-Y., et al., ACS Nano. **4,** 2955 (2010)
20. Rowell, M. W., Topinka, M. A., McGehee, M. D., et al., App Phys Lett. **88,** 233506 (2006)
21. Kang, M. G., Guo, L. J., Adv Mater. **19,** 1391 (2007)
22. Yugang, S., Byron, G., Brian, M., et al., Nano Letters. **2,** (2002)
23. Rathmell, A. R., Bergin, S. M., Hua, Y. L., et al., Adv Mater. **22,** 3558 (2010)
24. Rathmell, A. R., Wiley, B. J., Adv Mater. **23,** 4798 (2011)
25. Zhang, D., Wang, R., Wen, M., et al., J Am Chem Soc. **134,** 14283 (2012)
26. Kim, U. J., Lee, I., Bae, J. J., et al., Adv Mater. **23,** 3809 (2011)
27. Rathmell, A. R., Nguyen, M., Chi, M., et al., Nano Lett. **12,** 3193 (2012)
28. Ozbay, E., Science (New York, N.Y.). **311,** 189 (2006)
29. Wang, W., Yang, Q., Fan, F., et al., Nano Lett. **11,** 1603 (2011)
30. Sanders, A. W., Routenberg, D. A., Wiley, B. J., et al., Nano Lett. **6,** 1822 (2006)
31. Guo, X., Qiu, M., Bao, J., et al., Nano Lett. **9,** 4515 (2009)
32. Fang, Y., Li, Z., Huang, Y., et al., Nano Lett. **10,** 1950 (2010)
33. Oliver, B., Nature. **480,** (2011)
34. Ahn, B. Y., Duoss, E. B., Motala, M. J., et al., Science. **323,** 1590 (2009)

FSC
www.fsc.org
MIX
Papier aus verantwortungsvollen Quellen
Paper from responsible sources
FSC® C105338